27.93

D1171921

Analog and Computer Electronics for Scientists

Analog and Computer Electronics for Scientists

Fourth Edition

Basil H. Vassos   Galen W. Ewing

A Wiley-Interscience Publication

John Wiley & Sons, Inc.

New York / Chichester / Brisbane / Toronto / Singapore

Copyright © 1993 by John Wiley & Sons, Inc.

Library of Congress Cataloging in Publication Data:
Vassos, Basil H.
     Analog and computer electronics for scientists / Basil H. Vassos.
Galen W. Ewing. — 4th ed.
          p.   cm.
     "A Wiley-Interscience publication."
     Rev. ed. of Analog and digital electronics for scientists.
     Includes index.
     ISBN 0-471-54559-7
     1. Electronic instruments.   2. Analog electronic systems.
3. Digital electronics.   I. Ewing, Galen Wood, 1914–
II. Title.
TK7878.4.V37   1993
621.3815—dc20   92-18330
                                                           CIP

Printed in the United States of America

10 9 8 7 6 5 4 3 2 1

Preface

This fourth edition of *Analog and Digital Electronics for Scientists* differs from its predecessors in that it includes a greatly expanded treatment of microcomputers as the ultimate electronic components.

The first two parts of the book cover essentially the same ground as the whole of each of the previous editions, but with suitable up-dating and rearrangement. We have reduced the dependence on phasors, which we feel are not essential to our intended audience. These and some other important mathematical tools, such as Fourier and Laplace transforms and Boolean algebra, are treated in the final chapter as a useful reference.

Part III gives an entirely new treatment of microcomputers and makes use of the principles of circuitry set forth in the preceding two parts. Discussions of computer peripherals, communications, and networking are included. No significant treatment of computer languages and programming is attempted.

Much has been written lately about the future of computers, and we have contributed our personal assessments. The general consensus seems to be that no great changes in computer electronics are imminent. There will be significant improvements, no doubt, in such areas as speed, miniaturization, intercommunication between computers, and general convenience of operation. We describe the existing electronic equipment and its principles of operation which are not likely to change qualitatively within the projected lifetime of this textbook.

As with the previous editions, this text is intended for profesional scientists as well as undergraduate and graduate students in the sciences. It makes no claim of mathematical rigor such as would be needed by students of electronic engineering. An elementary knowledge of calculus and of complex variables is assumed, as is an introductory course in physics.

We acknowledge our indebtedness to the many students and colleagues over the years. We especially wish to express our gratitude to our respective wives for their understanding and help. We also thank Dr. Theodore P. Hoffman and others of the publishing staff at John Wiley & Sons, for their encouragement and assistance in our work.

This book has been composed by the authors on Word Perfect and transferred to Ventura Publisher with artwork in Corel DRAW! The typeface used is Baskerville by Bitstream, with display in ITC Korinna. Many of the figures involving waveforms and other mathematically defined curves were drawn from the original equations using BASIC with DesignCAD. The computer used was a 386/25 clone operating under DR DOS 6.0, and the printer was a Hewlett-Packard Laserjet III, used to prepare the camera-ready copy.

Basil H. Vassos
Galen W. Ewing

October 1992

Contents

Contents

Part III Computer Electronics

Contents

Appendixes

Analog and Computer Electronics
for Scientists

Part I

Analog Electronics

Chapter 1

Introduction

That electronics is of great significance in today's world is almost a truism. Electronics is inescapable. Think of satellite television, VCR recorders and video cameras, personal computers, and automotive cruise controls, as examples of some of the more visible contributions to our life style that depend on electronics. Any person with a scientific or technological bent must learn something about this discipline.

This book is intended to bring such a knowledge to scientists in various areas. It is not intended for electronic engineers; for them a series of many textbooks would be needed, each twice as big as this one.

Scientists and students, who are expected to be the chief users of this book, are generally interested in laboratory experiments; for them electronics is merely a means to an end. So we will commence with a brief discussion of typical laboratory environments.

Analog and Digital Domains

For the majority of experiments, the first stage is carried out in the **analog domain**, which means that measurable quantities describing the phenomena under study can have any value within a continuum extending over a wide range. (The only exceptions are those experiments which consist of counting, such as found in radioactivity measurements.) A modern computer, however, operates on signals that are described only by two permissible voltage levels, often zero and +5

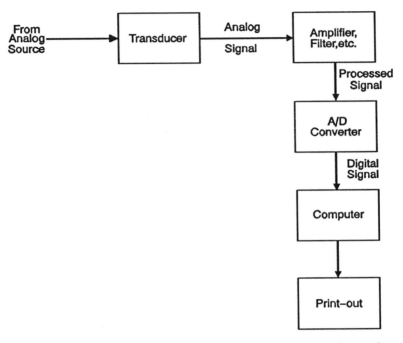

Figure 1-1 A chart showing the flow of information in a typical case, from the analog source to the print-out from a computer.

volts, but usually referred to as "high" and "low." These signals are said to be in the **digital domain.** Hence we must treat both areas of electronics, analog and digital, with more or less equal weight.

If we wish to manipulate continuously variable signals with a computer, we must supply some device that can convert analog signals to their digital counterparts, as will be described in Chapter 13. A simple analog-to-digital conversion, however, is rarely sufficient, and a number of successive steps are necessary. Often the first step takes the information from the "real world" and converts it into an analog electrical signal. The device to do this is a **transducer.** Examples of such transducers include thermocouples, photocells, and strain gauges. The signal produced by the transducer may need further alteration before conversion to the digital domain; for instance, a signal in the form of a current may have to be converted into a corresponding voltage. Only after all these steps have been completed can the signal finally be translated into the digital domain.

Once injected into the computer, the digital signal can be further manipulated in many different ways. Among others, it can be filtered to remove noise, and it can be compared to similar signals represent-

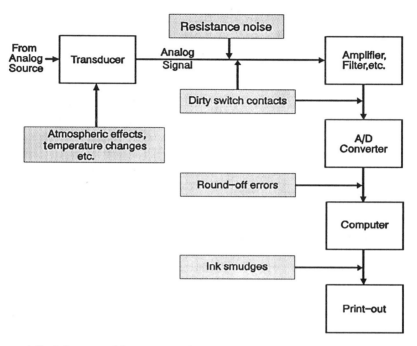

Figure 1-2 A few possible sources of noise intruding upon the information-flow system shown in Figure 1-1.

ing standard data through subtraction or ratioing. The processed data may then be stored in the computer's electronic memory for future use. The data may be reconverted to the analog form if it is to be utilized for recording or control.

The activity of the computer is controlled principally through the use of programs introduced by the operator by means of a keyboard. Figure 1-1 shows in schematic form the relationships between these several steps.

Noise

An electrical current consists essentially of electrons in motion. At ordinary temperatures there is considerable chaotic, random motion of electrons on which is superimposed the directed flow that we call a **current**. We often have need to measure very small voltages, and whenever we do, we are almost certain to find interference from the random electron motion. One very familiar form of this interference occurs in a radio receiver if the signal to be observed is too weak or if it disappears altogether; the receiver contains an automatic volume control that turns up the gain to its maximum trying to get a useful response, with the result that all we hear is an annoying hiss. The sound produced carries no information beyond calling to our attention that something is wrong. It consists of all sound frequencies over the region to which our ears are sensitive, and perhaps beyond.

This effect is called **electronic noise**, and is universally present in all electronic circuits including those that do not involve sound. Comparable deleterious effects are often seen in nonelectronic systems. Everyone knows that it is more difficult to understand a speaker in a crowded hall if the audience is noisy than if it is quiet. Figure 1-2 will suggest some ways in which noise can intrude itself into a system including both electronic and nonelectronic parts. In a later chapter we will explore the nature of noise and discuss techniques for minimizing its effects.

Computer Control

If it is desired to have an experiment controlled by a computer, it must be supplied with provision for reconverting signals from the digital to the analog domain, conceptually the reverse of the procedure described above. Here a transducer is needed that can produce a physical effect on its environment. Examples are light bulbs, valves, and heaters.

From the standpoint of the information flow through a system, a computer can be considered a single electronic circuit component, albeit a highly complex and powerful one. One of the objectives of this book is to exploit this concept and demonstrate that in order to use the computer to best advantage in connection with laboratory instruments, consideration must be given to the system as a whole.

First, however, we must lay some foundations in terms of the basic circuits assembled together to form an electronic system. This task we will begin in the next chapter.

Chapter 2

Circuits and Schematics

Circuits

Any electrical instrument, from the simplest to the most complex, must contain one or more **circuits**, defined as a closed loop of electrically conducting material through which a current flows. A basic circuit is made up of **conductors**, typically metallic wires, that offer only negligible hindrance to the flow of electrons. That means that all parts of a conductor can be considered to be at the same electrical potential. In addition to conductors, any useful circuit must contain a number of components that *do* affect the electron flow; hence they are generically called **impedances**.

In general a circuit will contain a source of voltage that causes the electrons to move, conductors to let them move freely, and impedances to introduce restrictions to the flow. Ohm's law gives the fundamental relation between these quantities, serving as a definition of impedance, Z:

$$Z = E/I, \quad \text{or} \quad E = ZI \tag{2-1}$$

where E is the source potential and I is the current produced. If E is in volts and I in amperes, then Z will be in ohms. An example of a typical circuit is shown in Figure 2-1.

7

Current can only flow in a circuit if it is driven by some sort of **power supply**. The simplest kind of power supply to describe is the battery. This is a device that converts chemical potential energy into electrical energy. It produces a characteristic potential (voltage) across its two terminals, about 1.5 volts in one common variety. When the switch is turned on, this voltage drives a current through the circuit in proportion to the total impedance, following Ohm's law.

Ohm's Law is perhaps the most basic of all concepts in the entire field of electricity and electronics. If the impedance consists of just a resistor[1], the symbol Z can be replaced by R, the resistance. In this case Eq. (2-1) reduces to the conventional form of Ohm's law:

$$R = E/I \quad \text{and} \quad E = IR \qquad (2\text{-}2)$$

In the simple circuit of Figure 2-1, it is realistic to consider the resistance of the wires to be negligible, so the R-value to enter into the equation is the combined resistance of the battery and the load (i.e., the impedance[2] of Figure 2-1). Often the internal resistance of the power supply is much smaller than the resistance of the load, and need not be considered. In some cases, as we will see later, it *is* important, and then can conveniently be represented separately in the circuit diagram, as in Figure 2-2. A circuit drawn in this way, to show the effective values of each component, is called an **equivalent circuit**. This equivalent-circuit technique will be used many times in this book.

The electronics of a typical laboratory instrument consists of many interconnected circuits. Each circuit element is characterized by its impedance, which measures the opposition to the current flow. For a simple resistor, the impedance is the same as the resistance. If the total impedance in the circuit of Figure 2-1 were zero, the power supply would be short-circuited, and an effectively unlimited current would

---

[1] Terms such as resistor, that end in -*or* refer to the physical object or component, whereas the corresponding terms ending -*ance* refer to the magnitude of the physical property.

[2] The term "impedance" is used in two senses: it describes an object, and also The term "impedance" is used in two senses: it describes an object, and also it describes a quantity denoted by Z. The word is also used in connection with certain fields other than electronics, notably acoustics and electrochemistry. Note that the term "impedor" does not exist.

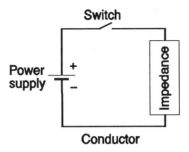

Figure 2-1 An elementary electric circuit. The power supply, here indicated by the conventional battery symbol, may in fact be a rectifier, that converts line AC to DC. The conductors can be simple wires, or can be metallic traces on an insulating board. The unit marked "impedance" is ordinarily the portion of the circuit that is driven by the power supply acting through conductors, in order to perform some useful function.

flow that would immediately burn something up, probably the power supply itself.

The load is usually the interesting part of the circuit. It is commonly constructed of combinations of **components**, each of which has its own impedance. The impedances of the individual components can be combined, following computational rules that will be detailed later, to give the total impedance of the circuit. Before we can describe circuits meaningfully, we must examine the principal types of components of which they are made.

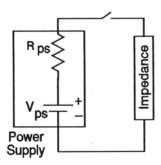

Figure 2-2 The same circuit as in Figure 2-1, but emphasizing that the power supply can be considered to consist of a voltage source V_{ps} and a resistance R_{ps} in series with it.

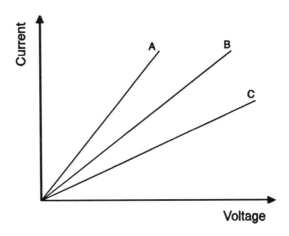

Figure 2-3 Current-voltage curves for three different linear impedances at a specific frequency.

Active and Passive Components

Active elements are devices that must be supplied with operating power. They are thus able to increase signals by amplification. Prime examples of active devices are the various types of transistors that will be treated in later chapters. Passive components, on the other hand, are those that do not require any external source of power other than the signal itself and only interact with impressed signals by diminishing the voltage or current level. This effect can include the selective removing of specific frequencies. The most familiar passive elements are resistors, capacitors and inductors.

Passive Components

An important property of these components is linearity. A component is said to be **linear** if the current passing through it is proportional to the driving potential. Figure 2-3 will exemplify this important concept.

To describe passive components more fully, we must take account of their impedance, their response to potentials and currents of varying frequency, and their ability to dissipate power.

Resistors

An ideal resistor is characterized by only a single operating parameter, its resistance, R. It is equally effective in AC and DC circuits. The resistance required in a particular place in a circuit can often be calculated through a direct application of Ohm's law.

Resistors are manufactured in a large selection of types, varying with respect to precision, stability, and physical form. For applications where the value of the resistor must be changed from time to time, variable resistors are available, usually called **potentiometers**, or "pots" for short. Most pots are provided with three external connections, one at each end of the resistor, the third connected to a sliding contact called a "wiper," so that any resistance value from zero to the maximum can be picked off with ease. Some precision pots are in the form of large boxes with calibrated dials, while others, such as those used as radio volume controls, are not calibrated.

The only parameter other than the resistance value, which must be considered in this context is the power rating. The power, in watts, dissipated by a resistor[3] is the product of the potential between its terminals, measured in volts, and the current flowing through it, in amperes:

$$P = VI = RI^2 = V^2/R \qquad\qquad (2\text{-}3)$$

Thus if it is desired to produce a potential drop of 5.00 V with a current of 30.0 mA (0.0300 A), a resistor with a value of $V/I = 5.00/0.0300 = 167\ \Omega$ is needed. The power dissipated will be $P = VI = (5.00)(0.0300) = 0.150$ W. The next larger standard resistor[4] is rated at one-quarter watt, so this should be selected.

Capacitors

A capacitor consists of two conductors separated by a thin insulating layer (the dielectric). There are three principal types: (1) **solid dielectric**, in which the separating layer may be made of various plastic films, mica, paper, or ceramic; (2) **electrolytic**, composed of aluminum or tantalum foil covered by a thin film of oxide produced electrolytically; and (3) **air-dielectric**, best known as the variable capacitors used in the tuning circuits of radio receivers.

---

[3] It is interesting to note that only in resistors is electrical energy transformed directly into heat. In (ideal) capacitors and inductors no heat is generated.

[4] Resistors are normally available in specified power ratings such as 1/8, 1/4, 1/5, 1, 2, 5, and 10-watt. The resistance values are also standardized, as detailed in the Appendix.

Capacitors are rated by their **capacitance**, measured in farads, or more often, in microfarads[5] (μF) or picofarads (pF). Fundamentally, the capacitance can be defined in terms of the area of the conducting plates and the dielectric constant of the nonconducting spacer, but this relation is seldom of practical use in electronics. More important for our applications is the relation that gives the value of the capacitance C in terms of the charge Q on each plate resulting from an applied voltage V

$$C = \frac{Q}{V} \qquad\qquad (2\text{-}4)$$

The several types of capacitors mentioned above differ in their secondary characteristics, which may be of great practical importance. For instance, an electrolytic unit (unless it is marked "NP" or "nonpolarized") must be connected in such a way as to maintain a DC voltage of proper polarity across it; wrong polarity may well result in explosive failure. These units can be obtained with very large capacitance at low cost, and are especially useful in power-supply circuits. Electrolytic capacitors are less than ideal; they behave as though they had a resistor of a few megohms in parallel (the leakage resistance), and this may limit their areas of application. There are two types of electrolytic capacitors, aluminum, the common type, and tantalum, characterized by a smaller size and more favorable leakage characteristics.

Of the nonelectrolytic types, those with dielectric of polystyrene or polycarbonate are preferred over paper, since their capacitance varies little with time and their leakage resistance is very high. Ceramic disk capacitors are often more convenient than others, and of low cost.

Inductors

A simple inductor consists of a coil of wire containing a core of iron or ferrite,[6] or simply air. It is characterized by its **inductance** (symbol

---

[5] The lack of Greek letters on conventional typewriters has given rise to the unfortunate practice of designating microfarads by "uF."

[6] Ferrite is a ceramic-like material made by sintering together various heavy metal oxides. It has particularly favorable magnetic properties and extremely high electrical resistance.

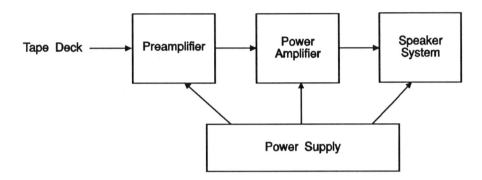

Figure 2-4 A typical block diagram: an audio amplifier for a tape player.

L), expressed in henries[7] (H). Inductors are frequency dependent. The iron-core types, called "chokes," have large inductance and are used principally in high-current applications. Ferrite inductors are important in high-frequency equipment, such as the antenna coil in a radio receiver. Transformers are inductors with two or more windings. They are useful in power supplies, and they are also used to couple two AC circuits without direct connection.

Real Components

Actual components, such as one might find in laboratory instruments, nearly always include features characteristic of all three classes, but with one predominating. Thus a resistor, particularly if it is made of a coil of resistance wire, can be expected to have a certain amount of both inductance and capacitance, which may effect its high frequency applications. Inductors, being made of wire, have by necessity a finite resistance. Capacitors made by coiling strips of metal foil together with a film dielectric, have finite inductance; because of this effect, it is considered good practice to bypass large capacitors with small ceramic units.

Circuit Diagrams

There are several distinct types of circuit diagrams; each has its own area of utility. First, we can consider a **block diagram**, which does not contain information about specific components but only functional segments (an example is seen in Figure 2-4). Next is a type of diagram

---

[7] The henry is defined as the inductance of a coil across which one volt is induced by a current changing at a constant rate of one ampere per second.

usually called a **schematic**, designed to show the flow of information (or of signals) from one component to another. It may have little or no resemblance to the actual layout of the parts on a circuit board. Each component, either active or passive, is given a characteristic symbol to represent it in the schematic, the symbols being interconnected by lines indicating the wires in the real circuit. Figure 2-5 shows the usual symbols for various types of components. Both block diagrams and schematics are used by a designer in planning the circuits, and they are the easiest for a neophyte to understand. All the circuit diagrams used so far in this book, and the majority of those yet to come, are of these types.

Finally, there is a kind of diagram that uses boxes (usually rectangles or triangles) to represent integrated circuit components, often with individual pin designations. This type of diagram is a useful aid in thinking through a proposed system that combines a number of components, but it presupposes some familiarity with what is inside the boxes.

Another class of diagrams is used primarily in the manufacture and repair of electronic instruments. It shows individual components, laid out as they are in the actual circuit, with the connecting wires shown in detail. It is difficult to understand the action of a circuit by examining this type of diagram.

Drawing good schematic diagrams, that will be easily understood by others, takes considerable skill and practice, but the art is worth cultivating. It is well to arrange components so that the flow of information is more or less linear, from left to right, with subsidiary circuit segments, such as the power supply, above or below. Lines representing wires should be either horizontal or vertical, with sharp angles as needed. Crossing over of two lines should be avoided where possible, but where crossovers are necessary, it should be made perfectly clear whether or not the wires make electrical connection. It is recommended that the student give careful attention to the many circuits of various types used in this and other books, noting techniques used in their drawing.

The diagram of a fairly complicated electronic instrument can often be broken down into parts, each of which can be understood readily enough. Then these parts will generally fall into place, permitting a comprehensive understanding of the whole instrument.

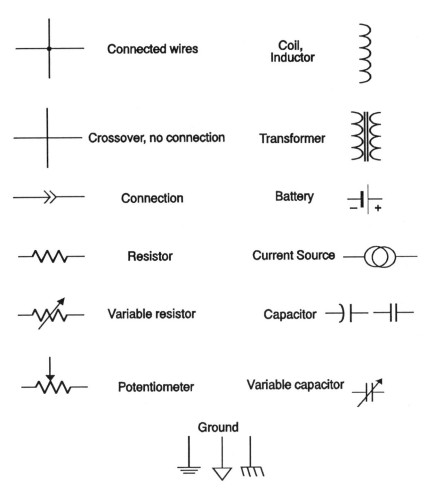

Figure 2-5 Some common symbols used in electronic circuitry. Whichever symbol is chosen for ground, it should be used consistently. Symbols for active devices will be introduced later.

Circuit Ground

Special comments are needed with respect to the concept of ground.[8] It is usual practice to measure all DC voltages in a group of circuits

---

[8] The name derives from the early days of radio when circuits were literally grounded, attached to a metal rod driven deeply into the earth. This type of grounding is still important in electrical power transmission, but is only encountered in laboratory electronics in connection with safety precautions.

with respect to a common zero, and this zero level is designated as the **ground** of the system. There are often many points in the circuit that are established at ground by direct interconnection, and to avoid confusion, each such point in the schematic is given the ground symbol, rather than showing a profusion of wires connecting them. The current passing through a grounding wire may cause a small IR drop in potential, so that various "ground" points may not actually be at exactly the same potential. Precautions must be taken to avoid connecting two points nominally at ground potential but actually differing by even a small amount, because large currents may flow, due to the small resistance of the wires.

Direct and Alternating Current Circuits

When a potential is applied abruptly to a circuit, it takes a certain length of time for the resulting current to achieve its steady state. During this time, the voltages and currents in various parts of the circuit are changing, an effect called a **transient.** This transient can be very short if the circuit is made up entirely of resistors, but can be quite long if inductors or capacitors are present. After the transient has died away, the current and voltage remain constant until some change is made in the circuit connections. Such constant conditions are referred to as **direct current** (DC).[9]

On the other hand, **alternating currents** and voltages require three parameters for complete description: the amplitude A, the period T (or its reciprocal, the **frequency** f), and the phase ϕ. Figure 2-6a shows the definitions of amplitude (in volts or amperes) and period (in seconds) of a sine wave. The **period** is defined as the elapsed time between two equivalent points in the wave (i.e. two crests or two zero-crossings). The wave in Figure 2-6a has a period of 1/2 s. The frequency has the units of reciprocal seconds (hertz), 2 Hz in the example. For reasons that will be made clear shortly, it is convenient to define another symbol for frequency, denoted by ω, with the units radians per second. The relation between these two is given by the equation

$$\omega = 2\pi f \qquad (2\text{-}5)$$

---

[9] The phrase "DC voltage" is strictly a misnomer, since it stands for "direct current voltage."

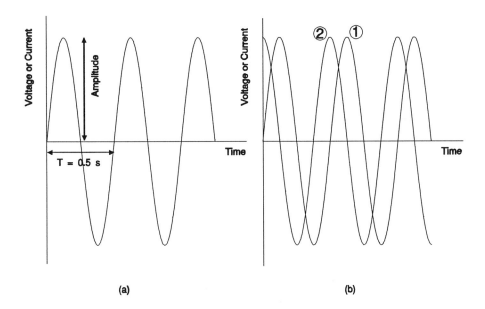

Figure 2-6 (a) An example of an AC signal; the frequency is 2 Hz, the period, one-half second. (b) Two signals of the same frequency and amplitude. The time-delay between the two signals generates a difference in phase.

Figure 2-6*b* shows that two signals of the same frequency and amplitude can have different values at any given instant; they are said to differ in **phase.** Signal 2 in the figure reaches any particular point earlier in time than does 1; hence 2 is said to **lead** and 1 to **lag** in phase.

Since a sinusoidal signal has different amplitudes at different moments, it is customary to define some type of average. The arithmetic mean of a sine wave is zero; hence the conventional average is useless. In its place a type of average called the **root-mean-square** (RMS) voltage (or current) is used, given by

$$V_{RMS} = \sqrt{\overline{V^2}} \tag{2-6}$$

where the bar in $\overline{V^2}$ indicates the time average taken over an integral number of cycles. For sine waves, the RMS value is $1/\sqrt{2}$ or 70.7% of the peak value A. The line voltage is conventionally expressed in terms of RMS quantities: thus "115 V RMS" describes a wave with an amplitude of $115\sqrt{2} = 163$ V. The excursions are alternately positive

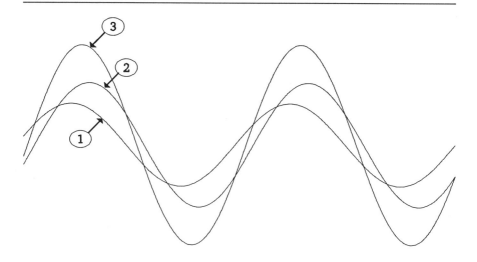

Figure 2-7 Demonstration of the addition of two sine waves to give a third. Curves 1 and 2 are sine waves of the same frequency but different amplitude and 30° apart in phase. Curve 3 is the sum of the first two; it has the same frequency but is different from either of the others in amplitude and phase.

and negative with respect to ground, giving 326 V peak-to-peak. For alternating currents other than sine waves, the factor varies from $\sqrt{2}$. An operational definition says that the RMS current can be defined as that value of a DC current that produces an equal heating effect in a resistor.

Sinusoidal currents or voltages have a number of unique properties. For instance, any sum of sine waves *of the same frequency* reduces to a single sine wave with phase and amplitude determined by its components (see Figure 2-7). Another interesting property is that all integrals and derivatives of sine waves are also sine waves, differing only in amplitude and phase from the original.

Phase is not an absolute quantity, but must always be referred to some arbitrary standard, the only measurable quantity being the difference between the phases of two voltages or currents. Capacitors and inductors in a circuit affect the phase of an alternating current passing through them in various ways, hence it is convenient to assign a zero phase angle to the *source* voltage, and to describe the phases at any point in the circuit with respect to this as a standard. The voltage at any point can be described by an equation of the type

$$V = A \sin{(2\pi ft + \phi)} = A \sin{(\omega t + \phi)} \tag{2-7}$$

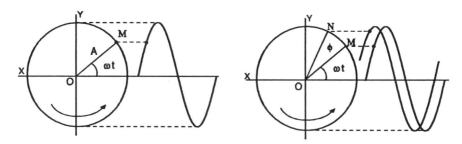

Figure 2-8 (a) Generation of a sine wave from circular motion. The radius OM (of magnitude A) can be considered to be a vector rotating in the counterclockwise direction. The position of point M, projected on the vertical axis, becomes a sinewave when represented as a function of time. $y = A \sin \omega t$. (b) Two sinewaves differing in phase, $A \sin \omega t$ and $A \sin (\omega t + \phi)$.

where A is the amplitude, t the time, and ϕ the phase difference with respect to the reference, in radians or degrees. (Strictly speaking, ϕ should be given only in radians, but degrees are frequently substituted for convenience.)

A sine wave can be interpreted in terms of a rotating vector. Consider vector **OM** in Figure 2-8a, rotating counterclockwise around the point O. The tip of the vector, at M, moves up and down from the center line with simple harmonic motion so that its vertical position traces a sinewave as a function of time, in accordance with the relation $y = \sin \omega t$.

If now we consider two such vectors, displaced from each other by an angle ϕ, as in Figure 2-8b, rotating at the same rate, it can be seen that two sinewaves result, separated by a constant time difference which can be expressed in terms of the phase angle ϕ. If the first curve, generated by the vector **OM**, is $y = \sin \omega t$, then the curve generated by the vector **ON** gives $y = \sin (\omega t + \phi)$. If $\phi = 180°$ (π radians), the two waves are exactly opposite each other. The waves shown previously in Figure 2-6b differ by 90° or $\pi/2$ radians.

Impedance

As we have seen, impedance can be defined as the ratio of the voltage across a component or circuit to the current passing through it. The current and voltage can be either AC or DC (note that DC can be considered to be the limit of AC as the frequency approaches zero).

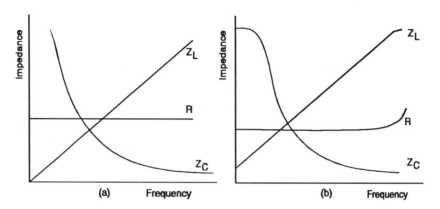

Figure 2-9 (a) The frequency dependence of impedance for ideal resistors, inductors and capacitors. (b) The behavior of real components, showing deviations at zero frequency (DC) and at very high frequencies. The exact shapes in (b) depend on the physical construction.

For pure resistances, the impedance is a constant identical with the resistance, but for capacitances and inductances the response is frequency dependent. Any relationship concerning them must involve the frequency. The additional information that is required can best be handled in terms of complex numbers. The notation includes the symbol j for the imaginary operator, the square root of minus one.[10] In this notation, the impedance of a capacitor is given by

$$Z_C = X_C = \frac{1}{j\omega C} = -\frac{j}{\omega C} \qquad (2\text{-}8)$$

and that of an inductor by

$$Z_L = X_L = j\omega L \qquad (2\text{-}9)$$

In these relations, the symbol X represents a quantity called the **reactance**, a special class of impedance that refers only to capacitance and inductance. The strong dependence on the frequency is illustrated in Figure 2-9.

[10] Mathematicians usually express the imaginary operator by the letter i. In electronics, however, this is likely to be confused with the symbol for the current, so j is substituted.

A table in the Appendix gives the impedance of capacitors at a number of frequencies, in a form that is convenient for calculations. Since inductors are seldom used in the low-frequency circuitry that we are mostly dealing with, a comparable table for the impedances of inductors is not included.

Since the current in a capacitor is given by $I = C(dE/dt)$, it follows that a sine-wave voltage results in a cosine-wave current. Thus if $E = A \sin \omega t$, then $I = A\omega \cos \omega t = A\omega \sin (\omega t + 90°)$. Thus in a capacitor, the current **leads** the voltage by 90° ($\pi/2$ radians), whereas for an inductor the opposite is true and the current **lags** by 90°.

Reactance

The reactance is that portion of the impedance that is due to the nonresistive properties of the component and is described by an imaginary number. If the component also has resistance, then its total impedance is the combination of the reactance and resistance, both of which have the same units (ohms). Consequently the impedance is generally a complex quantity, of which the pure resistance is the real part, while the reactance is the imaginary part. For example, a particular impedance might be represented by the quantity $Z = 50 + j60$, which would mean that the resistance is 50 Ω, and the reactance 60 Ω. A positive imaginary part indicates an inductive component, the negative sign a capacitive component. An ohmmeter measures only the real component (except if capacitors are present in series, thus preventing any DC current flow).

Series Combinations

The net resistance of two resistors connected in series is equal to the sum of the individual values. This can be visualized readily by means of an equivalent circuit, a concept introduced earlier. Figure 2-10 shows two circuit fragments with their simplified equivalents. In Figure 2-10a, the total resistance of the combination of two resistances is merely equal to their arithmetic sum, $R_T = R_1 + R_2$. In Figure 2-10b, the comparable combination involves combining R and X_C, the capacitive reactance. The general relation for a series combination of impedances gives the total value, Z_T, as

$$Z_T = Z_1 + Z_2 + Z_3 + \dots \qquad (2\text{-}10)$$

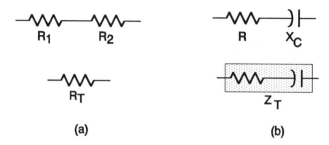

(a)   (b)

Figure 2-10 (a) A series-connected pair of resistors R_1 and R_2, together with their one-component equivalent R_T. (b) A similar pair of components, one a resistor the other a capacitor, with their equivalent impedance Z_T.

Note that this treatment requires that the equivalent impedance Z_T has only *two* terminals; this means that no connection to any other component, power supply or ground, can be made to any point between the members of the series string without invalidating the equivalence to Z_T.

The combined impedance shown in Figure 2-10*b* is given by the complex number Z_T

$$Z_T = R + X_C = R + 1/(j\omega C) \qquad (2\text{-}11)$$

These relations are best considered as the lengths of vectors in the complex plane, for which the coordinates are shown in Figure 2-11. By convention, inductive impedance is plotted upward and capacitive downward on the imaginary axis.

As with all vectors, impedances must be combined vectorially. To obtain the length of Z_T, we can apply the Pythagorean formula, as in the following example. Consider a series combination of a 10-μF capacitor and a 100-Ω resistor, energized at a frequency of 1000 Hz. The total impedance is given by[11]

$$Z = \sqrt{R^2 + 1/(2\pi f C)^2}$$

$$= \sqrt{(100)^2 + (6280 \times 10^{-5})^{-2}} = 101.3 \ \Omega \qquad (2\text{-}12)$$

---

[11] This can be interpreted as a conversion from Cartesian to polar coordinates.

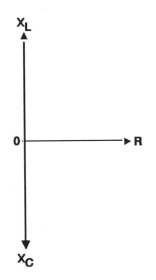

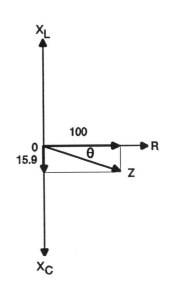

Figure 2-11 The coordinate systems for the depiction of complex impedances. The horizontal axis measures the real component, whereas the vertical axis shows the imaginary.

Figure 2-12 Method of determinig the impedance of the RC combination. The phase angle θ can be calculated by $\theta = \tan^{-1} \dfrac{X}{R}$; in this case $\theta = -9°$.

The presence of the operator j prevents us from adding the two terms directly. If simple arithmetic summing rather than vectorial addition were attempted, the calculation would yield an incorrect answer:

$$Z \neq R + 1/(2\pi f C)$$
$$\neq 100 + 1/(6280 \times 10^{-5}) = 115.9 \ \Omega \qquad (2\text{-}13)$$

Figure 2-12 shows how the two components of impedance combine to give a single resultant vector, Z_T, which is neither purely real nor purely imaginary. The angle θ is equal to the phase difference between voltage and current. The resultant impedance lies in the fourth quadrant because the sign of X_C is negative. This does not confer a negative sign upon the impedance, however, because the *length* of a vector can never be negative.

An important application of the series connection of impedances is found in the **voltage divider.** Clearly the current through a string of components must be the same for all of them, since the electrons

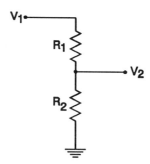

V_1
R_1
V_2
R_2

Figure 2-13 A voltage divider. The resistors shown can be replaced by reactive impedances in which case the response is dependent on the frequency.

have nowhere else to go. In contrast, the voltages developed across the components will in general be different, each governed by Ohm's law.

Making use of this fact permits a very convenient way of producing a desired potential, as illustrated in Figure 2-13. It is readily shown by application of Ohm's law that the voltage V_2 is given by the following relation, called the voltage-divider equation:

$$V_2 = V_1 \frac{R_2}{R_1 + R_2} \qquad (2\text{-}14)$$

This is a most important relation that will be invoked many times in this book.

A frequent application of a voltage divider is to adapt a circuit to operate an analog recorder that is too sensitive. Figure 2-14 shows, as an example, a circuit that gives a 15-V full-scale output driving a recorder with a full-scale deflection of only 10-mV. Substituting these voltages in Eq. (2-14) gives

$$0.010 = 15 \frac{R_2}{R_1 + R_2}$$

We can arbitrarily choose a value for one of the resistors and calculate the other. Inspection shows that R_1 must be much larger than R_2, so we can select a large value for it, say, 100 kΩ. Calculation then gives $R_2 = 70$ Ω. Figure 2-14b shows a potentiometer replacing the smaller resistor, to make a variable voltage divider. This circuit has the advantage that it can be adjusted to give the exact 10 mV output. The total resistance of the potentiometer is not critical; 100 Ω would be

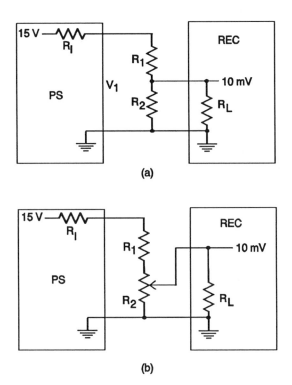

(a)

(b)

Figure 2-14 The use of a voltage divider to interface a 15 V source (PS) with a 10 mV recorder (REC). It is assumed that the internal resistance of the source (R_i) is negligible compared to R_1, and that the load resistance R_L is much larger than R_2. In (a) the resistors are fixed, whereas in (b) they are made variable to permit fine adjustments of the ratio.

satisfactory in this example, giving sufficient leeway on both sides of the calculated 70 ohms.

Power in Series Combinations

In a practical application of the relations described above, the power-dissipating capability of the components must be given due consideration. For example, suppose two resistors R_1 and R_2, each of 1 kΩ, are connected in series with a power supply giving 20 volts. The current will be $I = V/R_T = (20 \text{ V})/(2000 \text{ }\Omega) = 0.010$ A. The power dissipated as heat in each resistor will be $P = VI = RI^2 = (1000 \text{ }\Omega)(0.01 \text{ A})^2 = 0.1$ $W = 100$ mW for a total of 200 mW. If the two resistors were 100 Ω and 1900 Ω, respectively, the total resistance would still be 2 kΩ, and the current would still be 0.01 A, but the power dissipated in R_1 would

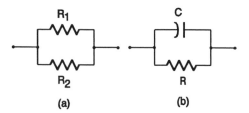

Figure 2-15 Impedances in parallel: (a) two resistances; (b) one resistance and one capacitance.

be $P_1 = (100)(0.01)^2 = 0.01$ W $= 10$ mW, whereas the power in R_2 would be $P_2 = (1900)(0.01)^2 = 0.19$ W $= 190$ mW. Hence the larger resistor would produce more heat than the smaller by a ratio of 19 to 1, but the total power would remain unchanged.

Parallel Combinations

When impedances are combined in parallel, as in Figure 2-15, the arithmetic becomes a little less direct. In the case of two resistances (Figure 2-15a), application of Ohm's law tells us that a voltage V impressed across the circuit will give currents through the two resistors of $I_1 = V/R_1$ and $I_2 = V/R_2$, respectively. The total current I_T is given by $I_1 + I_2 = V(1/R_1 + 1/R_2)$. The overall resistance, R_T, is given by V/I_T, which is equal to $1/(1/R_1 + 1/R_2)$, so that the desired parallel resistance is

$$R_T = \frac{R_1 R_2}{R_1 + R_2} \qquad (2\text{-}15)$$

In the case of the parallel RC circuit of Figure 2-15b, the same arithmetic holds, if we replace one of the R's by $1/j\omega C$:

$$Z_T = \frac{\dfrac{R}{j\omega C}}{R + \dfrac{1}{j\omega C}} = \frac{R}{j\omega RC + 1} \qquad (2\text{-}16)$$

For the analogous case of an inductor and resistor in parallel, similar considerations show that

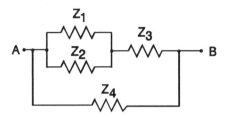

Figure 2-16 A network of impedances.

$$Z_T = j\omega \frac{RL}{R + j\omega L} \qquad (2\text{-}17)$$

The general formula for parallel combinations of impedances is

$$\frac{1}{Z_T} = \frac{1}{Z_1} + \frac{1}{Z_2} + \cdots \qquad (2\text{-}18)$$

In the case of only two components, this reduces to

$$Z_T = \frac{Z_1 Z_2}{Z_1 + Z_2} \qquad (2\text{-}19)$$

In more involved circuits, such as that shown in Figure 2-16, the equivalent impedance between points A and B can be found most readily by combining the elements in pairs. For instance, first calculate the impedance of Z_1 and Z_2 in parallel, then add to this the value of Z_3, and finally determine the impedance of the Z_1, Z_2, Z_3 group in parallel with Z_4. Any of these Z's could be resistances or reactances.

Power in AC Circuits

In a DC or an AC circuit that contains only resistors, the power dissipated as heat is merely the product of the voltage across the circuit and the current flowing through it

$$P = VI \qquad (2\text{-}20)$$

where all symbols refer to DC or RMS quantities.

With AC circuits including reactances, however, this definition must be modified. The *VI* product, called the **volt-ampere power**, now represents the *maximum* attainable power dissipation.

The actual power (the **effective power**) must be determined by vector treatment, and involves the phase angle between the current and voltage

$$P_{(actual)} = V_{RMS} \times I_{RMS} \times \cos\theta \qquad (2\text{-}21)$$

The quantity $\cos\theta$ is defined as the **power factor**. Its value can be found by taking the ratio of the actual power to the volt-ampere power, or theoretically, from the ratio of reactance to resistance, $\theta = \tan^{-1}(X/R)$. The expression for power dissipated is commonly given as

$$P = VI\cos\theta \qquad (2\text{-}22)$$

in which RMS values are implicitly assumed.

Since the phase angle between voltage and current in a capacitor is 90°, and the cosine of 90° is zero, it follows, as expected, that the power dissipated in an ideal capacitor or inductor is also zero.

Equivalent Circuits

An equivalent circuit is a substitute assembly that has the same current-voltage characteristics as the original circuit when connected between the same contact points. For example, two resistors in series can be replaced by a single resistor with a value equal to their sum. It is impossible to distinguish between the two equivalents on the basis of total current and voltage measurements, even though they may differ in size and number of components.

The analysis of passive electronic circuits is greatly simplified by the use of such equivalents. A network consisting of a number of resistors, some in series and others in parallel, can be replaced by a single resistor, as shown earlier in connection with Figure 2-16. Similarly a circuit containing reactive components along with resistors can be replaced by one resistor and one reactive component to form an equivalent with the same electrical characteristics as the original.

Given such a combination of components, the total impedance can be obtained by the series and parallel rules. The resulting expression can be quite complicated. Nevertheless, after separating the real and imaginary parts, an equivalent circuit can be made by using a resistor with the value of the real part in series with a reactance that has the value of the imaginary part.

It is important to be able to interpret equivalent circuits in which some components have much higher impedance than others. In series combinations the smaller value can be neglected, whereas in parallel combinations the larger can be ignored. The criterion depends on the acceptable error of measurement. For example, if the acceptable error is 10% for a series combination of 330 Ω and 33 Ω, it can be replaced by the 330-Ω resistor alone. If the permitted error is 1 percent, then only a 3.3-Ω (or lower) resistor can be neglected. It turns out that in parallel combinations, the ratios are the same as in the series case, but the *larger* resistor can be ignored. The same procedure applies equally to combinations of impedances in general. It must be remembered that reactances are frequency-dependent, so that a series capacitor can be neglected at high frequencies, while a parallel capacitor can be neglected at low frequencies.

Thévenin and Norton Equivalents

The concept of equivalent circuit can be extended to the case where energy sources are also present. In the material that follows, two types of ideal sources will be dealt with: (1) the **voltage source,** which must have a *low* internal resistance (ideally zero), and (2) the **current source**, with a *high* internal resistance (ideally infinite). A real source such as a battery represents an intermediate case, much closer to a voltage than a current source. It may have, typically, a fraction of an ohm internal resistance. It is essential that the relations between these kinds of sources be thoroughly understood.

There are two types of equivalents involving electrical energy sources, known respectively as the Thévenin and the Norton equivalents, the former given here in a slightly modified form:

> **The Thévenin Theorem:** Any two-terminal linear electronic network that contains voltage and/or current sources of a single frequency (or DC) can be replaced by an equivalent circuit consisting of a single voltage source and series impedance.

Note the restriction that the circuit has only two terminals. According to this theorem, the functioning of any combination of

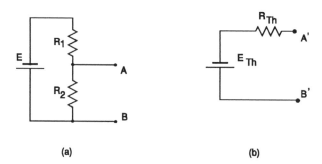

Figure 2-17 Example of Thévenin equivalent.

resistors, capacitors, and inductors, together with voltage or current sources (of a single frequency), as observed between two points A and B, can be represented by only one complex impedance in series with a single ideal voltage source (AC or DC). Thévenin equivalent circuits can be written for any linear network with the limitations noted above. They are most often used to describe energy sources, such as power supplies and oscillators.

The general method used in finding the Thévenin equivalents depends upon the fact that the equivalent circuit must be valid no matter what is connected to the output terminals, including conditions of open-circuit and short-circuit.

Let us consider first the application of this theorem to the purely resistive voltage divider, shown in Figure 2-17a. We now draw Figure 2-17b and define it as the Thévenin equivalent of (a). Then any measurement made at terminals A and B must be reproduced exactly at A' and B'. If we measure the open-circuit voltage between A and B with a high-impedance voltmeter, the reading will be given by the voltage-divider equation

$$E_{AB} = E \frac{R_2}{R_1 + R_2}$$

The equivalent measurement of the circuit in Figure 2-17b will give

$$E_{A'B'} = E_{Th} - IR_{Th}$$

The last term is negligible and, since the two circuits are equivalent, it follows that

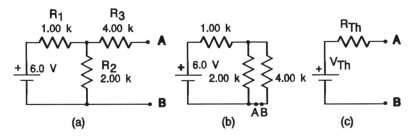

Figure 2-18 An example of Thévenin analysis. The original circuit is shown in (a); (b) is the equivalent to (a) when the output is short-circuited; (c) is the final equivalent circuit.

$$E_{Th} = E\frac{R_2}{R_1 + R_2}$$

Thus open circuit measurements give directly the Thévenin equivalent voltage.

On the other hand, we can determine the short-circuit current by connecting a low-impedance *ammeter* between A and B. This effectively removes R_2 from the circuit, and the current is given by

$$I_{short\ circuit} = E/R_1$$

Transfering this to the circuit in Figure 2-17*b*, should give the same short circuit current, and the Thévenin resistance[12] is merely the ratio of voltage to current

$$R_{Th} = \frac{E_{Th}}{I_{shortcircuit}} = \frac{R_1 R_2}{R_1 + R_2}$$

Now let us consider a further example, as in Figure 2-18*a*. The actual circuit can be simplified as follows: First, notice that resistances R_1 and R_2 form a voltage divider. In open circuit, the potential at the node where the resistances join is

[12] An alternative method for determining the Thévenin resistance is to calculate the output resistance with the circuit simplified by replacing voltage sources by short circuits and current sources by open circuits. This can be verified by application to the circuit of Figure 2-17.

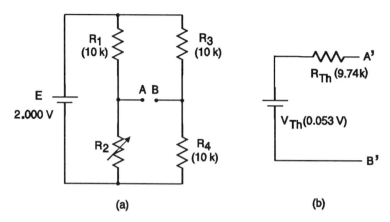

Figure 2-19   (a) A Wheatstone bridge and (b) its Thévenin equivalent.

$$V_1 \frac{R_2}{R_1 + R_2} = 4.0 \text{ V.}$$

While the voltage between points A and B is being measured with our high-impedance voltmeter, there will be no potential drop across R_3, so the measured voltage will be that across R_2, which becomes the Thévenin voltage, $V_{Th} = 4.0$ V. Now we must determine the short-circuit current. With the output shorted, the circuit takes the form of Figure 2-18b, and it is convenient to treat resistors R_2 and R_3 as one, using the formula for parallel impedances, $1/R_{parallel} = 1/2000 + 1/4000$, which gives $R_{parallel} = 1.33$ kΩ. The current that will flow from the battery is therefore given by $I = V/(R_1 + R_{parallel}) = 6.0/2.3$ k = 2.6 mA. The voltage across the combination of R_2 and R_3 is 2.6 mA × 1.33 kΩ = 3.45 V. The output short circuit current is 3.45/4000 = 0.86 mA. The value of R_{Th} is then calculated to be $V_{Th}/I_{Th} = 4/0.86 = 4.6$ kΩ.

The example just treated is an easy one, for illustrative purposes. Let us now consider a Wheatstone bridge circuit, Figure 2-19a. When the bridge is open-circuited, with nothing connected to points A and B, it acts as a pair of voltage dividers. If we suppose that R_2 is originally set at 9000 Ω, the voltages will be

$$V_A = 2.00 \frac{9000}{9000 + 10,000} = 0.947 \text{ V} \qquad (2\text{-}23)$$

$$V_B = 2.00 \frac{10{,}000}{10{,}000 + 10{,}000} = 1.000 \text{ V} \qquad (2\text{-}24)$$

Hence the voltage between A and B is $V_A - V_B = -0.053$ V. Since the open-circuit *current* is zero, there can be no drop of voltage across the Thévenin resistance, hence V_{Th} must be equal to $V_A - V_B$.

The value of R_{Th} can be calculated by considering the bridge with the battery replaced by a direct connection. This places R_3 and R_4 in parallel, giving an equivalent resistance of 5.00 kΩ. Similarly R_1 and R_2 give a combined resistance of 4.74 kΩ. The Thévenin resistance is the sum of $5.00 + 4.74 = 9.74$ kΩ. The complete equivalent circuit is shown in Figure 2-19*b*. As the bridge is brought to balance by increasing R_2, the Thévenin equivalent values tend toward their limits of $V_{Th} = 0$ V and $R_{Th} = 10$ kΩ.

The values of the Thévenin impedance and voltage of a physical unit can be determined experimentally through external observations. The Thévenin voltage is determined directly by measuring the voltage appearing at the terminals under no-load conditions. The Thévenin impedance can be found by attaching a variable resistance to the output terminals, and adjusting it until the potential observed is exactly halved; the resistance required then becomes numerically equal to R_{Th}.

Note that there can be an infinity of actual circuits all of which have the same Thévenin equivalent. As a consequence, Thévenin equivalents give little information about the actual implementation of the circuit.

Norton Equivalents

The Norton Theorem: Any two-terminal linear electronic network containing voltage and current sources of only one frequency (or DC) can be replaced by an equivalent circuit consisting of a single current source and an impedance in parallel.

This theorem is the counterpart of the previous one but with a current source instead of the voltage source, the single impedance being placed in parallel with the source rather than in series with it. The relation between the two equivalents is illustrated in Figure 2-20.

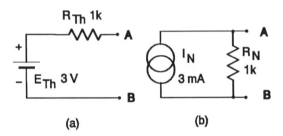

Figure 2-20 (a) A Thévenin circuit, and (b) its Norton counterpart. R_{Th} and I_N are related through Ohm's law.

It can be easily ascertained that any measurements at points A and B would detect no difference between the two circuits.

Some actual circuits are best considered in terms of voltage sources, others as sources of currents. The most appropriate of the two equivalent circuits can be chosen to best fit the system being studied.

Kirchhoff's Laws

In this section we discuss two procedures for calculating the various voltages and currents in a **network**, defined as a set of interconnected components involving voltage and current sources. Only linear components will be considered.

A network may contain one or more closed **loops,** each of which can be thought of as a circuit. Figure 2-21 is a diagram of a network containing two inner loops in addition to the major outer loop. The points where loops join are designated as **nodes.**

As seen above, equivalent circuit methods can reduce a two-terminal network to only two components. However all current, voltage,

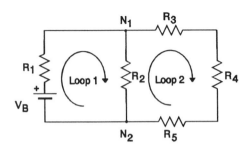

Figure 2-21 A typical network with two inner loops. The outer circuit, containing all components except R_2, constitutes the third loop. The points marked N_1 and N_2 are nodes.

and power information about individual components is lost. A more elaborate treatment is needed to fully explore the properties of a network.

A set of components connected in loops can be analyzed in either of two ways, in terms of voltages (**Kirchhoff's voltage law**, KVL) or currents (**Kirchhoff's current law**, KCL). According to the first of these laws, the sum of all voltage sources and potential drops within a loop must sum to zero: $\Sigma V = 0$. The second law states that the sum of all currents entering a node must be zero, $\Sigma I = 0$.

The voltage law can be clarified by noting that voltage is actually a difference of potential, independent of path. For instance, in a loop ABCD, the potential difference between A and D can be obtained directly or, equally well, by going through B and C.

Similarly the current law can be rationalized as a consequence of the conservation of charge, recalling that current measures the number of charges flowing per unit time. The number of charges leaving a given node must equal the number entering. This null balance applies equally to currents, hence the Kirchhoff current law follows. As in all electronic applications positive current is taken by convention as the direction of flow of positive charges, the opposite to the flow of electrons.

To apply the first law (KVL), we must start at the voltage source and proceed around the loop in the direction that *positive* current would flow[13], taking the voltage drops across passive components as negative. Voltage sources are taken as positive if the current crosses them from minus to plus and negative in the opposite case.

Applying this to loop 1 in the figure, we can write

$$V_B - R_1 I_1 - R_2 I_1 + R_2 I_2 = 0 \qquad\qquad (2\text{-}25)$$

where I_1 is the current flowing through loop 1 and I_2 that through loop 2. The contributions of the two currents in R_2 are considered independently, even though physically only a single current is flowing. For the second loop, the voltage drop across R_2 produced by current I_1 is taken to be the source of voltage driving the loop. The KVL then gives us

---

[13] If this this is not known beforehand, either direction may be chosen; a resulting negative sign indicates that the "wrong" direction was chosen.

$$R_2I_1 - I_2(R_2 + R_3 + R_4 + R_5) = 0 \qquad (2\text{-}26)$$

In applying this law it is necessary to be careful in assigning algebraic signs to the various quantities. We can also apply the KVL to the major loop, getting

$$V_B - I_1R_1 - I_2(R_3 + R_4 + R_5) = 0 \qquad (2\text{-}27)$$

Any two of these three equations can be used to solve for the currents I_1 and I_2, if the resistances and voltages are known. For example, let $R_1 = 2000$, $R_2 = 3000$, $R_3 = 4000$, $R_4 = 5000$, and $R_5 = 6000$ Ω, while $V_B = 3.0$ V. Then by Eq. (2-25)

$$3 - (2000 + 3000)\,I_1 + 3000\,I_2 = 0$$

and by Eq. (2-26),

$$3000\,I_1 - (3000 + 4000 + 5000 + 6000)\,I_2 = 0$$

These two equations can be solved simultaneously to give $I_1 = 0.667$ mA and $I_2 = 0.111$ mA.

Similar derivations can be made for reactive circuits with AC excitation, but the mathematics involved become quite extensive.

The Superposition Principle

Earlier in this chapter linearity was described as a proportionality between the current in a circuit and its driving potential. One might inquire about the relation between current and voltage in a circuit in which there is more than one driving potential.

Consider, for example, the circuit shown in Figure 2-22. If we eliminate, for the moment, the AC source (i.e., replace it by a short circuit), then we can show that the DC voltage developed across the capacitor (neglecting the transient) is

$$E_{DC} = V_B \frac{R_2}{R_1 + R_2}$$

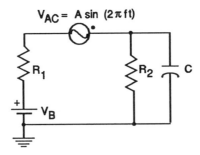

Figure 2-22 A circuit containing two separate sources of potential, one DC the other AC of frequency f Hz. The dot at the AC source serves as a reference for establishing the sign of the curents.

Now let us replace the battery by a short circuit and reinstate the AC source. The voltage across the capacitor can now be calculated using the voltage divider equation:

$$E_{AC} = \frac{R_2\, Z_C/\, (R_2\, +\, Z_C)}{[R_2\, Z_C/\, (R_2\, +\, Z_C\,)]\, +\, R_1} \times V_{AC} \qquad (2\text{-}28)$$

The numerical calculation consists of replacing R_1 and R_2 by their resistance values and Z_C by $1/j\omega C$. If we make an actual measurement with a meter that can respond to both AC and DC (e.g., an oscilloscope), we will find that both calculations still hold true without interfering with each other. The voltage observed is the sum of the two contributions.

The preceding example illustrates the **superposition principle** which is valid for linear circuits. Stated generally, this principle says that if two potentials applied separately produce voltages E_1 and E_2, respectively, they will produce the voltages $E_1 + E_2$. when they are applied together. This principle can be stated equally well in terms of currents. It applies to combinations of DC as well as AC sources of any frequency and voltage, as long as the circuit remains linear.

Sources of Noise

Electronic noise was mentioned in Chapter 1. We will now treat the subject in more depth. Some forms of noise can be eliminated at the source, but most are unavoidable and can only be minimized. The basic difficulty in getting rid of it is that noise is observed by the same techniques as the desired signal.

There are four main categories of noise of importance to us: **resistance noise, shot noise, current noise, impulse noise,** and **environmental noise.** The first two are inherent in electronic systems, the third results from properties of particular components, and the last two produced by external phenomena.

In general, one type of noise is dominant in a system. Thus resistive sources are limited by Johnson noise, whereas photomultiplier tubes give mostly shot noise. Most instruments show increasing noise at low frequencies. Noise is usually observed at the output of an amplifier, but for a quantitative measure, it is conveniently referred to the input by means of a parameter called the **noise equivalent power** (NEP). This is the output noise power divided by the amplifier gain.

Resistance Noise

Resistance noise, also called **Johnson noise,** results directly from the random thermal motion of electrons in resistive components. The voltage that it produces, squared and averaged over a period of time, is given by the expression

$$\overline{e^2} = 4kTR\Delta f \tag{2-29}$$

$$e_{\text{RMS}} = \sqrt{\overline{e^2}} = \sqrt{4kT}\sqrt{R}\sqrt{\Delta f} \tag{2-30}$$

in volts. The quantity k is the Boltzmann constant (1.38×10^{-23} J/K), T is the Kelvin temperature, R is the source resistance in which the noise originates, and Δf represents the span of frequencies over which the measurement is made (the **bandwidth**). In other words, within a particular frequency span, say, 100 Hz, the resistance noise will be the same, regardless of what part of the frequency spectrum is selected. This is an example of **white noise** which is characterized by an even distribution of all frequencies—in a manner similar to white light that contains all visible frequencies.

Resistance noise is specified in the units $V/\sqrt{\text{Hz}}$ which comes directly from Eq. (2-30) as the units of the factor $\sqrt{4kTR}$. Experimentally, if a device shows 1 μV (RMS) of noise over the frequency range of 150 to 250 Hz, the noise (at 25°C) would be expressed as $10^{-6}/\sqrt{100} = 10^{-7}$ V/$\sqrt{\text{Hz}}$.

For a given resistance value, the square of the voltage is proportional to power; hence we can rewrite Eq. (2-30) as

$$\overline{P} = \frac{\overline{e^2}}{R} = 4kT\Delta f \qquad (2\text{-}31)$$

which gives the rather surprising result that the power associated with resistance noise is independent of the resistance.

It must be emphasized that resistance noise cannot be eliminated, as it is due to a fundamental fluctuation that occurs in all devices that dissipate power (ideal capacitors and inductors are free of it). It represents the irreducible minimum amount of noise under a given set of circumstances. It can only be decreased by changes in instrumental parameters, such as by lowering the resistance, the temperature, or the bandwidth.

Shot Noise

Shot noise, also called **Schottky noise**, is observed whenever a current passes through some component. It reflects the fact that electricity is quantized and can flow only in units of single electrons. Shot noise is usually specified as a random fluctuation i of the current I, again squared and averaged over a time interval:

$$\overline{i^2} = 2I\varepsilon\Delta f \qquad (2\text{-}32)$$

$$\overline{P} = \overline{i^2}R = 2RI\varepsilon\Delta f \qquad (2\text{-}33)$$

Here ε is the electronic charge, $1.602 \times 10^{-19}\,C$, and R refers to the load resistance through which the current $(I \pm i)$ is flowing. Power in this type of noise is indeed proportional to the resistance. Shot noise is also white and dependent on the bandwidth Δf.

Current Noise

Current noise, or **flicker noise**, originates by a rather involved mechanism in those resistors and other components that are granular in structure. It is not white, but has an inverse dependence on frequency:

$$\overline{e^2} \propto \frac{\Delta f}{f} \tag{2-34}$$

$$\overline{P} \propto \frac{\Delta f}{Rf} \tag{2-35}$$

Current noise is found to some extent in all components, but it is especially evident in transistors, carbon-paste resistors, and photocells. Metallic film and wire-wound resistors show the effect to a lesser degree. The $1/f$ dependence suggests that this type of noise can be important at very low frequencies, including DC.[14] This is indeed true, and means that it is desirable to avoid DC measurements when working with very small signals.

One of the most annoying phenomena encountered with low-level DC signals is that of **drift**. This is a gradual change in output caused by some minor progressive variation in the circuit. It is most often the result of slow changes in temperature caused by self-heating, but may also represent the effects of aging of components. It is a form of low-frequency $1/f$ noise.

Impulse Noise

Impulse noise in analog electronics often results from the switching transients in nearby digital circuits, especially if both systems are powered from the same supply. Digital circuits are characterized by sudden shifts in voltage from one level to another, often synchronized over a whole system. Unless design precautions are taken, these transients can be picked up and amplified in the analog circuits. They can be reduced or eliminated by suitable shielding and by filtering of power-supply connections.

Another variety of impulse noise, sometimes called **popcorn noise**, or burst noise, arises within the semiconductor devices themselves. Specially processed units may have little of this type of noise.

---

[14] The fact that the expressions in Eq. (2-34) as well as the one in Eq.(2-35) become infinite for $f = 0$ should not disturb us. Zero frequency implies an infinite period, not realizable experimentally.

Environmental Noise

Most other types of noise can be lumped together under the umbrella of environmental noise or interference. Below about 10 Hz such noise appears to show a $1/f$ frequency response, similar to that of current noise.

A prevalent source arises in the coupling of signals from one system to another by means of AC fields. This type of noise, sometimes designated as **electromagnetic interference** (EMI), can originate in neighboring devices that utilize radio frequencies. However, an electronic instrument can also be a source of EMI if it contains fast switching circuits such as those in computers. Power lines can conduct such interference from one instrument to another if plugged into the same branch.

Power Line Effects

A special case that can be particularly troublesome is pickup of spurious signals at the power-line frequency itself and its first few harmonics. This pickup can occur through a direct conductive path (e.g., leakage through a cracked insulator) or by magnetic or capacitive coupling. It can be controlled by a combination of carefully planned grounding and shielding. Recall that the purpose of grounding is to ensure that a particular conductor is maintained at 0 V. Shielding consists of surrounding a component or wire with a grounded conductor. The shield serves to intercept EMI, and drain it off harmlessly to ground.

Environmental noise is also present in the various disturbances of electrical distribution lines. Frequency components up to about 30 MHz are likely to be present. The line noise is often transmitted into instruments and causes an increase in the total noise.

In addition the power line can carry transients of very short duration and very high voltage that may represent major causes of instrument failure, or perhaps only generate erroneous digital data in computers. The line transients can be clasified as follows:

1. Type A originating from remote lightning strikes (about 6 kV peak and 1 μs duration), illustrated in Figure 2-23. They can cause catastrophic failures.

2. Type B originating from remote fuse blowing, circuit breakers, or any major power-off (perhaps 1 kV for 50 μs). These might occur

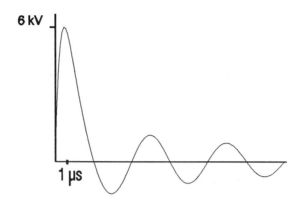

Figure 2-23 A typical transient caused by a remote lightning strike. The IEEE 587 standard recommends a similar waveform to test equipment for immunity to type A transients.

several times in a single day, thousands of transients during the lifetime of a piece of electronic equipment.

3. Type C transients originating from switching motors on and off (in the range of 500 V for 0.5 ms). They may be quite frequent, and carry more total energy than the other types but are neutralized by the power circuitry of most instruments.

Commercial protection devices can eliminate most types of line noise but some types are slowly damaged by type A transients and should be replaced periodically.

Chapter 3

Operational Amplifiers–I

Amplifiers unquestionably constitute the most important class of active electronic devices. There are many types of amplifiers, each designed or optimized for a particular application, a well-known example being the high-fidelity amplifier for the reproduction of music. Of greatest significance to us is a class known as **operational amplifiers**, commonly known as "op amps."

The first op amps were bulky items built with vacuum tubes. With the introduction of semiconductor technology, "transistorized" op amps were designed in which the old vacuum tubes were replaced with discrete transistors. More recently, however, even these have been almost totally replaced by **integrated circuits** (ICs) in which a complete amplifier is created on a tiny wafer of silicon. Most are fabricated as DIPs (dual inline packages) containing one, two, or more amplifiers in a unit provided with 8 or 14 pins for insertion into sockets or circuit boards. These ICs are manufactured in large quantity, hence are inexpensive, typical units being available for less than a dollar.

Amplifiers

Briefly stated, an amplifier is a device that can be used to augment a signal, that is, it can increase the amplitude of the signal as required. The multiplying constant (the **gain** A) is defined as the ratio of output to input and can be as great as 10^5 or 10^6. The quantity to be amplified can be a voltage, a current, or a power, thus defining a voltage, current, or power gain. The power in a DC circuit is the product of current and voltage (the same is true of AC circuits if the power factor is unity). Hence it follows that the power gain of an amplifier is the product of its voltage and current gains:[1]

$$A_P = \frac{P_{\text{out}}}{P_{\text{in}}} = A_E \times A_I = \frac{E_{\text{out}}}{E_{\text{in}}} \times \frac{I_{\text{out}}}{I_{\text{in}}} \qquad (3\text{-}1)$$

It is quite possible for a particular amplifier to have a voltage (or current) gain of unity or even less, while still having a large power gain. Clearly, if the amplifier is to increase the power in a circuit, it must have access to an external source upon which it can draw. This may be supplied from batteries or from the AC line through a suitable **power supply**.

The multiplying factor in an amplifier can be less than unity, thus giving an output that is less than the input. This action, known as **attenuation**, can be thought of as dividing the input by a factor. In principle, an attenuator need not be an active device, as it has no inherent need for an external power supply. The familiar volume control, often consisting of a variable resistor, is an example of an attenuator.

The Decibel Scale

The gain and attenuation vary over many orders of magnitude therefore are commonly represented on a logarithmic scale, the **decibel** scale, dB. This scale is defined in terms of the output/input ratios, as follows:

$$dB_E = 20 \log (E_{\text{out}}/E_{\text{in}}) \qquad (3\text{-}2)$$

$$dB_I = 20 \log (I_{\text{out}}/I_{\text{in}}) \qquad (3\text{-}3)$$

For example, a gain of 10^6 (either voltage or current) corresponds to $20 \log (10^6) = 120$ dB. The decibel scale has the advantage of using small numbers to represent large ratios. The case of attenuation, the gain is less than one. Its logarithm is negative, and therefore the

---

[1] Voltages (or potential differences) are sometimes designated by E, sometimes by V. We will follow the convention that E refers to information-carrying signals V to power supplies.

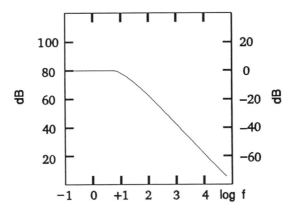

Figure 3-1 Frequency response curve (Bode plot) for an amplifier (left scale) or an attenuator (right scale). Many of the frequency-selective networks responsible for the sloping portions of the curve produce a "roll-off" of 20 dB per decade, as shown.

decibel number is also negative. Consecutive stages of amplification or attenuation, expressed in dB, lead to a *sum* (rather than a product) of individual gains. Consider as an example three cascaded amplifiers with voltage gains of 10, 100, and 1000. Multiplying these numbers together gives an overall gain of 1,000,000. The corresponding decibel gains from Eq. (3-2) are 20, 40, and 60, giving a sum total of 120 dB, which is also given directly as dB = 20 log (1,000,000) = 20 × 6 = 120 dB.

In addition to the current and voltage definitions of the decibel, a distinct formula is applicable to power amplification.[2] Recalling that the power dissipated in a resistor is $P = E^2/R$, we can substitute for E its equal $\sqrt{RP}$:

$$dB_E \;=\; 20 \log \sqrt{RP_{out}/RP_{in}} \;=\; 20 \log \sqrt{P_{out}/P_{in}} \qquad (3\text{-}4)$$

Assume now that $R_{in} = R_{out} = R$. This gives an equation with a numerical coefficient of 10 rather than 20:

---

[2] Power ratios are sometimes designated as dBm, taken in reference to 1 mW as standard. The decibel scale is also widely used in acoustics with a rather different meaning.

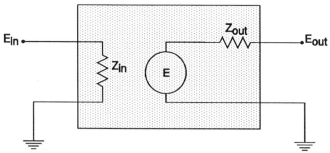

Figure 3-2 The equivalent circuit for an amplifier, showing the input and output impedances. The Thévenin voltage, shown as E, is established by the action of the amplifier operating on the input signal.

$$dB_P = 10 \log \left[\frac{P_{out}}{P_{in}} \right] \tag{3-5}$$

Amplifier gain (or attenuation) in decibels is often plotted as a function of frequency, with the latter also on a logarithmic scale, as in Figure 3-1. This is called a **Bode plot**, and is useful in representing the frequency response of any device. The use of logarithmic scales on both axes makes it possible to display very wide spans of frequencies and of gain ratios on a single graph. Note, however, that DC cannot be represented on a Bode plot, as it would fall at negative infinity [log $(0) = \infty$]. In many instances, as we shall see, the characteristic curve of an amplifier over a part of its range is represented on a Bode plot by a diagonal straight line as illustrated in the figure. Most commonly the slope corresponds to 20 dB per decade of frequency. This means a 1:1 slope, since both 20 dB and a decade describes a factor of 10.

Input and Output Impedances

In addition to gain, the most important properties of an amplifier are its input and output impedances. Figure 3-2 shows the conventional representation of these quantities. As a rule it is advantageous to have Z_{in} large and Z_{out} small, though there is a less important class of amplifiers where exactly the reverse is required.

Operational Amplifiers

An amplifier, to be termed **operational**, must be characterized by a very high gain, a very high input impedance, and very low output impedance. The basic op amp is represented in circuit diagrams by a triangle with two input and one output connections, as shown in

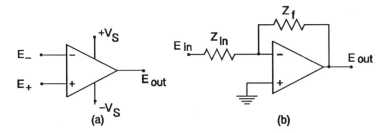

Figure 3-3 (a) The basic symbol for an operational amplifier; the leads marked
$+V_s$ and $-V_s$ are often omitted for clarity. (b) A typical configuration for an op amp;
Z_{in} is the input impedance, Z_f the feedback impedance.

Figure 3-3. Op amps are commonly powered by a bipolar power
supply, that is, one that provides both positive and negative voltages
relative to ground. These connections are indicated in Figure 3-3a by
vertical lines above and below the triangle but are usually omitted in
schematic diagrams to avoid unnecessary clutter. Some op amps can
use a single supply, generally the positive, with the opposite lead
connected to ground.

The (+) and (–) signs attached to the inputs signify that the
corresponding input voltages must be multiplied by the gains $+A$ or
$-A$, respectively, in determining what the output voltage will be.[3] The
output voltage from the op amp is thus proportional to the *difference*
between the potentials at the two inputs, following the relationship:

$$E_{out} = A \, (E_+ - E_-) \tag{3-6}$$

where A is the inherent gain of the amplifier, a large number, perhaps
10^5 or 10^6.

If there were no external components, such as those in Figure 3-3b,
the output would quickly reach a practical limit for anything but the
smallest input signals. For example, if $A = 10^6$, an input signal $(E_+ -
E_-)$ any greater than 12 μV would produce an output limited to 12
V, which is approximately the highest output voltage that can be
obtained using a 15-V supply. Figure 3-4 shows a transfer plot (i.e.,
output as a function of input) for such an op amp. It will be seen that
for any input greater than $+12$ μV the output will be $+12$ V, and
similarly any input less than -12 μV gives -12 V out. The amplifier

---

[3] These signs are sometimes left out of schematics, but it is best practice to
include them.

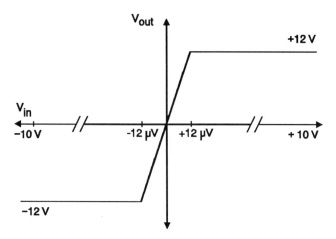

Figure 3-4 Transfer plot (output vs. input) of an op amp devoid of external circuitry. The slope of the midportion is the inherent gain A. Note the changes of scale on the horizontal axis.

operating at these limits is said to be **saturated.** Only in the very narrow region around the origin will the output be a linear function of the input.

The gain of an op amp is a function of frequency f as shown in the Bode diagram of Figure 3-5. It can be approximated by a straight horizontal line from DC (negative infinity) to some low frequency, perhaps 10 Hz, where it intersects with a diagonal line of 45° slope. Throughout the range covered by the sloping line (about 10 Hz to 10^7 Hz in the example), the product $A \times f$ is a constant, called the **gain-bandwidth product.** Note that in the figure, the gain becomes unity (i.e., 0 dB) at a frequency of 10 MHz, which means that the gain-bandwidth product is 10 MHz (remember that the gain is dimensionless). At some higher point on the sloping portion of the

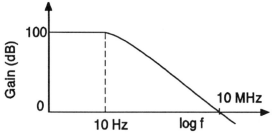

Figure 3-5 A Bode plot for a typical op amp. Note that DC cannot be represented on this graph as it would lie at minus infinity.

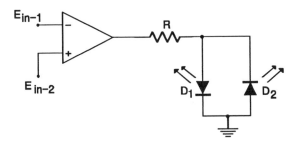

Figure 3-6   A comparator used to indicate the relative magnitudes of two volt-ages. If E_{in-1} is more positive than E_{in-2}, the output will be negative, and the light-emitting diode D_2 will light up, whereas the opposite situation will energize D_1. The resistor R serves to protect the diodes from excess current.

curve, the gain might be 10 and the bandwidth only 1 MHz, for the same product. The higher this product, the higher will be the fre-quency at which the amplifier can be operated. Hence it is a useful figure of merit for an op amp.

Comparators

An op amp, used directly without any feedback connection is called a **comparator,** since it compares the voltages E_+ and E_- and produces an output of corresponding sign at the saturation voltage. Figure 3-6 shows the use of a comparator to monitor two voltages, lighting either of two light-emitting diodes (LEDs)[4] according to which voltage is higher. If the (–) input is grounded, the comparator will indicate the sign of the input, by positive or negative saturation. The use of a comparator in converting analog to digital signals will be mentioned in a later chapter.

Linear Applications

Many more circuits are possible if the amplifier is forced to operate within the limits of its linear region (see Figure 3-4). This can be accomplished by incorporating **negative feedback** into the circuitry external to the amplifier itself. Consider the amplifier of Figure 3-7, with the noninverting input (i.e., E_+) at zero potential (grounded) and with resistors connected from the inverting input to both E_{in} and the

[4] An LED is a small device that generate light when current passes through it in the direction denoted by the arrowhead but blocks any current in the opposite direction.

output, as shown. R_f is called the **feedback resistor.** We know from Ohm's law that the current flowing through a resistor is given by the difference in potential between its terminals divided by its resistance. Therefore the input current I_{in} must equal $(E_{in} - E_{SJ})/R_{in}$, and the feedback current, I_f, must be given by $(E_{out} - E_{SJ})/R_f$. The negative sign results from the application of Kirchhoff's current law. Since the inherent input resistance of the amplifier[5] is very high, we can assume that no current is lost into the amplifier. Thus I_{in} and I_f must be equal, so we can write

$$\frac{E_{in} - E_{SJ}}{R_{in}} = \frac{E_{SJ} - E_{out}}{R_f} \tag{3-7}$$

We also know that in general $E_{out} = A(E_+ - E_-)$ which for the present circuit becomes $E_{out} = A(0 - E_{SJ})$. Combining these gives

$$E_{SJ} = -E_{out}/A,$$

and substituting this into both sides of Eq. (3-7), we obtain:

$$\frac{E_{in} + (E_{out}/A)}{R_{in}} = \frac{(-E_{out}/A) - E_{out}}{R_f} \tag{3-8}$$

This equation simplifies algebraically to

$$\frac{E_{in}}{E_{out}} = -\frac{R_{in}}{R_f} \bigg/ \left(1 + \frac{1}{\beta A} \right) \tag{3-9}$$

where

$$\beta = \frac{R_{in}}{R_{in} + R_f} = \frac{1}{(R_f/R_{in}) + 1} \tag{3-10}$$

---

[5] One must distinguish between the input resistance of the amplifier, and the input resistor R_{in}. The former is a property of the amplifer itself, whereas the latter depends on the external passive components.

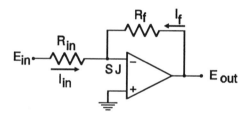

Figure 3-7 An amplifier with negative feedback. The SJ connection to the inverting input is the point where the two currents, I_{in} and I_f are said to be "summed". It is named the **summing junction.**

At higher frequencies, where either A or β is small, the complete equation must be used. Otherwise, if the quantity $1/\beta A$ is negligible (less than perhaps $1/10^4$), we can use the simplified form:

$$E_{out} = -\frac{R_f}{R_{in}} E_{in} \tag{3-11}$$

or, more generally,

$$E_{out} = -\frac{Z_f}{Z_{in}} E_{in} \tag{3-12}$$

The negative sign carries considerable significance; it means that an op amp connected basically as in Figure 3-7 always changes the algebraic sign. This of course ties in with the designation of the inverting and noninverting inputs of the amplifier.

Equation (3-12) is thus the basic relationship for such applications of the operational amplifier. Note the absence of A from this equation. This means that the internal nature of the amplifier is irrelevant, as long as the requirements are met (large A, high input impedance, not too small β). This points out a great advantage of op amps: they obey simple mathematical formulas with great precision, irrespective of the identity and parameters of the specific unit. The only errors that must normally be taken into account are those brought about by high-frequency operation, and certain offset errors discussed in the next chapter.

Op amp circuits can be analyzed most easily by observing two rules, both of which stem from the treatment just given:

1. The input and feedback currents are equal but of opposite sign: $I_{in} = -I_f$.
2. The potentials of the two input connections are very nearly equal: $E_+ \cong E_-$.

These rules will always apply, provided only that the amplifier is in control of the circuit, that is, provided it is operating in its linear region (nonsaturated). The input and feedback impedances can be networks of resistors, capacitors, and inductors. We will now describe a variety of simple op-amp circuits that can serve as building blocks for the more complex circuitry found in many analog instruments.

The Voltage Summer

In many circuits, op amps have their noninverting inputs grounded (as in Figure 3-7). They are said to be operating in the single-input mode. Because of the requirement of essential equality between the two inputs, the negative input in such circuits is very close to ground potential, a condition referred to as **virtual ground**. Another example of such operation is the circuit shown in Figure 3-8a which permits additive combination of several voltages, which may be of either sign. Assuming, as before, that no significant current is flowing into the amplifier, we write:

$$I_1 + I_2 + I_3 = I_f \tag{3-13}$$

It is because of this relation that the negative input is often called the **summing junction**. The currents in Eq. (3-13) can then be ex-

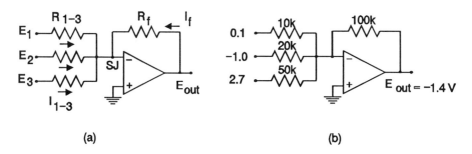

Figure 3-8 A voltage summer with three inputs: (a) the general circuit, (b) a specific implementation.

pressed in terms of the input voltages by substituting for each I the corresponding E/R ratio. Rearrangement gives

$$E_{out} = R_f I_f = -R_f \left[\frac{E_1}{R_1} + \frac{E_2}{R_2} + \frac{E_3}{R_3} \right] \qquad (3\text{-}14)$$

This circuit sums all the input voltages while multiplying them by constant coefficients with the usual sign inversion. This is called a **weighted summer**. If all the resistors are equal, the operation is the simple addition of voltages. The voltages may be time-dependent, whereas the $1/R$ multiplying factors usually are not. The reader should find no difficulty in verifying the voltage produced by the circuit of Figure 3-8b.

The equations derived in the preceding paragraphs have defined the circuit gain in terms of ratios of resistances. A given ratio of course can be implemented by various combinations. A ratio of R_f/R_{in} of 10, for example, could be made up of $R_f = 10$ kΩ and $R_{in} = 1$ kΩ, or 100 kΩ and 10 kΩ, or 1 MΩ and 100 kΩ, among many other choices. There are, however, some practical limitations. The maximum current in a typical op amp circuit is about 20 mA, therefore, assuming a maximum excursion of 12 V, the minimum values of resistors should be about 600 Ω, though in practice resistors less than 1 kΩ should be avoided. At the other extreme, resistors greater than about 10 MΩ should not be selected, because they would restrict the currents to less than 1 μA,

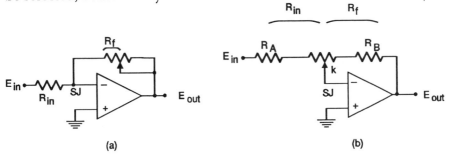

(a)   (b)

Figure 3-9 Variable-gain summers. In (a) the gain can be varied from zero to R_f/R_{in}. In (b) the gain could, in principle, be varied from zero to infinity ($0/R_{in}$ to $R_f/0$); if it were not for the series resistors R_A and R_B, inserted to keep the gain finite. Thus suppose that the variable resistance in (b) were 1 MΩ, and R_A and R_B both 100 Ω, the gain could be varied from the ratio $100/1{,}000{,}100$ ($\cong 10^{-4}$) to $1{,}000{,}100/100$ ($\cong 10^{+4}$).

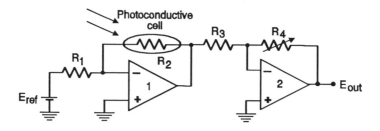

Figure 3-10 A possible light-intensity measurement system. The photoconduc-tive cell, R_2, decreases its resistance as a nearly logarithmic function of increasing illumination. The output is easily calculated to be $E_{out} = E_{ref} \dfrac{R_4}{R_1 R_3} \times R_2$. It should be clear to the reader why the customary negative sign does not appear.

where the effect of any stray currents and noise would be excessive. These restrictions apply to the common general-purpose op amps; special premium models are available that can operate outside these ranges.

It is often desirable to be able to adjust the gain of the amplifier. This can be done by changing either R_f or R_{in}, or both. If $R_f = R_{in}$ the action of the amplifier is to invert the sign without altering the value. The circuit is called an inverter. Other circuit arrangements are shown in Figure 3-9. (Note that we now refer to the gain of the *stage* R_f/R_{in}, rather than the inherent gain A of the amplifier.)

Amplifiers are often connected in a chain, with the overall gain distributed between the component amplifiers. Figure 3-10 shows an example in which a photoconductive cell is energized at constant current by amplifier 1, and the effective gain can be adjusted by a variable resistor in the feedback of amplifier 2.

The Current-to-Voltage Converter

Since the essential action of the op amp is to generate a current (I_f) to be equal to the input current, there is no need for an input resistor if the signal is already in the form of a current. Figure 3-11 shows a three-input **current summer**, also called a **current-to-voltage con-verter**. The output is readily given as

$$E_{out} = -R_f (I_1 + I_2 + I_3) \qquad (3\text{-}15)$$

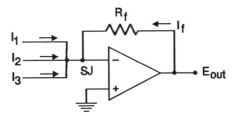

Figure 3-11   A multiple-input current summer or current-to-voltage converter.

The three currents to be summed meet at a point of virtual ground, so that changes in one current do not affect the other currents, often a valuable property.

As an example of this type of circuitry, consider Figure 3-12. A **photodiode** is a unit in which the *current* is generated by light and flows only in the direction pointed by the triangle. If the light intensity is the same on the two diodes, cross-connected as shown, the currents through them will be identical. This means that no current will pass through the feedback resistor R_f, so the output will be zero. If the light falling on one diode is greater than on the other, then a current will flow through R_f, and the output voltage will show a finite value, either positive or negative, according to the relative illumination. This circuit can be used in photometric measurement devices, when it is desired to compare the intensity of an experimental light beam against a standard light source. Another application is to monitor shifting positions in a beam of light.

The Integrator

Another operation that can be carried out with the aid of an op amp is integration with respect to time. This function is implemented by connecting a capacitor as Z_f (Figure 3-13). The capacitor is charged by

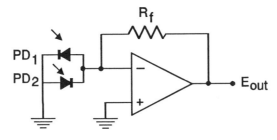

Figure 3-12   Two photodiode currents summed at the inverting input to an op amp.

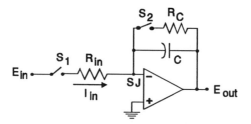

Figure 3-13 An analog integrator. Switch S_2 is used to reset (discharge) the capacitor through a resistor.

the feedback current, and hence the output voltage varies according to the relation describing the behavior of a capacitor:

$$C\frac{dE}{dt} = I_f \tag{3-16}$$

Setting I_f equal to the negative of the input current gives

$$\frac{E_{in}}{R_{in}} = -C\frac{dE_{out}}{dt} \tag{3-17}$$

Rearrangement and integration shows the output voltage to be the time-integral of the input:

$$E_{out} = -\frac{1}{R_{in}C}\int_0^t E_{in}\, dt \tag{3-18}$$

If the input signal is in the form of a current, the resistor R_{in} can be omitted, and the expression becomes

$$E_{out} = -\frac{1}{C}\int_0^t I_{in}\, dt \tag{3-19}$$

Zero time (t_o) refers to the moment when switch S_1 is closed, thus permitting current to flow, charging the capacitor. Charging is linear[6]

---

[6] Compare this with the connection of a capacitor and battery through a resistor (no op amp). In this case the voltage will change *logarithmically* from zero, approaching the battery voltage asymptotically.

with time until one of two events occurs: either the switch is opened, or the amplifier reaches saturation. In general, the capacitor must be discharged before another cycle of operation can start. This is accomplished by closing switch S_2. Thus the integrator has three modes: (1) **integrate**, while current flows into the capacitor, (2) **hold**, when both switches S_1 and S_2 are open, so that no current can flow either into or out of the capacitor, and (3) **reset,** when S_2 is closed.

As seen above, the integrator can be used to generate a DC ramp voltage simply by holding the input voltage constant. Equation (3-20) then becomes:

$$E_{out} = -\frac{E_{in}}{RC} \int_0^t dt = kt + E_0 \qquad (3\text{-}20)$$

where the constant of integration E_0 represents the initial voltage, and the constant k is the slope of the ramp. It is of interest to note the result of integrating a sine wave:

$$E_{in} = B \sin \omega t \qquad (3\text{-}21)$$

$$E_{out} = -\frac{1}{RC} \int B \sin \omega t\, dt = \frac{1}{\omega RC} B \cos \omega t \qquad (3\text{-}22)$$

$$= \frac{1}{\omega RC} B \sin (\omega t + \frac{\pi}{2})$$

The output is a sine wave of the same frequency ω but with the phase changed by $\pi/2$, the amplitude decreasing as the frequency increases. Its amplitude is equal to the original only for the special case where $\omega = 1/RC$. For example, suppose that $\omega = 1000$ rad/s, $R = 1$ MΩ, $C = 1$ μF, and $B = 1$ V. Since farads $\times$ ohms = seconds, the RC product in this example is 1 s, and the amplitude of E_{out} will be $1/(\omega RC) = 10^{-3}$ V, compared to 1 V for the input. On the other hand, if this integrator (with the same R and C), is excited at 100,000 rad/s, the output will be only 10^{-5} V. This property makes possible the use of an integrator to selectively amplify low frequencies, a type of low-pass filter.

Equation (3-22) implies that at zero frequency (at DC), the gain is infinite, a condition which, of course, is unattainable because the

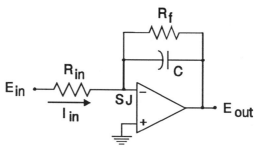

Figure 3-14 Modified integrator, for which the gain is $-R_f/R_{in}$. It can also be thought of as a low pass filter.

amplifier would saturate long before infinity is approached. This behavior can be modified by adding a resistor in parallel with the capacitor, as in Figure 3-14; at very low frequencies the reactance of the capacitor is so great that its effect on the feedback impedance is essentially nil, while at higher frequencies the effect of the capacitor outweighs the resistor.

The effect can be derived from the mathematical description of the impedances. First, write the op amp equation in the form:

$$\frac{E_{out}}{E_{in}} = -\frac{Z_f}{Z_{in}} \qquad (3\text{-}23)$$

then substituting for Z_f its equal, $Z_C R_f/(Z_C + R_f)$, and letting $R_f = R_{in} = R$, we obtain the relation:

$$\frac{E_{out}}{E_{in}} = -\frac{Z_C}{Z_C + R} \qquad (3\text{-}24)$$

Now we replace Z_C by $1/j\omega C$, to give

$$\frac{E_{out}}{E_{in}} = -\frac{\dfrac{1}{j\omega C}}{\dfrac{1}{j\omega C} + R} = -\frac{1}{1 + j\omega RC} \qquad (3\text{-}25)$$

Let us examine the results for the three following cases:

Case I. At low frequency, where $\omega \ll 1/RC$, the term $j\omega RC$ vanishes, and the gain is unity:

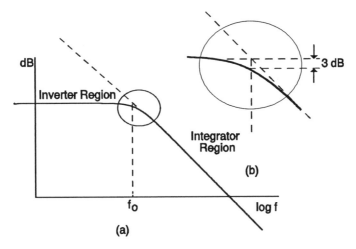

Figure 3-15 (a) Bode plot showing the behavior of the modified integrator circuit of Figure 3-14. The dashed diagonal line shows the plot for the unmodified integrator. The frequency f_0 corresponds to the break point, where $f_0 = 1/(2\pi RC)$ (i.e., 3 dB down) as shown in the inset (b).

$$\frac{E_{out}}{E_{in}} = -\frac{1}{1 + j\omega RC} = -1 \qquad (3\text{-}26)$$

Case II. At the frequency $\omega = 1/RC$:

$$\frac{E_{out}}{E_{in}} = -\frac{1}{2} + \frac{j}{2} \qquad (3\text{-}27)$$

The magnitude of the gain is

$$\sqrt{(-1/2)^2 + (1/2)^2} = 1/\sqrt{2}$$

This case corresponds to a gain of 3dB.

Case III. At high frequency, where $\omega >> 1/RC$, the gain becomes zero:

$$\frac{E_{out}}{E_{in}} = -\frac{1}{j\omega RC} = 0 \qquad (3\text{-}28)$$

The corresponding Bode plot is shown in Figure 3-15.

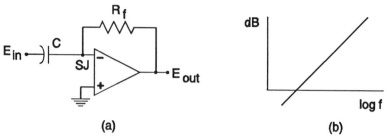

Figure 3-16 (a) A differentiator; (b) its Bode plot.

This frequency response is characteristic of a low-pass filter in that high frequencies are not amplified to the extent that low frequencies are. It is always wise to take advantage of this effect by placing a small capacitor across the feedback resistor in all summers, to attenuate high-frequency noise. As a rule of thumb, for DC or low-frequency operation, a capacitor that will give the product $R_fC = 1$ ms should be adequate.

The Differentiator

As the complementary function to integration, differentiation can also be carried out with the aid of an op amp. For this application, the capacitor is moved to the input circuit, as shown in Figure 3-16. For this circuit we can write, as before,

$$I_{in} = C(dE_{in}/dt) = -I_f = -E_{out}/R_f \qquad (3\text{-}29)$$

$$E_{out} = -R_f C(dE_{in}/dt) \qquad (3\text{-}30)$$

If $E_{in} = B$ sin ωt, then $dE_{in}/dt = \omega B$ cos ωt, and

$$E_{out} = \omega RCB\ cos\ \omega t = \omega RCB\ sin\ (\omega t + \pi/2) \qquad (3\text{-}31)$$

which is the working relation for the simple differentiator.

The difficulty with this circuit is that it tends to overemphasize the noise, much of which is at high frequency. The situation can be improved by adding a resistor in a manner analogous to the previous example, as shown in Figure 3-17. For this arrangement, it is evident that $Z_{in} = R_{in} + 1/j\omega C$. Again taking $R_{in} = R_f = R$,

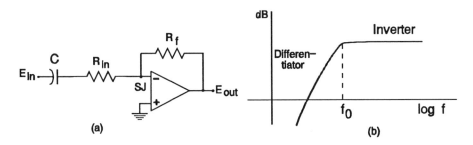

(a)   (b)

Figure 3-17 A modified differentiator with reduced high-frequency gain. The breakpoint is $f_0 = 1/(2\pi R_{in}C_{in})$.

$$Z_{in} = \frac{1 + j\omega RC}{j\omega C} \qquad\qquad (3\text{-}32)$$

and

$$\frac{E_{out}}{E_{in}} = -\frac{j\omega RC}{1 + j\omega RC} \qquad\qquad (3\text{-}33)$$

Again let us consider three cases:

Case I: For $\omega \ll 1/RC$, $E_{out}/E_{in} = -j\omega RC$.
Case II: For $\omega = 1/RC$, $E_{out}/E_{in} = -(1/2 + j/2)$.
Case III: For $\omega \gg 1/RC$, $E_{out}/E_{in} = -1$.

These three cases demonstrate that, in agreement with the Bode plot, the (negative) gain is unity above the f_0 point, 3 dB down at that point, and falls off linearly with decreasing frequency below f_0.

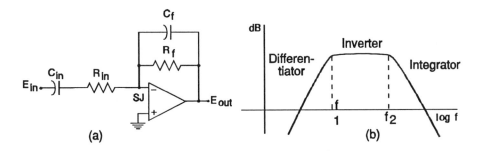

(a)   (b)

Figure 3-18 (a) A combination of differentiator and integrator; (b) its Bode plot.

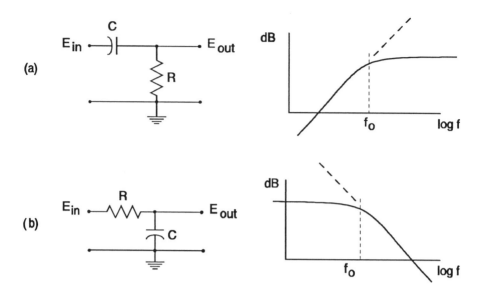

Figure 3-19 Passive *RC* filters and their Bode diagrams: (a) high pass, (b) low pass. In each, $f_0 = 1/(2\pi RC)$ Hz or $1/RC$ radians per second.

A useful circuit, combining the input circuitry of Figure 3-17 with the feedback network of Figure 3-14, is shown in Figure 3-18. This forms a band-pass filter in which the region between f_1 and f_2 is passed with fixed gain, whereas both higher and lower frequencies are rolled off at the usual 20 dB per decade. If the two *RC* products were made equal, the passed band would be narrow.

Filters

Circuits based on simple integrator and differentiators, such as those just discussed, are members of a much larger class of circuits called **filters.** A filter is defined as a device that attenuates selectively certain ranges of frequency. Distinction must be made between passive and active filters. A passive filter is a network made up of resistors, capacitors, and sometimes inductors, that is characterized by a se-lected shape of its frequency response curve (Bode plot). See Figure 3-19. These simple circuits change their frequency characteristics somewhat depending on the source and load impedances, but op amps can ameliorate this problem.

Filters can be made more powerful by incorporating op amps to provide gain. The integration and differentiation circuits described

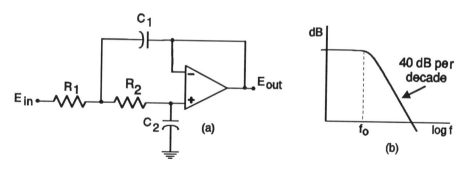

Figure 3-20 A second-order low-pass active filter and its Bode diagram. The break frequency is given by $f_0 = 1/(2\pi\sqrt{R_1 R_2 C_1 C_2})$.

above are examples. The Bode plots for these filters, are presented in Figures 3-15, 3-17, and 3-18; they show the same 20 dB/decade roll-off as similar plots for passive filters. These are called **first-order active filters.**

Another class, called **second-order active filters,** include more specialized op amp circuits; their Bode slopes are twice as steep, 40 dB/decade, allowing a more effective attenuation of unwanted frequencies. Examples are shown in Figures 3-20, 3-21, and 3-22.

The Twin-T Notch Filter

An important passive filter is the Twin-T, shown in Figure 3-23. This dual network can be considered functionally to be a combination of a high-pass with a low-pass filter, resulting in a very high impedance at the notch frequency, given by $f_0 = 1/(2\pi RC)$. The sharpness of the

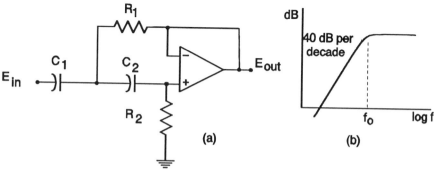

Figure 3-21 A second-order high-pass active filter and its Bode diagram. The break frequency is given by $f_0 = 1/(2\pi\sqrt{R_1 R_2 C_1 C_2})$.

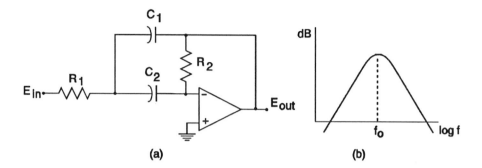

(a)   (b)

Figure 3-22 A second-order band-pass filter. Note that the peak in its Bode plot is much sharper than that of the first-order filter in Figure 3-18.

notch is dependent on the closeness of match of the resistors and capacitors. This can be used to reject a particular frequency, such as that of the power line, which otherwise might be picked up sufficiently to interfere with subsequent circuitry. There are a number of similar circuits with a less sharp notch, which may offer an advantage because of easier tuning to the specified frequency.

The Twin-T can be used in the feedback loop of an op amp to produce a sharply tuned amplifier, which will amplify AC signals of the desired frequency while strongly attenuating other frequencies (Figure 3-24). This works because of the high impedance of the Twin-T at frequency f_0, which makes the Z_f/Z_{in} gain very large. A practical difficulty may arise in that the gain at the tuned frequency may be so high that the amplifier will saturate. To prevent this effect, it is wise

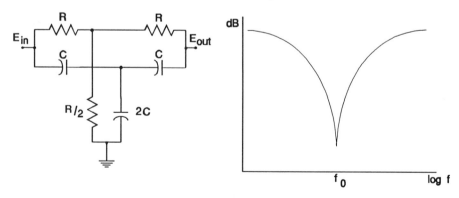

Figure 3-23 Twin -T notch filter. The minimum is at $f_0 = 1/(2\pi RC)$.

to shunt the Twin-T network with a large resistor R_S, as shown in the figure; the feedback impedance can then never exceed the value of this shunt resistor.

Applications of Filters

The most common application of a filter is in noise reduction. As detailed elsewhere in this book, much noise occurs at high frequencies, and in addition there is likely to be considerable noise present at the frequencies of the power line and its low harmonics.

Any unwanted signal that has a well-defined frequency can, in principle, be greatly attenuated by means of a notch filter. If the true signal has significant amplitude at the same frequency as the noise, the notch filter will remove some of the desired information along with the noise. Hence the notch should be made as narrow as possible, even though this makes it more difficult to tune with precision. One or more of the resistors in the notch circuit is often made variable to aid in fine tuning.

Broad-band high-frequency noise, such as that originating in resistors, is usually attenuated with a low-pass filter that has a cut-off beyond the highest frequency component of the signal. If the signal has a rather narrow bandwidth, then a band-pass filter is the best choice.

Isolation of a Selected Frequency

A good example of the use of filters to eliminate unwanted frequencies is encountered in a **beat-frequency sine-wave generator**. This is a device that is sometimes used to produce sine waves in the audio range by combining two waves of higher frequency. Suppose that two transistor oscillators are available that produce sine waves at 50.00 kHz and 50.10 kHz, respectively. If these two waves are combined in a *linear* circuit, the superposition principle tells us that there will be no interaction between them. However, if the combining circuit contains a nonlinear element, such as a diode, the two frequencies will interact in such a way that *four* frequencies will be found in the output: the two original frequencies, their sum, and their difference. Thus in the example above, we would find 0.10 kHz, 50.00 kHz, 50.10 kHz, and 100.1 kHz. For an output in the audio range, we would want the 0.10 kHz (100 Hz) wave free of the others. How would this be accomplished? It would be possible to construct a Twin-T filter tuned at 100 Hz and use this in the feedback of an op amp in order to pass the

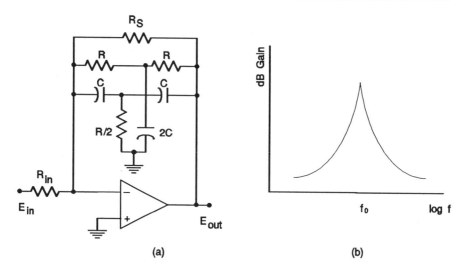

(a)   (b)

Figure 3-24 An op amp tuned to the frequency f_0 with a Twin-T network. The corresponding Bode plot is the inverse of that of Figure 3-23, with its maximum amplification at f_0.

desired frequency and exclude all others. This would have a major disadvantage in its rigidity making it inconvenient in variable-frequency application.

A better solution would be to choose a second order low-pass filter that would pass anything below, say, 2 kHz, and reject higher frequencies. The higher the order of the filter, the sharper will be its roll-off, It is convenient in the circuit of Figure 3-20 to let $C_1 = C_2 = C$ and also $R_1 = R_2 = R$. Then $f_0 = 1/(2\pi RC)$. Let us select two matched 0.01-μF capacitors, then solve for $R = 1/(2\pi \times 2000 \times 10^{-8}) = 7960$ Ω. (Remember that μF $\times$ MΩ = seconds.) Hence our circuit can be built up of two 0.01-μF capacitors and two 8000-Ω resistors. (A resistor of 7960 Ω is not normally available) resulting in a frequency f_0 of about 1.99 kHz.

The roll-off of 40 dB per decade means that at 50 kHz, the attenuation will be about 56 dB, which is fully adequate. The sum frequency (100.1 kHz) would, of course, be totally wiped out.

Chapter 4

Operational Amplifiers–II

Difference Amplifiers

Increased flexibility in the design of op amp circuits can be had if both inputs are actively used, rather than one being restricted to circuit ground. Configured in this way, the amplifier is referred to as a **difference amplifier,**[1] as it is the *difference* between the potentials of the two inputs that is forced to be zero, $E_+ - E_- = 0$, rather than the summing point being forced to virtual ground, $E_- = 0$.

One such design is the **voltage follower**, shown in Figure 4-1. A direct connection between the output and summing junction serves as the feedback loop, the signal being introduced through the noninverting input. Since the potentials of the two inputs are maintained at virtual equality, this connection ensures that the output is exactly equal to the input and of the same sign. The great merit of this circuit is that it provides impedance conversion. Since negligible current is drawn from the voltage source, the amplifier gives the exact value of the Thévenin voltage of the source. Without the follower, an error equal to IR_{Th} would appear. On the other hand, since the op amp has very small R_{Th}, current can be drawn from the output without affecting the voltage. The follower is widely used to *buffer* the output of a

---

[1] Or, less correctly, called a differential amplifier.

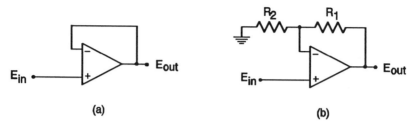

(a)   (b)

Figure 4-1   (a) A voltage follower; (b) a voltage follower with gain.

high-impedance source so that its voltage will be unchanged regardless of the current demands made upon it. In other words, the circuit
transforms a high-impedance voltage source into a very low-impedance source of exactly the same voltage.

An application of a buffer is shown in Figure 4-2. Calculation will
show that the voltage drop across the 5 kΩ load resistor in part (a) of
the figure is only 2.5 V, whereas in (b) it is the expected 5.0 V, which
will not vary as the load is changed over a wide range.

The modified follower circuit shown in Figure 4-1*b* is called a
follower with gain. The resistors R_1 and R_2 form a voltage divider
between the output of the amplifier and ground; hence the voltage
at the summing junction must be given by

$$E_{SJ} = E_{out} \frac{R_2}{R_1 + R_2}$$

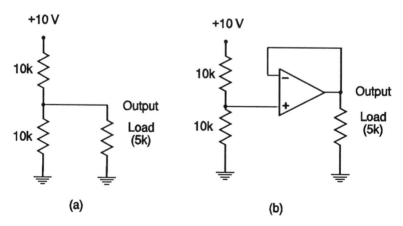

(a)   (b)

Figure 4-2   A voltage divider, (a) without buffering and (b) with a voltage follower
as buffer.

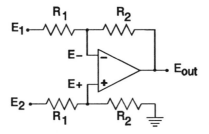

Figure 4-3 A difference amplifier, also called a subtractor. The equality of resistors on the top and bottom lines is not essential, but leads to simplified arithmetic.

Rearrangement of this equation, together with the virtual equality of E_{SJ} and E_{in} gives the governing expression

$$E_{out} = E_{in}\left[\frac{R_1 + R_2}{R_2}\right] = E_{in}\left[\frac{R_1}{R_2} + 1\right] \qquad (4\text{-}1)$$

Notice that this permits multiplication of the input by a fixed factor without change of sign. Since the expression includes the $+1$, the output can never be less than the input.

A more complex example of a difference amplifier, (Figure 4-3), shows two separate signals being sensed by the amplifier at the same time. To analyze this circuit, note that R_1 and R_2 form a voltage divider from E_2 to ground, so that E_+ is given by

$$E_+ = E_2\frac{R_2}{R_1 + R_2}$$

Similarly,

$$E_- = E_1 + \left[\frac{R_1}{R_1 + R_2}\right](E_{out} - E_1)$$

$$= \frac{E_1R_1 + E_1R_2 + E_{out}R_1 - E_1R_1}{R_1 + R_2}$$

but $E_- = E_+$, so we can write

$$\left[\frac{R_2}{R_1 + R_2}\right]E_2 = \frac{E_1R_2 + E_{out}R_1}{R_1 + R_2}$$

Multiplying both sides by $R_1 + R_2$ and rearranging gives

$$2R_2 - E_1 R_2 = E_{out} R_1$$

$$E_{out} = \frac{R_2}{R_1}(E_2 - E_1) \tag{4-2}$$

Thus we have an output proportional to the *difference* between the two inputs, with a gain given by the ratio of the resistors[2]

An important property of this configuration is its insensitivity to input changes that affect *both* inputs equally, a **common-mode** voltage. Consider what would happen if the two inputs of Figure 4-3 were tied together: the output would be zero, no matter what the common input voltage, up to the point where the amplifier saturates.

The Instrumentation Amplifier

The difficulty with the difference amplifier discussed above is its relatively low input impedance. It would be desirable to have a symmetrical difference amplifier with the high impedance of a voltage follower for both its inputs. This can be accomplished by placing followers in both two input lines, as shown in Figure 4-4a, but a more efficient arrangement is that of Figure 4-4b in the same figure.

Analysis of this circuit shows that the potentials at points marked A and B are:

$$E_A = \left(1 + \frac{R_3}{R_4}\right)E_1 - \frac{R_3}{R_4}E_2$$

$$E_A = \left(1 + \frac{R_3}{R_4}\right)E_2 - \frac{R_3}{R_4}E_1$$

so that, if $R_1 = R_2$

$$E_{out} = (E_2 - E_1)\left(1 + 2R_3/R_4\right)$$

Since R_3 is the equal value of two resistors, it is convenient to use R_4 to alter the gain. Thus R_4 often consists of a number of switch-

---

[2] The student can derive a similar expression for the case where all four resistors are different, but the exercise may not be worth the frustration.

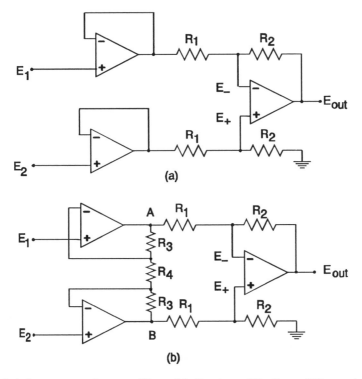

Figure 4-4 Instrumentation amplifiers: (a) direct modification of Figure 4-3; (b) a preferred circuit.

selectable resistors, called the **gain resistors**, to give convenient fixed gains, such as 10, 100, and 1000.

Instrumentation amplifiers are available as single ICs, and are widely employed where their combinations of high impedance, very low common-mode response, and exact gains are desirable.

The Sample-and-Hold Amplifier

Figure 4-5 shows a special purpose amplifier combination that is very useful in the collection of data, the **sample-and-hold** (S/H) amplifier. When the switch is closed, the signal E_{in} will appear unchanged at the output, and the capacitor will be charged to whatever voltage the signal has at the instant when the switch is opened. Since the capacitor is connected only to the high-impedance input of the second amplifier, it will hold its charge unchanged for a reasonable length of time, and the corresponding voltage will be present at the amplifier's output. Thus the unit **samples** the signal, then **holds** it until the switch is again

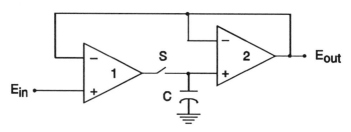

Figure 4-5 A sample-and-hold amplifier. The switch S is ordinarily a semiconductor unit .

closed (or the capacitor is shorted). This action is essential if a changing signal is to be converted to its digital equivalent, because the process of conversion takes appreciable time during which the signal must remain constant.

In a digital panel meter, for example, the meter reading must remain constant for a long enough time to be read by the user. Quickly changing digits are difficult or impossible to read. In many applications, a repetitive S/H operation with a period of something like half a second will allow the eye to scan the display painlessly even if the voltage being monitored is changing rapidly. The accuracy of a S/H circuit depends largely on the quality of the capacitor; if the latter has appreciable leakage resistance the "held" voltage will fall off ("droop"). The capacitance value is not critical, and 0.001 to 0.1 μF is usually suitable.

Diodes: A Preview

In the treatment of op amps we will have the occasion to include diodes as circuit elements. The full treatment of diodes will appear in a future chapter, but a short description is needed at this point.

A diode is a two-terminal passive semiconductor device that conducts current only in one direction. Its two terminals are labeled **anode** and **cathode**. When current passes through the diode, a voltage is present across it (about 0.7 V), called the **forward voltage drop**. Diodes are introduced in a circuit to discriminate between positive and negative polarities, and to convert AC into DC (rectification).

Sine-Wave Oscillators

Many electronic applications call for an AC source. Sometimes the power line can be used, but otherwise a sine-wave oscillator (or "generator") is required. Suitable units can be designed around op

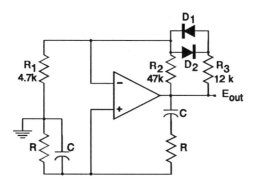

Figure 4-6 A sine-wave oscillator in which a Wien bridge determines the frequency, given by $f = 1/(2\pi RC)$. The values given for R_1, R_2, and R_3 are representative only.

amps in a number of ways. The frequency is determined by an RC network providing positive feedback at some selected frequency. Figures 4-6 and 4-7 show two such circuits.

In the circuit of Figure 4-6, called a **Wien-bridge oscillator,** the frequency is determined by the series and parallel RC networks connected to the noninverting input of the amplifier. The impedance of the parallel RC combination decreases with increasing frequency, whereas that of the series branch increases, and the circuit oscillates at the frequency where these two impedances are equal and the phases are such that the amplifier receives positive feedback. The ratio between the other two arms of the bridge must be adjusted to give sufficient feedback to maintain oscillation; $R_2/R_1 = 10$ is about right. If the two resistors marked R and the two capacitors C are equal, the frequency is $f = 1/(2\pi RC)$.

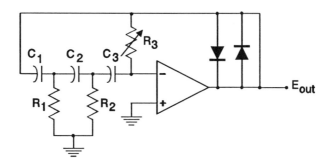

Figure 4-7 A phase-shift oscillator.

The pair of crossed diodes D_1 and D_2 serve to stabilize the amplitude of oscillations. If the output of the bridge is less than a few hundred millivolts, neither diode will pass current. When the output is fairly large, however, the two diodes will start to conduct, thus effectively placing R_3 in parallel with R_2 and reducing the amplifier gain. An equilibrium is quickly reached that results in a sine wave of very stable amplitude.

Another way of producing positive feedback is found in a circuit called a **phase-shift oscillator**, which incorporates a network that will return a portion of the output to the inverting input, but with its phase shifted by 180°. An oscillator designed this way is shown in Figure 4-7. The phase-shifting network is composed of three RC segments which may include a variable resistor to permit adjustment of the frequency. Each of the high-pass elements ($R_1 C_1$, $R_2 C_2$, $R_3 C_3$) shifts the phase by exactly 60° for some particular frequency, the three segments combining to give the 180° phase shift required for oscillations. If all resistors and all capacitors are made equal, the frequency is given by $f = 1/(2\sqrt{6}\,\pi RC)$. If these components are not all equal, the frequency must be calculated by the relation:

$$f = \frac{1}{2\pi\sqrt{6}\,\sqrt[3]{R_1 C_1 R_2 C_2 R_3 C_3}}$$

It should be realized that the maximum frequency obtainable from an op-amp oscillator is limited by the frequency characteristic of the amplifier. For frequencies above 100 kHz, it is better to use an oscillator designed around a single transistor, described in Chapter 6.

The Precision Rectifier

The circuit shown in Figure 4-8, known as a **precision rectifier** or **perfect rectifier**, utilizes cross-connected diodes with resistors in series. The connection to the second stage is not taken directly from the output of this amplifier, but rather two outputs are taken from points between the diodes and resistors, and then fed into a difference amplifier.

To follow how this works, consider what happens when the input is positive. Because of the inversion inherent in an op amp, the upper diode will now conduct and the lower one will be cut off. Because of the equality of resistors, the potential at X will be equal to the negative

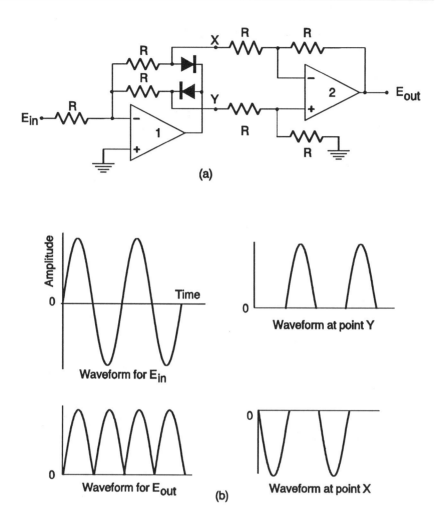

Figure 4-8 (a) A precision rectifier, sometimes known as a "perfect" rectifier. For convenient comparison, all resistors are assumed to be equal. (b) Waveforms are shown for key points in the circuit.

of the input voltage; during the same time Y will be effectively at ground. Hence the output of the second amplifier is $V_Y - V_X$ making E_{out} identical to the positive input. Likewise, if the input is negative, only the lower diode will conduct, and the potential at V_Y will be equal to the input but of opposite sign while $V_X = 0$. Now, $V_Y - V_X$ is again positive, and equal in magnitude to the negative input. The circuit generates the absolute value of the input, since it inverts only the negative component. On the other hand, if the input is given an AC signal, the output will be full-wave rectified. A low-pass filter with cutoff frequency about 1/10 of the input frequency will convert this to a steady DC.

Op-Amp Specifications

Up to this point we have presented op amps with the assumption that they are "perfect," that is to say, that they obey the simple formulas derived for the various configurations. Actually very few things in this world are truly perfect, and op amps are no exception, though they can be close to perfection if operated at moderate gain, frequency and output current. Op amps can only be completely specified by a long list of properties, and the circuit designer must often choose between different models by trading-off one property against another.

We will now discuss the major descriptives, and tabulate them for a number of commercial amplifiers that have been optimized for various desirable features or applications. Then we will explore ways for measuring and minimizing the residual errors.

The Specification Sheet

We will start by a discussion of the model 741, a general-purpose op amp that is used so widely that it serves as a convenient basis for comparison for other types. Figure 4-9 is a reproduction of the specification sheet for the 741. Its format is typical of most spec sheets, not only for op amps but for many other integrated circuits as well (there is some variation in style from one manufacturer to another, but the same data must be given). Let us look at each feature in turn.

The first item is the heading. Note that the "741" is here designated as μA741M and μA741C. The prefix μA indicates that this type was originally devised by Fairchild, though now manufactured by many firms[3] The letters "M" and "C" are the designations of Texas Instruments (TI) for different temperature ranges as will discussed below. Sometimes other letters are appended designating different shapes of the finished product such as dual-in-line (DIP) plastic, ceramic, or cylindrical can housing, but these vary between manufacturers.

---

[3] Some of the manufacturers' prefixes are "AM" (AMD Corp.), "μA" (Fairchild Semiconductor), "MC" (Motorola), "LM" (National Semiconductor), "OP" (Precision Monolithics), "RC" (Raytheon), "NE" (Signetics), "SG" Silicon General, "TLC" (Texas Instruments).

Beneath the heading is a list of salient features and a short description, intended to tell the circuit designer whether or not this model is appropriate for the intended application. Also on this page are pin-out diagrams of the several styles in which the 741 is available (the 8-pin DIP package is most convenient for small-scale laboratory use; the others may be more appropriate for use in large instruments, such as computers.)

On the next page is given an equivalent schematic of the entire circuit; this one shows no less than 22 transistor units, 1 diode, 1 capacitor, and 11 resistors, all of which are fabricated on a single chip of silicon. Also on this page is found a table of "absolute maximum ratings" which should not be exceeded for fear of damage to the unit. At this point, we find two columns of figures, headed μA741M and μA741C. Most op amps are specified in more than one rated type, often more than the two seen here; these differ in certain data, showing that one type has more relaxed specifications than another (priced accordingly). In the present case, the chief difference is the maximum permitted power supply voltages and operating temperature.

On the third page is a long table of electrical characteristics (we explain each of these below). Note that the data are listed under "MIN," "TYP," and "MAX." To take the first entry as an example, the input offset voltage is typically 1 mV but is guaranteed never to be greater than the MAX limit, 5 mV. Some items, such as this one, should be as small as possible, so only a MAX value is given, whereas others, such as "common-mode input voltage range" should be as large as possible, so only the MIN limit is guaranteed. For rough calculations, one can use the TYP values, but to plan for the "worst-case" situation, the extreme values should be used.

Underneath is a similar shorter table of "operating characteristics," which have to do with dynamic properties of the amplifier, usually of importance only in high-speed or high-frequency work. In the next page is given a circuit to be used for measuring these dynamic properties. Under "typical application data" is shown the usual triangular symbol for the amplifier with a 10-kΩ nulling potentiometer, often called a "balance pot." This is an optional addition that can be used to bring the output to exactly zero, in the absence of signals. There follow several pages of graphical data showing the effects of a number of variables on the action of the amplifier. Figures 3 and 4 in the spec sheet, for instance, show that the input offset current and bias current (to be defined later) both fall off with increased temperature.

- Short-Circuit Protection
- Offset-Voltage Null Capability
- Large Common-Mode and Differential Voltage Ranges
- No Frequency Compensation Required
- Low Power Consumption
- No Latch-up
- Designed to be Interchangeable with Fairchild μA741M, μA741C

description

The uA741 is a general-purpose operational amplifier featuring offset-voltage null capability.

The high common-mode input voltage range and the absence of latch-up make the amplifier ideal for voltage-follower applications. The device is short-circuit protected and the internal frequency compensation ensures stability without external components. A low potentiometer may be connected between the offset null inputs to null out the offset voltage as shown in Figure 2.

The uA741M is characterized for operation over the full military temperature range of −55°C to 125°C; the uA741C is characterized for operation from 0°C to 70°C.

symbol

NONINVERTING INPUT IN +

INVERTING INPUT IN −

OUTPUT

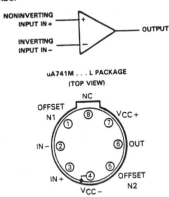

uA741M . . . L PACKAGE
(TOP VIEW)

PIN 4 IS IN ELECTRICAL CONTACT WITH THE CASE

uA741M . . . J PACKAGE
(TOP VIEW)

NC	1	14 NC
NC	2	13 NC
OFFSET N1	3	12 NC
IN −	4	11 VCC +
IN +	5	10 OUT
VCC −	6	9 OFFSET N2
NC	7	8 NC

uA741M . . . JG PACKAGE
uA741C . . . D, JG, OR P PACKAGE
(TOP VIEW)

OFFSET N1	1	8 NC
IN −	2	7 VCC +
IN +	3	6 OUT
VCC −	4	5 OFFSET N2

uA741M . . . U FLAT PACKAGE
(TOP VIEW)

NC	1	10 NC
OFFSET N1	2	9 NC
IN −	3	8 VCC +
IN +	4	7 OUT
VCC −	5	6 OFFSET N2

uA741M . . . FK PACKAGE
(TOP VIEW)

NC — No internal connection

TEXAS INSTRUMENTS

POST OFFICE BOX 655012 • DALLAS, TEXAS 75265

Figure 4-9 Typical op-amp specification sheet. See continuation in the next few pages. Courtesy of Texas Instrument Inc.

schematic

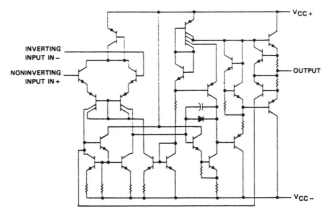

INVERTING
INPUT IN −

NONINVERTING
INPUT IN +

V$_{CC+}$

OUTPUT

V$_{CC-}$

absolute maximum ratings over operating free-air temperature range (unless otherwise noted)

		uA741M	uA741C	UNIT
Supply voltage V$_{CC+}$ (see Note 1)		22	18	V
Supply voltage V$_{CC-}$ (see Note 1)		−22	−18	V
Differential input voltage (see Note 2)		±30	±30	V
Input voltage any input (see Notes 1 and 3)		±15	±15	V
Voltage between either offset null terminal (N1/N2) and V$_{CC-}$		±0.5	±0.5	V
Duration of output short-circuit (see Note 4)		unlimited	unlimited	
Continuous total power dissipation		See Dissipation Rating Table		
Operating free-air temperature range		−55 to 125	0 to 70	°C
Storage temperature range		−65 to 150	−65 to 150	°C
Case temperature for 60 seconds	FK package	260		°C
Lead temperature 1.6 mm (1/16 inch) from case for 60 seconds	J, JG, or U package	300	300	°C
Lead temperature 1.6 mm (1/16 inch) from case for 10 seconds	D or P package		260	°C
Lead temperature 1.6 mm (1/16 inch) from case for 10 seconds	L package	300		°C

NOTES: 1. All voltage values, unless otherwise noted, are with respect to the midpoint between V$_{CC+}$ and V$_{CC-}$.
2. Differential voltages are at the noninverting input terminal with respect to the inverting input terminal.
3. The magnitude of the input voltage must never exceed the magnitude of the supply voltage or 15 V whichever is less.
4. The output may be shorted to ground or either power supply. For the uA741M only, the unlimited duration of the short-circuit applies at (or below) 125 °C case temperature or 75 °C free-air temperature.

DISSIPATION RATING TABLE

PACKAGE	T$_A$ ≤ 25°C POWER RATING	DERATING FACTOR	DERATING ABOVE T$_A$	T$_A$ = 70°C POWER RATING	T$_A$ = 125°C POWER RATING
D	500 mW	5.8 mW/°C	64°C	464 mW	N/A
FK	500 mW	11.0 mW/°C	105°C	500 mW	275 mW
J (uA741M)	500 mW	11.0 mW/°C	105°C	500 mW	275 mW
JG (uA741M)	500 mW	8.4 mW/°C	90°C	500 mW	210 mW
JG (all others)	500 mW	N/A	N/A	500 mW	N/A
L	500 mW	6.7 mW/°C	75°C	500 mW	167 mW
P	500 mW	N/A	N/A	500 mW	N/A
U	500 mW	5.4 mW/°C	57°C	432 mW	135 mW

TEXAS
INSTRUMENTS
POST OFFICE BOX 655012 • DALLAS, TEXAS 75265

Figure 4-9-continuation. Typical op-amp specification sheet. Courtesy of Texas Instrument Inc.

electrical characteristics at specified free-air temperature, V_{CC+} = 15 V, V_{CC-} = –15 V

PARAMETER		TEST CONDITIONS†		uA741M			uA741C			UNIT
				MIN	TYP	MAX	MIN	TYP	MAX	
V_{IO}	Input offset voltage	V_O = 0	25°C		1	5		1	6	mV
			Full range			6			7.5	
$\Delta V_{IO(adj)}$	Offset voltage adjust range	V_O = 0	25°C	±15			±15			mV
I_{IO}	Input offset current	V_O = 0	25°C		20	200		20	200	nA
			Full range			500			300	
I_{IB}	Input bias current	V_O = 0	25°C		80	500		80	500	nA
			Full range			1500			800	
V_{ICR}	Common-mode input voltage range		25°C	±12	±13		±12	±13		V
			Full range	±12			±12			
V_{OM}	Maximum peak output voltage swing	R_L = 10 kΩ	25°C	±12	±14		±12	±14		V
		R_L ≥ 10 kΩ	Full range	±12			±12			
		R_L = 2 kΩ	25°C	±10	±13		±10	±13		
		R_L ≥ 2 kΩ	Full range	±10			±10			
A_{VD}	Large-signal differential voltage amplification	R_L ≥ 2 kΩ	25°C	50	200		20	200		V/mV
		V_O = ±10 V	Full range	25			15			
r_i	Input resistance		25°C	0.3	2		0.3	2		MΩ
r_o	Output resistance	V_O = 0, See Note 5	25°C		75			75		Ω
C_i	Input capacitance		25°C		1.4			1.4		pF
CMRR	Common-mode rejection ratio	V_{IC} = V_{ICR} min	25°C	70	90		70	90		dB
			Full range	70			70			
k_{SVS}	Supply voltage sensitivity ($\Delta V_{IO}/\Delta V_{CC}$)	V_{CC} = ±9 V to ±15 V	25°C		30	150		30	150	μV/V
			Full range			150			150	
I_{OS}	Short-circuit output current		25°C	±25	±40		±25	±40		mA
I_{CC}	Supply current	No load, V_O = 0	25°C		1.7	2.8		1.7	2.8	mA
			Full range			3.3			3.3	
P_D	Total power dissipation	No load, V_O = 0	25°C		50	85		50	85	mW
			Full range			100			100	

† All characteristics are measured under open-loop conditions with zero common-mode input voltage unless otherwise specified. Full range for uA741M is –55°C to 125°C and for uA741C is 0°C to 70°C.
NOTE 5: This typical value applies only at frequencies above a few hundred hertz because of the effects of drift and thermal feedback.

operating characteristics, V_{CC+} = 15 V, V_{CC-} = –15 V, T_A = 25°C

PARAMETER		TEST CONDITIONS		uA741M			uA741C			UNIT
				MIN	TYP	MAX	MIN	TYP	MAX	
t_r	Rise time	V_I = 20 mV,	R_L = 2 kΩ,		0.3			0.3		μs
	Overshoot factor	C_L = 100 pF,	See Figure 1		5%			5%		
SR	Slew rate at unity gain	V_I = 10 V,	R_L = 2 kΩ,		0.5			0.5		V/μs
		C_L = 100 pF,	See Figure 1							

Texas Instruments
POST OFFICE BOX 655012 • DALLAS, TEXAS 75265

Figure 4-9-continuation. Typical op-amp specification sheet. Courtesy of Texas Instrument Inc.

PARAMETER MEASUREMENT INFORMATION

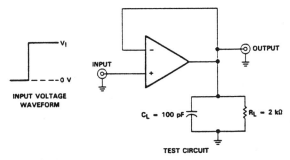

FIGURE 1. RISE TIME, OVERSHOOT, AND SLEW RATE

TYPICAL APPLICATION DATA

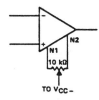

FIGURE 2. INPUT OFFSET VOLTAGE NULL CIRCUIT

Texas
Instruments
POST OFFICE BOX 655012 • DALLAS, TEXAS 75265

Figure 4-9-continuation. Typical op-amp specification sheet. Courtesy of Texas Instrument Inc.

TYPICAL CHARACTERISTICS

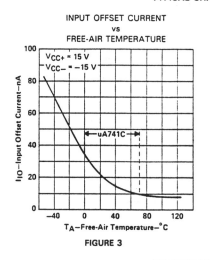

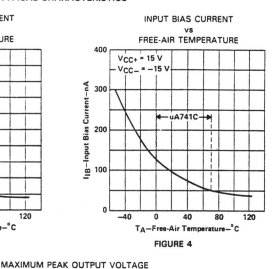

Figure 4-9-continuation. Typical op-amp specification sheet. Courtesy of Texas Instrument Inc.

TYPICAL CHARACTERISTICS

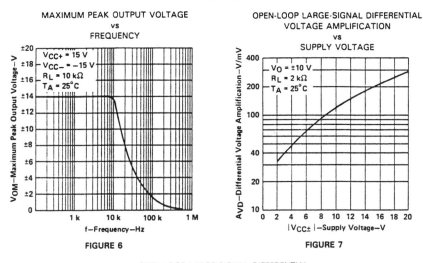

MAXIMUM PEAK OUTPUT VOLTAGE
vs
FREQUENCY

FIGURE 6

OPEN-LOOP LARGE-SIGNAL DIFFERENTIAL
VOLTAGE AMPLIFICATION
vs
SUPPLY VOLTAGE

FIGURE 7

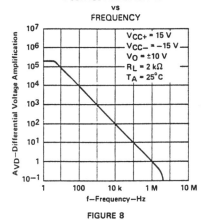

OPEN-LOOP LARGE-SIGNAL DIFFERENTIAL
VOLTAGE AMPLIFICATION
vs
FREQUENCY

FIGURE 8

TEXAS
INSTRUMENTS
POST OFFICE BOX 655012 • DALLAS. TEXAS 75265

Figure 4-9-continuation. Typical op-amp specification sheet. Courtesy of Texas Instrument Inc.

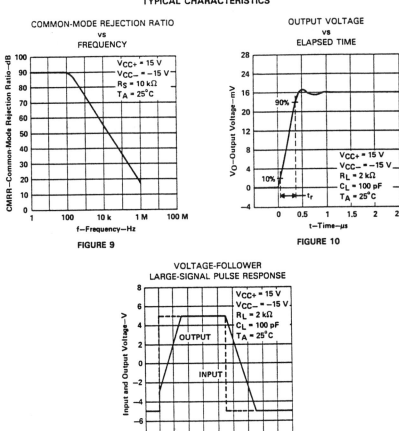

TYPICAL CHARACTERISTICS

FIGURE 9

FIGURE 10

FIGURE 11

TEXAS
INSTRUMENTS
POST OFFICE BOX 655012 • DALLAS. TEXAS 75285

Figure 4-9-continuation. Typical op-amp specification sheet. Courtesy of Texas Instrument Inc.

Figures 8 and 9 are Bode plots taken under specified conditions. Some spec sheets, particularly for less familiar amplifier types, include a few pages of circuit diagrams for typical applications. We will now discuss the specific entries in the tables of Figure 4-9.

Maximum Ratings

Note the distinction between "supply voltage," designated by V_{CC}, which is the voltage delivered by the power supply, and "input voltage," which is the potential at any input. In general, the input voltage should never be allowed to be as great as the supply voltage. The "voltage between either offset null terminal and V_{CC-}" refers to the nulling potentiometer (shown in Figure 2 of the spec sheet) of which the sliding contact goes to V_{CC-}; this maximum specification is to prevents damage to the amplifier from excessive voltage at this point.

Some amplifiers would be damaged by a continuous short circuit of the output to ground or to the power supply terminals; the listing shows that this is not true of the 741. The power dissipation can be important, especially if many ICs are mounted in close quarters where ventilation is poor. This ties in with the operating temperature range, as shown in Note 5. Storage temperature ranges are always less restrictive than operating temperatures.

Electrical Characteristics

The **input offset voltage** (V_{IO}) is an undesirable, internally-generated, input voltage; it can be regarded as the voltage that must be applied to the input terminals to result in zero output. The spec sheet shows it at 25°C and over the full temperature range. The entry for "$\Delta V_{IO(adj)}$" gives the range over which the offset voltage can be adjusted by the nulling pot.

The **input bias current** (I_{IB}) is the minute current (measured in nA) flowing into either amplifier input. The **input offset current** is the difference between the two bias currents when the output voltage is zero.

The **common-mode input voltage range** (V_{ICR}): The term "common-mode" refers to a voltage applied in common to the two inputs; ideally it should have no effect on the output, which should only respond to the difference between the two inputs. The table entry shows that V_{ICR} can be as great as ± 12 V guaranteed. The **common-mode rejection ratio** (CMRR) is shown as typically 90 dB. This is the ratio of V_{ICR} to the output voltage, V_O when V_{ICR} is applied to both inputs. (Note that 90 dB corresponds to a voltage ratio of about 31,000.)

The **maximum peak output voltage swing,** V_{OM}, tells us that the 741 will saturate at (typically) $V_O = \pm 13$ V if the load is 2 kΩ, a little greater for larger load resistances. This can be interpreted to mean that saturation will occur at about 2 V below the power supply (V_{CC}); thus if the power supply is ± 12 V, saturation can be expected at about ± 10 V.

The **large-signal voltage amplification** (A_{VD}) is the ratio of the output voltage swing to the input voltage swing, when the output is driven to a specified value, given in the table as $V_O = \pm 10$ V with a load resistance not less than 2 kΩ. This is also called the **open-loop gain**, and should not be confused with circuit gain, which is ordinarily given by $-Z_f/Z_{\text{in}}$. Note that the typical value, given as 200 V/mV, is approximately 100 dB.

The **input resistance** (r_i) is the resistance seen at either input if the other input is grounded. It is desirable for this to be as high as possible to minimize loss of current to the input. The 2-MΩ value is typical of the 741. If the inputs are at 10 V, as might be the case for a follower, a maximum current of 10 V / 2 MΩ = 5 μA will flow into the amplifier; this must be kept in mind in designing external circuitry, so that this 5 μA will always be negligible. The **output resistance** (r_o) on the other hand, should be as low as possible; it is the internal resistance through which the output current must flow. The **input capacitance,** (C^i) is only of significance at high frequencies, where it contributes to the phase shift observed in passing a signal through the amplifier. The relation between some of these quantities is represented in Figure 4-10.

The **supply-voltage sensitivity** (k_{SVS}), also known as the **power-supply rejection ratio** (PSRR), is the ratio, often expressed in decibels, of a change in offset voltage V_{IO} to the change in V_{CC} causing it. This is seldom significant if a regulated power supply is used.

The **short-circuit current** (I_{OS}) is the greatest current that the amplifier can deliver into a low resistance load. The **supply current** (I_{CC}) is the current that the amplifier will draw from the power supply when no output current is flowing. The **total power dissipation** (P_D) represents the power (in mW) that the amplifier will convert into heat when $I_o = 0$.

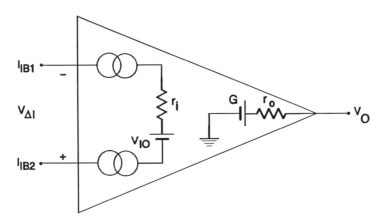

Figure 4-10 The equivalent circuit for an operational amplifier, showing the relations between the input currents (I_{IB1} and I_{IB2}), the differential input voltage ($V_{\Delta I}$), the input resistance (r_i), the input offset voltage (V_{IO}), the output resistance (r_O), and the output voltage (V_O). The G represents a voltage source yielding the product of $V_{\Delta I}$ and the open-loop gain.

Operating Characteristics

The dynamic properties of op amps are best explained with the aid of Figure 4-11. When a step voltage is applied suddenly to the input of an op amp, the response of the output will rise from zero to its final value following the curve shown. The **rise time** is defined as the time required for the response to rise from 10% to 90% of the final value, these points are chosen arbitrarily to avoid the curvature at both

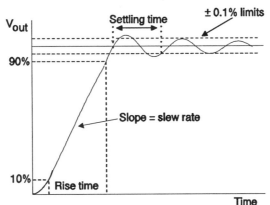

Figure 4-11   Illustration of the dynamic behavior of an op amp connected as a summer. The curve represents the output response to a step input. The oscillations around the final value, and the marked limits, are exaggerated for clarity.

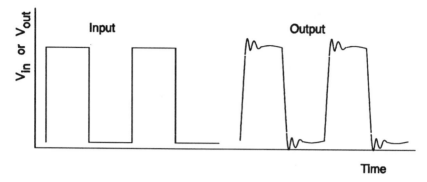

Figure 4-12 Distortion of a square wave input by an amplifier with insufficiently fast rise and settling times. The oscillations are called "ringing."

ends. The response generally overshoots the equilibrium output, oscillating about that value for a short time. The **settling time** is defined as the elapsed time after the response curve first crosses the final value until the oscillations quiet to within a specified band, shown here to be $\pm 0.1\%$. The **slew rate** is the slope of the nearly linear portion of the rising curve. These dynamic properties are significant not only when high frequencies are to be amplified, but also when square waves or square pulses are to be handled. For example, if the slew rate is too low and the settling time too long, a square wave input would give a distorted output, as shown in Figure 4-12.

Examples of Op Amps

The 741 is a general-purpose amplifier, particularly useful because it is inexpensive and ubiquitous but nevertheless of high quality. For a specific application, however, it may not be the best choice. Amplifiers are available that have been optimized in manufacture for specific kinds of circuits. These, as one would expect, generally command higher prices. A selection of op amps appear in Table 4-1 for easy comparison. For example, if a high current is needed, the LM675 can provide it, but if high frequency response is important, AD509 is a better choice.

Noise

As we have seen in Chapter 2 all electrical circuits contain noise, consisting of unwanted variations in the signal. Noise can be random, as is the case with resistance noise arising from fluctuations in the motion of electrons in resistors, or it can be repetitive, as with pick-up

Table 4-1

Characteristics of selected operational amplifiers (typical values)

Paramaters	741	LM308A	OP-07	CA3130B	AD509	LF400C	LM675	OP A 128	Units
Special features	[note 1]		[note2]	[note 3]	[note 4]	[Fast Settling]	[Power]	[note 5]	
			[Low offset]	[FET input]	[High Speed]			[Electrometer]	
Open-loop gain	100	300	500	320	10	300	32	2500	V/mV
Input impedance	2	40	80	1.5E06	50	100 000	--	1E09	megohms
Output impedance	75	500	60	--	--	75	--	100	ohms
Offset voltage	5	0.3	0.025	1	5	0.5	1	0.26	mV
Offset temp. coeff.	15	2	0.01	5	20	--	25	20	μV/K
Bias current	30	1.5	0.7	0.005	100	0.2	200	150 fA	nA
Offset current	200	0.2	0.3	0.0005	25	0.4	50	30 fA	nA
CMRR	90	110	125	100	80	100	90	118	dB
PSRR	90	110	130	90	--	100	90	--	dB
Bandwidth	1.5	1	0.3	15	20	16	5.5	1	MHz
Short circuit curr.	25	6	15	22	5	45	1190	10	mA
Slew rate	0.5	0.2	0.3	10	120	70	8	3	V/μs
Settling time	--	--	--	--	200	200	--	5000	ns [note7]
Noise voltage	--	--	0.01	23 μV [note 6]	0.03	0.023	--	<0.1	μV/SQRT(Hz)

Notes

1. Models 108,208, and 308, differ principally in the permitted temperature ranges.
2. Precision Monolithics Inc.
3. RCA Corporation.
4. Analog Devices.
5. Burr-Brown Corp.
6. For bandwidth = 0.2 MHz.
7. To 0.1% of the final value.

89

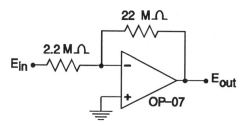

Figure 4-13 Circuit to illustrate calculations of ultimate sensitivities. The current noise referred to the input is taken to be 0.1 pA /$\sqrt{Hz}$.

at the power line frequency resulting from unshielded transformers or fluorescent light ballasts.

Noise is observed at the output of an amplifier but is referred to the input by means of the noise-equivalent power (NEP). Thus an amplifier with a power gain of 1000 and 10 mW of noise at the output has a NEP of 10 μW. Noise is generated in all stages of the amplifier, but is most significant in the input circuit, because its effect is then amplified. Such noise is usually specified in terms of voltage and current rather than power. Thus at 1000 Hz, an OP07 premium amplifier exhibits about 0.01 μV/$\sqrt{Hz}$ and 0.1 pA/$\sqrt{Hz}$ of noise. The ratio of these two quantities is sometimes referred to as the **characteristic noise resistance** R_n; in the above example, $R_n = (10^{-8}\ V)/(10^{-13}A) = 10^5\ \Omega$. If the parallel combination of all resistances connected to the summing junction, R_T, is larger than R_n, current noise predominates, whereas below that point the voltage noise is larger.

Let us determine the minimum amplitude of a 1-kHz signal that can be handled by the circuit shown in Figure 4-13, in terms of the S/N ratio. Let us assume the bandwidth to be 100 Hz, and the minimum S/N ratio that is acceptable to be 10. First, calculate the parallel combination of resistors connected to the summing junction: $R_T = 2.2 || 22 = 2\ M\Omega$. Since this is larger than R_n as calculated in the previous paragraph, we can conclude that current noise predominates. Now we can calculate the voltage resulting when the specified current passes through this resistance, as

$$e_n = \left(0.1\ \frac{pA}{\sqrt{Hz}} \right) \left(\sqrt{100\ Hz} \right) (2 \times 10^6\ \Omega) = 2\ \mu V\ \text{(RMS)}$$

$$(4\text{-}4)$$

Consequently, if the S/N ratio is to be greater than 10, the input signal must be larger than 20 μV.

Note that the overall gain of the circuit does not affect the S/N ratio, since both signal and noise are amplified by the same amount. The above calculation does not take into account the noise contribution of the two resistors. If we consider them together as the equivalent 2-MΩ resistor, the Johnson noise can be calculated as

$$e_{RMS} = 1.3 \times 10^{-10} \sqrt{2 \times 10^6} \sqrt{100} = 1.8 \mu V \quad (4\text{-}5)$$

The amplifier noise and the resistance noise do not add together directly but as the square-root of the sum of their squares (this is because they are independent random processes):

$$e_{total} = \sqrt{e^2 \text{(amplifier)} + e^2 \text{(resistance)}} \quad (4\text{-}6)$$

It should be noted that this calculation only gives a lower limit. Other types of noise, such as electromagnetic interference, are apt to be present and further increase the total noise.

Offset and Bias Errors

The drift of the offset voltage and bias current also constitute a kind of noise because the exact values and the signs they will take cannot be predicted beforehand. In contrast, the *initial* offset and bias can be measured, and hence should not be combined in RMS fashion. If we again make use of the symbol R_T for the parallel combination of input and feedback resistors in an inverting circuit, we can write for the total error, referred to the input:

$$e_{total} = V_{OS} + I_B R_T + \sqrt{\Sigma(\text{drifts})^2 + \Sigma(\text{noise})^2} \quad (4\text{-}7)$$

where the drifts are determined by time fluctuations in the offset voltage V_{OS}, bias current, I_B and by random temperature variations.

Errors Caused by Finite Gain

The fundamental magic by which op amp circuits manage to keep errors negligible is ascribable to the negative feedback used in connection with a basic amplifier of very high open-loop gain A reduced

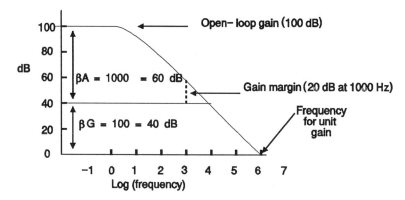

Figure 4-14 The relationship between various gain concepts. The circuit closed loop gain is specified as $G = 100$, which corresponds to 40 dB. The gain margin is constant at 60 dB up to a few Hz, then rolls off at 20 dB per decade.

to a smaller closed-loop gain G. It has been shown previously that the behavior of a summing amplifier can be described by the relation

$$\frac{E_{in}}{E_{out}} = \frac{R_{in}}{R_f}(1 + \frac{1}{\beta A}) \qquad (4\text{-}8)$$

so that the relative deviation from ideal behavior is of the order of $1/\beta A$. The constant β is given in terms of G by the equation

$$\beta = \frac{1}{\dfrac{R_f}{R_{in}} + 1} = \frac{1}{G + 1} \qquad (4\text{-}9)$$

The quantity β is thus approximately equal to $1/G$, i.e. the reciprocal of the overall gain of the circuit. The product βA, called the **gain margin**, is important in determining circuit errors, as will be shown below. When βA is large, the term $1/\beta A$ in Eq. (4-8) becomes negligible and the equation reduces to the familiar form

$$E_{out} = -E_{in}\frac{R_f}{R_{in}} = -GE_{in} \qquad (4\text{-}10)$$

The larger the quantity βA, the closer the agreement with this "ideal" equation. For example, if the open-loop gain A is 100,000, and the ratio R_f/R_{in} is 100 (the closed loop gain, $G = 100$), the value of β

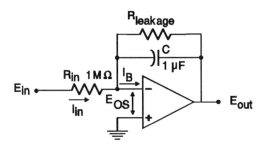

Figure 4-15 An integrator circuit showing the sources of error.

becomes 1/101 and the term $(1 + 1/\beta A)$ in Eq (4-8) can be calculated to be $1 + 101/100,000 = 1.001$. The deviation from ideality in this case is a negligible 0.1%, and Eq. (4-10) can be used with confidence. We say that the operational amplifier is behaving ideally.

The situation changes at higher over-all gains (β small) and at higher frequencies. Figure 4-14 illustrates the effect of increased frequency on the summing amplifier. The gain margin βA is constant at 1000 (or 60 dB) from DC to about 2 Hz. Beyond that frequency the gain margin decreases together with the open-loop gain at a rate of 20 dB per decade. It can be seen from the figure that the gain margin becomes 100 at 100 Hz and only 10 at 1 kHz. At this latter frequency the deviation from ideality becomes 10%. The effect is not as bad as it might seem, since "deviation" really means only that the gain no longer obeys the simple Eq. (4-10), but follows Eq. (4-8) instead.

The gain margin is more important in its effect on other errors, an important one being **distortion**. No amplifiers *exactly* reproduce the input waveform at the output. This effect, called distortion (D) has a value of perhaps 2% for an open-loop op amp. With feedback, D can be shown to decrease by a factor equal to the gain margin:

$$D_{\text{feedback}} = \frac{D_{\text{open loop}}}{\beta A} \qquad (4\text{-}11)$$

Thus a gain margin of about 20 would lower the distortion to an acceptable 0.1%. In the example of Figure 4-14 this gain margin is met by any frequency below about 500 Hz. From such data one can conclude that a general purpose op amp is not suited for high-fidelity (Hi-Fi) application, which allows no more than 0.05% distortion. The reader can verify, however, that an op amp with a gain-bandwidth

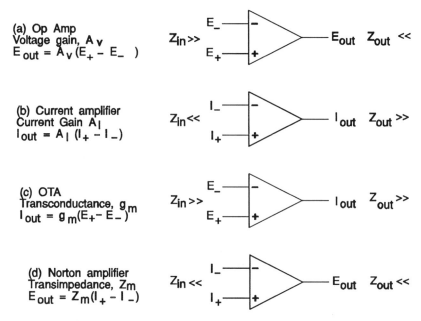

Figure 4-16 The four possible configurations of operational amplifiers: (a) The conventional "op amp." (b) A current gain amplifier, seldom used. (c) Operational transconductance amplifier (OTA). (d) The Norton transimpedance amplifier. In each the symbol "<<" means "very small," and ">>" signifies "very large."

product of about 20 MHz can be useful in this application, if the gain per amplifier is not too large. This points out an important caveat. One cannot have high closed-loop gain and wide frequency range simultaneously, without entailing considerable error.

Integrator Errors

If an integrator is to be used effectively, it is essential to compute the extent of offset and bias errors, as indicated in Figure 4-15. For high quality capacitors suitable for use in integrators, the leakage resistance is about 10^{11} $\Omega/\mu F$. Let us assume that a good quality amplifier, the OP07, is selected, for which $E_{OS} = 0.025$ mV and the bias current I_B = 0.7 nA. The operation of the circuit requires that

$$I_{in} = -C\frac{dE_{out}}{dt} \tag{4-12}$$

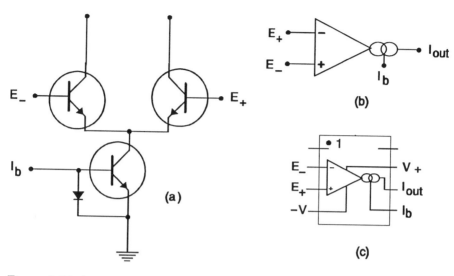

Figure 4-17 An operational transconductance amplifier. (a) The input circuitry, showing the bias control. (b) The schematic representation. (c) Pinout diagram for the type 3080 OTA.

In addition there exists an error current that also charges the capacitor:

$$I_{error} = I_B + \frac{E_{OS}}{R_{in}} + \frac{E_{out}}{R_{leakage}} \qquad (4\text{-}13)$$

For the data given, with $E_{out} = 10$ V, the error current becomes

$$I_{error} = (0.7 \times 10^{-9}) + \frac{0.025 \times 10^{-3}}{1 \times 10^6} + \frac{10}{10^{11}}$$
$$\approx 1 \times 10^{-9}\,A = 1nA \qquad (4\text{-}14)$$

This current will charge the capacitor at the rate of

$$(1 \times 10^{-9}) / (1 \times 10^{-6}) = 1 \text{ mV/s}$$

independently of the signal being integrated. In practice, this means that for a maximum error of 1 percent, the least permissible rate of integration is 100 mV/s or 6 V/min. Thus integrations over long periods of time (hours) are seldom feasible with op amps.

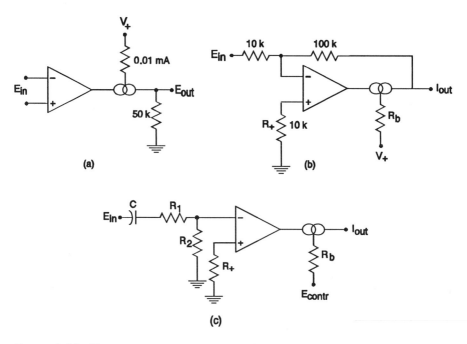

Figure 4-18 Three examples of the use of OTAs.

Other Classes of Operational Amplifiers

Up to this point we have assumed implicitly that the primary function of the amplifier is to manipulate voltages. This accounts for the mandatory high impedance input and low impedance output. However, this is not the only possibility, since information can be carried by either current or voltage. There are thus four alternative configurations, as depicted in Figure 4-16, corresponding to various combinations of large and small input and output impedances.

Operational Transconductance Amplifiers

The **operational transconductance amplifier** (OTA) responds to input voltages, producing a current output. The principal distinguishing feature is that the gain, with the units of a conductance, is controllable by a bias current I_b. In fact, not only the gain but also the values of the input and output impedances, the power requirements, and voltage swings, are all controlled by the bias current. As will be seen in Figure 4-17, an OTA is provided with a separate pin to accept an

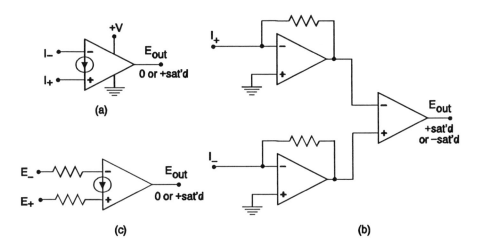

Figure 4-19 The Norton transconductance amplifier: (a) The circuit symbol. (b) An equivalent circuit using conventional op amps. (c) A comparator using a Norton amplifier.

externally adjustable bias current, I_b. The output is given by the formula

$$I_{out} = kI_b\,(E_+ - E_-) \qquad\qquad (4\text{-}15)$$

The quantity kI_b is known as the transconductance, with units of reciprocal ohms (siemens). For output currents in milliamperes and potentials in volts, the proportionality factor, k, is of the order of 10 V^{-1}. The circuits involving OTA are somewhat similar to those involving op-amps, with the addition of the bias control. A few examples are given in Figure 4-18. The example in Figure 4-18a is a comparator, somewhat more complicated than usual, Figure 4-18b shows an example of inverter, while in (c) is a filter, a type of application where one is more likely to encounter an OTA.

Norton Amplifiers

Somewhat more common than the OTA is the Norton amplifier, characterized by a current input and a voltage output. In contrast to the op amp, the input impedance is desired to be very low. Norton amplifiers are useful when the signal is carried by currents, as shown in Figure 4-19. In general, the circuits are similar to the ones used with conventional op-amps.

Chapter 5

Semiconductors

Most modern semiconductor devices are based on elemental silicon. For a few special purposes, however, germanium, or compounds of elements from the third and fifth columns of the periodic table, such as GaAs, are utilized. The great majority of applications that we will discuss are based on silicon.

Silicon in its pure crystalline form exists as a three-dimensional array of identical atoms covalently bonded to each other. This is conveniently diagrammed as a two-dimensional analog (Figure 5-1). The silicon atom has four valence electrons. In the silicon crystal structure each atom is joined to the four neighboring atoms by a bond consisting of two electrons, one from each participant. Thus each atom is surrounded by a stable complement of eight bonding electrons. As these electrons are effectively immobilized, the crystal is an extremely poor conductor of electricity.

Silicon becomes useful in electronic devices only when certain specified impurities are present in extremely small amounts, to render the crystal more conducting. One appropriate type of additive is an element that has five valence electrons, such as arsenic, antimony, or phosphorus. Figure 5-2 shows the effect on the silicon crystal of the addition of a small amount of arsenic. Each added arsenic atom is accompanied by one available electron in excess of the number needed for covalent bonding. These extra electrons are relatively free

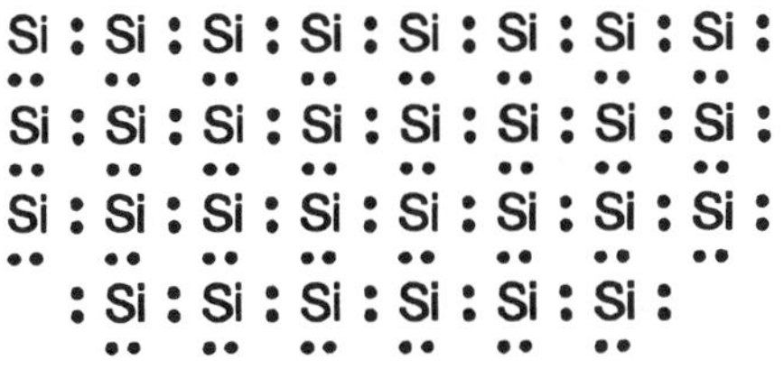

Si : Si : Si : Si : Si : Si : Si : Si :
Si : Si : Si : Si : Si : Si : Si : Si :
Si : Si : Si : Si : Si : Si : Si : Si :
 : Si : Si : Si : Si : Si : Si :

Figure 5-2 A two-dimensional analog of the interior of a pure silicon crystal. Each double-dot represents a pair of electrons in the covalent structure.

to move around in the crystal, leaving positively charged As^+ ions embedded in the lattice.

On the other hand, impurity atoms with only three valence electrons (aluminum, boron, or gallium), when present in silicon, give structures deficient in electrons (Figure 5-3). In this case, for each added atom a "hole" results that could accept an electron to form a complete covalent bond. It is always possible, as the result of thermal energy, that an electron from a neighboring Si−Si bond will free itself and jump into the hole, leaving a new hole behind. By this means, the positively charged hole can effectively move step by step through the crystalline lattice.

Materials that conduct electricity by this mechanism are examples of **semiconductors**. Pure (undoped) silicon is referred to as an **intrinsic** semiconductor. Semiconductors are intermediate in their ability to conduct electricity between true insulators on the one hand and metallic conductors on the other. Metals generally have conductivities in the range of 10^4 to 10^6 S cm^{-1}, while the values for insulators may be from 10^{-10} to 10^{-22}, leaving a vast region in between. This is where semiconductors lie, with conductivity values between about 10^3 and 10^{-9}. Both electrons and holes contribute to electrical conductivity.

The controlled addition of impurities to the pure substrate material is called **doping**. Additives with excess electrons give rise to

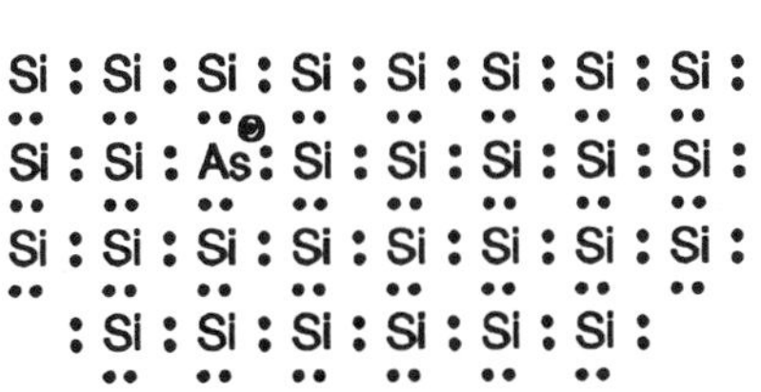

Si : Si : Si : Si : Si : Si : Si : Si :
Si : Si : As: Si : Si : Si : Si : Si :
Si : Si : Si : Si : Si : Si : Si : Si :
 : Si : Si : Si : Si : Si : Si :

Figure 5-1 A silicon crystal with an atom of arsenic substituted for one of silicon in the lattice. The encircled electron is mobile, conferring on the crystal a degree of electrical conductivity.

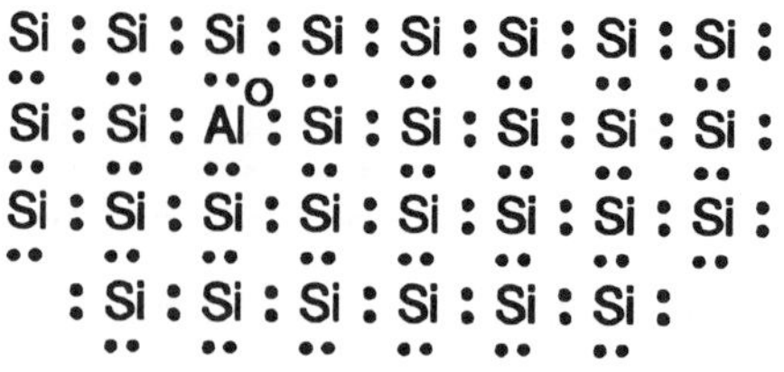

```
Si : Si : Si : Si : Si : Si : Si : Si :
••    ••   ••o ••   ••   ••   ••   ••
Si : Si : Al : Si : Si : Si : Si : Si :
••    ••   ••   ••   ••   ••   ••   ••
Si : Si : Si : Si : Si : Si : Si : Si :
••    ••   ••   ••   ••   ••   ••   ••
   : Si : Si : Si : Si : Si : Si :
      ••    ••   ••   ••   ••   ••
```

Figure 5-3 A silicon crystal with an atom of aluminum replacing one of silicon. The circle represents a "hole" where one electron is missing.

"n-doped" silicon, whereas those deficient in electrons produce "p-doped" material. The concentration of dopant is always very low, much less than one part per million. Appropriate combinations of n- and p-doped semiconductors has led to the development of diodes and transistors, the basic building blocks of all the important integrated electronic devices in use today.

Diodes

The direct junction between n- and p-doped silicon has unique characteristics that lead to a multitude of applications. A pn- junction can be diagramed as in Figure 5-4, in which electrons are designated by dots and holes by circles. There will be a tendency for electrons to cross the boundary from right to left and holes to cross in the opposite direction, as a result of coulombic attraction between opposite charges. This quickly produces a narrow region around the boundary, called the **depletion region**, that is free of both mobile electrons and holes. The result of this charge diffusion is that the p-region becomes negatively charged, the n-region positively. The difference in charge produces a potential across the boundary (the barrier potential) of about 0.7 V for silicon, or 0.5 V for germanium.

Now suppose that we provide the crystal with metallic electrical connections, as shown in Figure 5-5. Once these connections are made, the unit can be incorporated into an electrical circuit, and is called a **diode**. If an external voltage greater than the barrier potential is impressed upon the diode, with the p-material positive as in Figure 5-5a, the added field will drive electrons toward the depletion region from the right, where there are excess electrons available, at the same time that it will drive holes in the opposite direction from the left. These holes and electrons combine to neutralize each other where they meet at the pn-junction. The net result is a valence electron which is non-mobile. The supply of electrons in the n-region is maintained

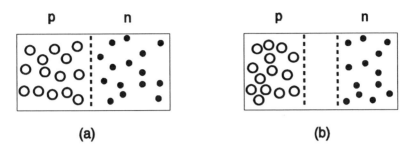

Figure 5-4 A *pn*-junction: (a) before charge diffusion occurs, and (b) following diffusion, showing the depletion region. Circles represent holes and dots electrons.

by the injection of more electrons from the external circuit, while electrons are pulled away on the opposite side, creating additional holes. The result is that current flows readily and the net voltage drop across the unit is essentially the barrier potential, 0.7 V. In this condition the diode is said to be **forward biased.**

If the applied potential is reversed as in Figure 5-5*b*, the situation will be quite different. Now both the electrons in the *n*-region and the holes in the *p*-region are pulled *away* from the depletion area, thus effectively preventing any current from flowing (the **reverse biased** condition). Thus the chief area of application of diodes becomes evident, namely the ability to conduct current in one direction while preventing its reverse flow. Figure 5-6 shows the current-voltage curves for silicon and germanium diodes.

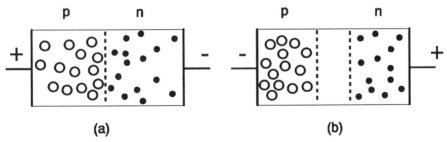

Figure 5-5   A *pn*-junction diode with electrical connections: (a) forward biased, with both electrons and holes driven toward the depletion region, and (b) reverse biased, with both electrons and holes withdrawn from the depletion region. The metal-to-silicon junctions (called **ohmic junctions**) do not form depletion regions but serve to conduct electrons in and out.

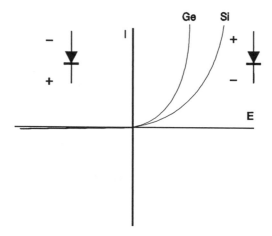

Figure 5-6 The current-voltage curves characteristic of silicon and germanium diodes. The circuit symbol is shown with both the forward biased condition (on the right) and the reverse biased on the left.

Zener Diodes

If the voltage across a silicon reverse-biased diode is increased, a point will eventually be reached at which the diode will break into conduction, as shown at V_z in Figure 5-7. This occurs when the electric field becomes large enough to produce new electron-hole pairs in the

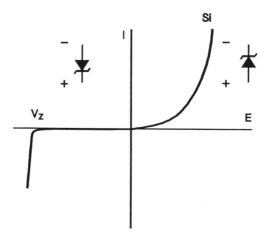

Figure 5-7 The current-voltage curves characteristic of silicon Zener diodes. The circuit symbol is shown with the forward-biased condition on the right and the reverse-biased on the left.

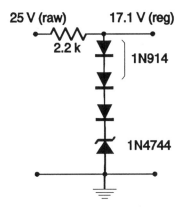

Figure 5-8 A temperature-compensated Zener voltage regulator. The "raw" voltage, 25 V or more, is dropped to a regulated value of 17.1 V, given by the 15-V Zener plus the forward voltage drop of the three diodes.

neighborhood of the depletion region. For heavily doped diodes the depletion zone is very narrow, and breakdown occurs in the range of 2 to 6 V. If the diode is less heavily doped, breakdown takes place only at higher voltages. Diodes optimized to make use of either of these breakdown effects are called **Zener diodes.** (Germanium diodes do not show an abrupt Zener potential.)

The Zener voltage has a temperature coefficient, and for high precision some kind of thermostatic compensation may be needed. The coefficient is nearly zero for a Zener voltage of about 5 or 6 V, increasingly negative below this and positive above. For a 10-V Zener, the coefficient is about +6 mV/K.

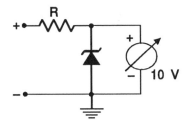

Figure 5-9 A 12-V Zener diode serving as protection for a l0-V DC voltmeter. If a higher voltage, say, 15 V should be connected across the input, current would be forced to flow through the resistor R and the Zener, with a voltage drop across R, resulting in not more than 12 V across the diode and meter. If a voltage of reversed sign were connected, it would send current through the Zener, which would be forward biased, with only very little voltage appearing across the meter.

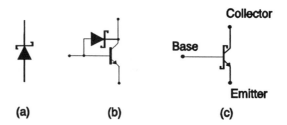

Figure 5-10 The symbol for (a) a Schottky diode, (b) a Schottky transistor, comprising an *npn* transistor and a Schottky diode, and (c) the conventional symbol.

The forward voltage drop of a silicon diode has a negative temperature coefficient, hence it is often possible to combine one or more forward diodes with a positive temperature coefficient Zener to produce a compensated reference voltage. For example, three type 1N914 diodes in series with a 1N4744 Zener give a combined voltage drop of about 17.1 V with a temperature coefficient of only 0.5 mV/K. Such a combination is illustrated in Figure 5-8.

A major field of application is in voltage regulating circuits. These will be described in a later chapter on power supplies. Zener diodes are also useful for the protection of various electronic components against excessive voltage. For example, a 10-V Zener in series with a small resistor, mounted across the terminals of a 10-V DC meter, as in Figure 5-9, will prevent damage from higher voltages of either polarity that might accidentally be applied.

Schottky diodes are fabricated with a metal-to-silicon junction rather than a junction depending on oppositely doped silicon (Figure 5-10*a*). Its forward voltage drop at normal currents is about 0.4 V, compared to the 0.7 V typical of silicon *pn*-junctions. They are similar in operation to germanium diodes, but can be fabricated directly on silicon substrates. A principal use is in the "LS" series of integrated circuits, (to be discussed at length in a later chapter), in the form of Schottky transistors, shown in Figure 5-10*b*.

Transistors

A transistor is a semiconductor device with three (or more) electrical contacts with other circuit components. The several types of transistors will be discussed in turn. The first to be developed were the so-called bipolar transistors. In the 1970s and 1980s these began to be supplemented by other devices with improved characteristics, permit-

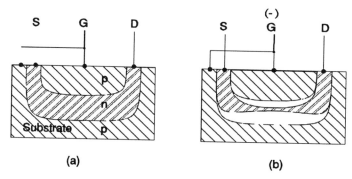

Figure 5-11 (a) Schematic of an *n*-channel JFET. (b) A negative charge on the gate terminal forms an electric field that causes the free electrons to withdraw from the *p*-regions, hence increasing the channel resistance. The letters S, G, and D refer to the source, gate, and drain. Actual size, about 25 μm.

ting greater speed of operation or greater impedance. As usual with new developments, both the old and the new devices continued to coexist and are widely used today, each in applications for which its properties are best suited. We shall begin with the simplest transistor to describe, the field effect type.

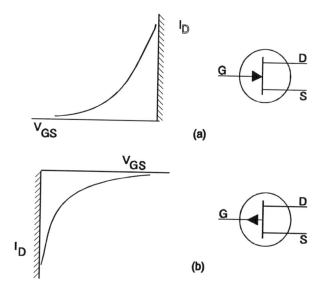

Figure 5-12 Current-voltage characteristics of JFETs: (a) *n*-channel, (b) *p*-channel. The shaded areas cannot be used, as they would imply forward bias.

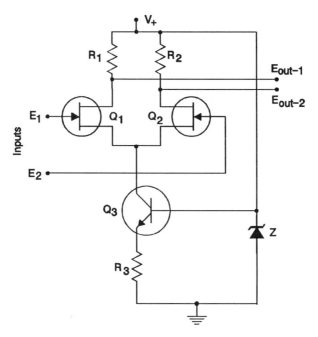

Figure 5-13 The JFET input stage of an op amp. The transistor Q_3, together with resistor R_3 and the Zener diode constitute a temperature-compensated constant current supply for the pair of FETs.

Junction Field-Effect Transistors

The **junction field-effect transistor** (JFET) is directly based on the mobility of electrons or holes for its operation. Figure 5-11*a* shows a cross section of a typical JFET, composed of two *p*-doped regions separated by a narrow channel of *n*-material, its ends designated by S (the **source**) and D (the **drain**). A reverse bias must be maintained across the *pn*-junctions. As a result current can pass between S and D along the *n*-channel without being diverted into the *p*-regions. When an increased negative charge is applied to the *p*-area, called the **gate** (G), electrons in the channel are repelled, forming a much more restricted channel, which therefore has increased resistance. Hence the channel resistance can be controlled by a voltage at the gate. There are two forms of JFETs, provided with *n*- and *p*-type channels, respectively. Their current-voltage characteristics and symbols are shown in Figure 5-12.

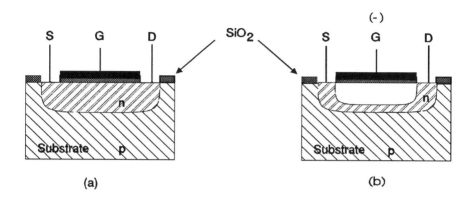

Figure 5-14 (a) Simplified structure of a depletion MOSFET with an *n*-type channel. (b)The effect of a negative gate potential, which restricts the width of the conductive channel.

Junction FETs are widely used in analog circuits, especially as the input stages to op amps. A typical circuit for a FET input stage is given in Figure 5-13. The prime advantage of JFETs as inputs to op amps lies in their very high impedance, with correspondingly small input and bias currents, though offset voltages may not be particularly low.

The MOSFETs

Another type of field-effect transistor is the MOSFET (for metal-oxide-semiconductor FET). It differs from the JFET in that a thin layer of

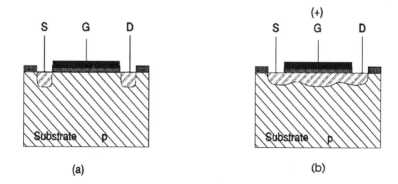

Figure 5-15 An *n*-channel enhancement MOSFET: (a) without an applied potential, and (b) with a positive voltage on the gate, creating a conductive channel. The shading is consistent with Figure 5-14.

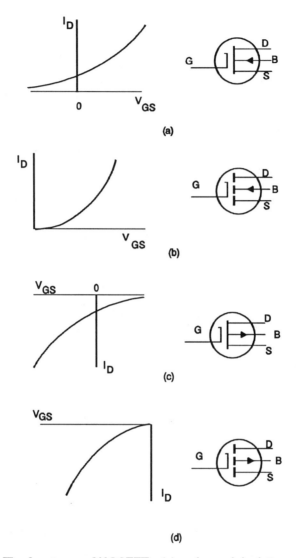

Figure 5-16 The four types of MOSFETs: (a) n-channel depletion, (b) n-channel enhancement, (c) p-channel depletion, and (d) p-channel enhancement. V_{GS}: gate-source voltage; I_D: drain current; G: gate; S: source; D: drain; B: bulk substrate connection.

SiO_2 covers the surface, insulating the gate from the transistor proper. Figure 5-14a is a section view of a typical MOSFET. A region at the surface of the p-type silicon (the **substrate**) is n-doped, forming the channel between the two contacts to the source and drain. The gate

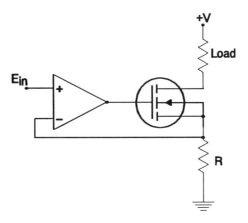

Figure 5-17   A MOSFET used as a current booster for an op amp. Note that the input voltage is impressed across the resistor R, but the corresponding current, E_{in}/R, is supplied by the power transistor, not by the op amp.

consists of a metallic coating between (but insulated from) the source and drain contacts. A negative potential applied to the gate produces an electric field within the channel. This gate field repels the mobile electrons, effectively decreasing the cross-sectional area of the channel thus increasing its resistance. The result is that the flow of electrons from source to drain can be controlled by means of the gate potential, just as in the JFET. This effect is diagrammed in Figure 5-14b. Note that the MOSFET can be operated with the gate either positive or negative.

The unit just described is known as a **depletion MOSFET** since it is operated by depleting the channel of its majority carriers. Another modification, called an **enhancement MOSFET**, has no channel at all until one is created by a positive voltage on the gate, attracting electrons to the adjacent area (Figure 5-15).

Both depletion- and enhancement-mode MOSFETs also exist in p-channel forms, the exact duplicate of those described above except for interchanging the two types of doped silicon, and the signs of the applied potentials. In each of the four kinds of MOSFET the channel current is dependent on the gate potential. The functional relationships are shown qualitatively in Figure 5-16, which also includes the conventional symbols for these circuit elements.

MOSFETs find their greatest applications in circuits that must handle considerable power, and also in digital logic gates (to be considered in a later chapter). Figure 5-17 shows a typical circuit in

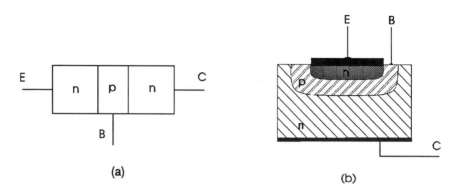

Figure 5-18   (a) Schematic representation of an *npn* transistor; (b) Section through one implementation. The letters C, B, and E designate the leads to the collector, base, and emitter, respectively.

which a MOSFET is used as a power booster or output stage in connection with a voltage follower. Only positive outputs can be accomodated.

Bipolar Transistors

The word "transistor," used without a modifier, generally refers to a bipolar transistor which we will now describe. The bipolar transistor consists of three layers of silicon, sequentially doped *n-p-n* or *p-n-p*. Let us consider first the *npn* configuration, depicted schematically in Figure 5-18*a*. One actual implementation is shown in a sectional view in Figure 5-18*b*. Suppose now that we connect two batteries as in Figure 5-19*a*. The currents designated by arrows are considered to

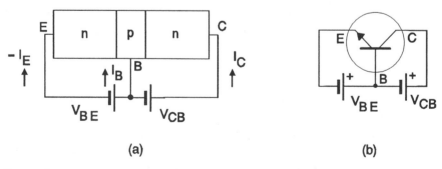

Figure 5-19 (a) An *npn* transistor shown with typical connections, and (b) the corresponding circuit diagram. The arrow indicates the direction of conventional positive current; the elecrons move in the opposite direction.

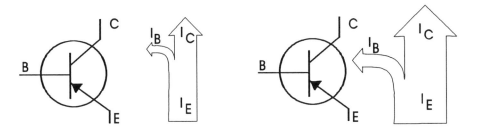

Figure 5-20 Symbol for an *pnp* transistor (that for the *npn* variety is the same except for the direction of the arrowhead). The division of current between base and collector is shown for two magnitudes of current. The ratio of I_C/I_B is the same for both and $I_E = I_B + I_C$.

be positive for the conventional (positive) current *entering* the transistor. Consequently I_E is negative, while I_B and I_C are both positive.

The principal current inside the transistor consists of a flow of electrons from emitter to collector, but the electrons are subject to interaction with holes in the base area. The base is so thin and so sparsely populated with holes (lightly *p*-doped) that most of the electrons pass through it without hindrance, though a small fraction are captured. The base current I_B withdraws electrons from the base (in the polarity shown), leaving excess holes. The added holes will of course capture more electrons, but this effect is greatly outweighed by the accelerating effect on electrons due to the increased electric field caused by the positive holes. Hence the overall effect is to increase the emitter-collector current. Figure 5-20 indicates the relation between the two currents.

The transistor is characterized by a factor β, the ratio of a change in collector current ΔI_C to the change in base current ΔI_B causing it:

$$\beta = \frac{\Delta I_C}{\Delta I_B}$$

The β factor for small-signal transistors is usually within the range of 50 to 200. As a practical application, suppose a transistor with $\beta = 100$ is required to supply 250 mA, for example, to operate a relay. The required base current must be 250/100 = 2.5 mA, and whatever device is providing the input to the transistor must be able to provide this much current.

Chapter 6

Diodes and Transistors as Circuit Elements

Now that we have introduced the various types of semiconductors, we will examine some of the basic circuits that employ them. As an introduction, consider the circuit in Figure 6-1a. This represents the conventional connections for a **voltage divider**, a circuit that reduces a voltage by the ratio of resistors. In Figure 6-1b, one resistor has been replaced by a diode, which as we have seen can conduct current in one direction only. With the orientation shown, current can pass freely through the diode if the polarity is positive, but not if it is negative. If no current can pass through the diode, its resistance is effectively infinite; hence the entire voltage drop must appear across the series resistor, and E_{out} will be equal to E_{in}. On the other hand, if the diode is in the conducting state, its resistance is nearly zero, and the potential at E_{out} will be equal to the forward voltage drop (≈ 0.7 V). If the diode were connected in the opposite sense (arrow head pointing upward), the results would be exactly the opposite.

If an AC voltage is presented to the input, the result will be a wave form like that of Figure 6-1c: in one polarity, only the forward voltage drop will be observed while in the other polarity one half of the sine-wave will be seen. This is an example of **half-wave rectification**, discussed in more detail in the chapter on power supplies. The forward voltage drop in this circuit precludes the effective rectification of small

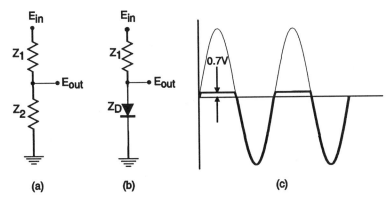

Figure 6-1   (a) A simple resistive voltage divider; the output is given by the equation $E_{out} = E_{in} \times [Z_2/(Z_1 + Z_2)]$. (b) A voltage divider including a junction diode as one arm; the output E_{out} is equal to E_{in} for negative values of E_{in}, and nearly zero for positive values. (c) The result of applying a sine wave at E_{in}.

signals. The difficulty can be overcome by the precision rectifier described in Chapter 4.

A circuit that shows a switching action based on a bipolar transistor is shown in Figure 6-2. If the input connection is tied to the positive supply line, as in (b), the transistor will be turned on, i.e., placed in its low-resistance state, which effectively grounds the output. Conversely, if the input is grounded (Figure 6-2c), the transistor will be

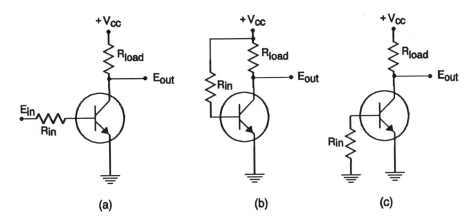

Figure 6-2   (a) An *npn* transistor acting as a switch. (b) If R_{in} is connected to the positive supply, E_{out} will be at zero; (c) grounding R_{in} will result in E_{out} being equal to the positive supply line.

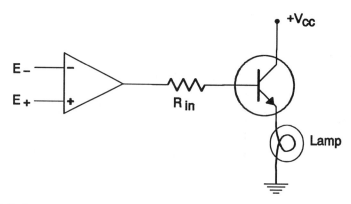

Figure 6-3 A transistor circuit used to turn a lamp on and off in accordance with the relative magnitudes of two inputs.

turned off, going to its high-resistance state so that no current can pass through it to ground. This means that no current will be passing through the load resistor, and there will be no potential drop across the resistor, so that the output will be at the power supply voltage V_{CC}. This circuit is often useful in digital electronics, where it is called an **inverter**. With a 5-V supply, the voltage is "inverted"; zero in gives +5 out, and +5 in gives zero out.

Figure 6-3 shows an application of this switching circuit. The lamp is switched by the output of a comparator. The lamp itself serves to limit the current, so that no resistor is needed in the collector lead.

Transistors in Linear Circuits

The transistor switches discussed above are on/off devices; either the transistors are in their nonconducting state or they are fully on (i.e., saturated). In the previous chapters we have seen that operational amplifiers can also function in this way. Now, just as op amps can be forced by suitable feedback circuitry to stay in their linear region, so can transistors be made to operate as linear devices.

To understand and design linear transistor circuits, we must be aware of the relationships between the several variables involved for any given type. Figure 6-4 shows two ways by which the action of a typical bipolar transistor, the 2N4074, can be described: the collector current I_C plotted either as a function of the collector-emitter voltage V_{CE} or as a function of the base-emitter voltage, V_{BE}. These curves show that a change of 20 μA in base current results in a change of

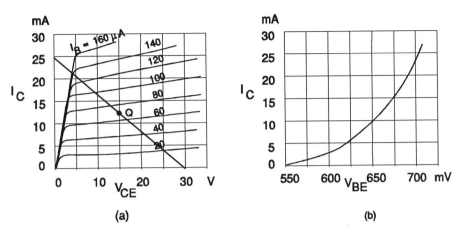

Figure 6-4 (a) The collector current as a function of collector-emitter voltage for a number of values of base current, for the 2N4079 *npn* transistor. (b) Collector current as a function of base-emitter voltage.

about 3.5 mA in collector current. The ratio of these two indicates a β of 170.

Consider now the simple amplifier circuit of Figure 6-5, where the same transistor is shown with suitable input and load resistors. Because of the relationships between currents described previously, we can write

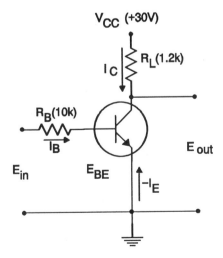

Figure 6-5 Basic circuit of a transistor amplifier using the 2N4074.

$$-I_E = I_B + I_C \qquad\qquad (6\text{-}2)$$

where the minus sign indicates that the current I_E is flowing *against* the arrow. The current flowing into the base is given by

$$I_B = \frac{E_{in} - V_{BE}}{R_B} \qquad\qquad (6\text{-}3)$$

Suppose that various voltages are applied to the input and corresponding outputs are measured. (Note that E_{CE} and E_{out} are identical.) Two pairs of data are:

E_{in} (mV)	E_{out} (V)
600	18.0
1800	12.0

For the first case, the current through the load, and hence the collector current is given by

$$I_C = \frac{V_{CC} - E_{out}}{R_L} = \frac{30 - 18}{1200} = 10 \text{ mA}$$

Figure 6-4b shows this to correspond to $V_{BE} = 650$ mV (as closely as one can read these graphs). Figure 6-4a shows us that $I_C = 10$ mA and $V_{CE} = 18$ V corresponds to a base current of $I_B = 0.06$ mA. Similar calculation for the second case yields values of $I_C = 15$ mA and $I_B = 0.09$ mA. Thus a difference of $\Delta I_C = 15 - 10 = 5$ mA corresponds to $\Delta I_B = 0.09 - 0.06 = 0.03$ mA, and the ratio gives

$$\beta = \Delta I_C / \Delta I_B = 5/0.03 = 170$$

which is in agreement with our original estimate of 180.

The voltage gain for this circuit can be calculated as

$$A_V = \frac{\Delta E_{out}}{\Delta E_{in}} = -\frac{18 - 12}{0.60 - 1.80} = -\frac{6}{-1.2} = 5 \qquad (6\text{-}5)$$

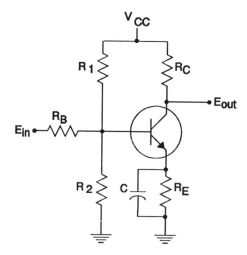

Figure 6-6 A transistor amplifier with bias circuitry shown.

It is sometimes convenient to regard the amplifier as converting a signal in the form of a current to a corresponding voltage, for which a useful parameter is the **transimpedance,** given by

$$Z_{tr} = \frac{\Delta E_{out}}{\Delta I_B} = \frac{6}{0.03} = 200 \text{ k}\Omega \tag{6-6}$$

The diagonal line superimposed on the curves of Figure 6-4a is termed the **load line.** Its ends correspond to the open-circuit voltage (i.e., 30 V at zero current) and the short-circuit current (25 mA with the output at ground potential). The ratio of these two quantities gives, by Ohm's law, the value 1.2 kΩ for the load resistor, R_L. Such a line can be drawn for any proposed load, and can be used to determine the result of current and voltage changes. With given currents and voltages present in the circuit an operating point is defined that must lie on the load line.

For optimum linear operation the transistor must be kept within an appropriate portion of its characteristics. This requires a **bias circuit** to provide the proper current to the base, the control element of the transistor. It is convenient to select a particular operating point, such as Q in Figure 6-4a, to be the **quiescent point,** which describes the currents through the transistor when no signal is present. The required bias can be supplied by an RC-network such as that shown in Figure 6-6. The voltage divider R_1R_2 serves to establish the Q point. The emitter resistor R_E, typically a few hundred ohms, increases the

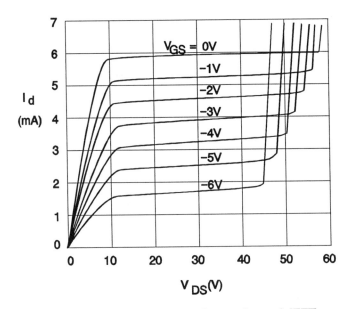

Figure 6-7 Current-voltage characteristics of an *n*-channel JFET.

stability of the circuit by providing some negative feedback. The capacitor across R_E reduces noise if the signal to be amplified is DC; it also establishes different operating conditions for AC and DC signals.

FET Circuits

The characteristics of field effect transistors are quite different from those of bipolars. Consider first the curve marked $V_{GS} = 0$ in Figure 6-7. For applied potentials (V_{DS}) less than about 10V the device acts as a simple resistor, obeying Ohm's law. However, beyond this point, when the current exceeds about 6 mA the voltage drop through the resistance of the channel becomes sufficient to produce a depletion region. This restricts the flow of current, which levels off into a region of saturation, the **pinch-off** effect. Eventually, if V_{DS} is increased further, breakdown will occur, similar to that of a Zener diode. For linear amplification, the FET is operated in the saturation region, where the changes in drain current are nearly proportional to the gate-source voltage as shown in the figure.

A unique property is the existence of a point where the channel resistance is independent of temperature (Figure 6-8). This suggests the utility of FETs in circuits where the effect of thermal drift must be minimized.

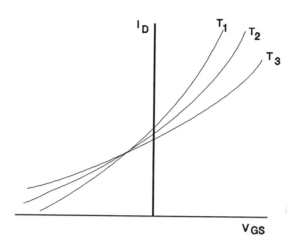

Figure 6-8 The drain current I_D as a function of the gate-source voltage V_{GS} shown here for a depletion MOSFET. The three curves correspond to various temperatures, where $T_1 < T_2 < T_3$. The point of zero temperature coefficient lies between 100 and 500 μA.

Input Circuitry of Op Amps

Considerations such as those discussed above may clarify the basis for some of the various op amps that were described in Table 4-1. The input stage of op amps usually consists of a pair of transistors connected as a **long-tailed pair** (Figure 6-9). The strong points of the bipolar input are its low noise, and its offset voltage which is easily trimmed to a very low value. This can be done by adjutment of the offset trimpot which equalizes the two bias currents and, in turn, nulls the offset voltage.

On the other hand, JFET op amps shown in Figure 6-9*b*, have extremely low bias currents at room temperature, but it is necessary to trim offset voltage and temperature drift separately. Drift is controlled by adjusting the variable resistor shown at the top of the figure. JFET amplifiers tend to be somewhat more noisy than their bipolar counterparts.

Another common configuration of a transistor amplifier is the **emitter-follower** circuit shown in Figure 6-10. The gain of this circuit is nearly unity, without sign inversion, and it is thus similar to the voltage-follower connection for an op amp. The collector is now connected directly to the power supply (V_{CC}) without any intervening resistor, and the load is taken from the emitter to ground. The current

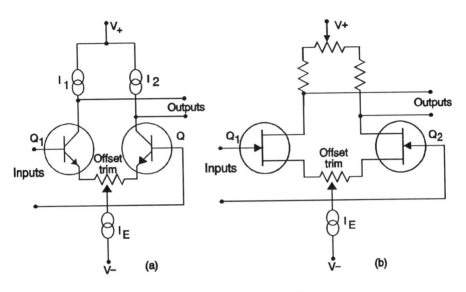

Figure 6-9 Input circuitry of (a) bipolar and (b) JFET op amps.

gain in this amplifier A_I, can be found by the following considerations The current in the load, I_L, is taken as flowing toward ground. This gives:

$$I_L = -I_E = I_B + \beta I_B = I_B(\beta + 1) \qquad (6\text{-}7)$$

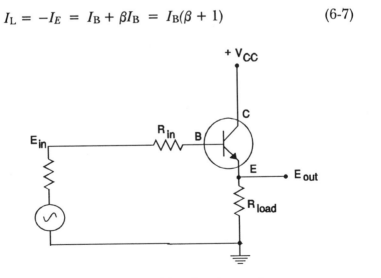

Figure 6-10 An emitter-follower amplifier showing a voltage source.

and

$$A_I = I_L/I_B = 1 + \beta \qquad (6\text{-}8)$$

This result should be compared to the voltage gain A_E which is practically unity.

Simple Current Regulator

It is often desirable to be able to maintain the current in a circuit at a stable value. This can be done with a single transistor that can be an integral part of the controlled circuit itself. Figure 6-11 gives two circuits, using a bipolar transistor and a FET, respectively. In both, R_L is the effective resistance of the circuit to be controlled, the "load." In Figure 6-11a, the application of Kirchhoff's voltage law to the emitter loop gives

$$IR_3 + V_{BE} = V_D + (V_{EE} - V_D)\frac{R_2}{R_1 + R_2} \qquad (6\text{-}9)$$

where V_{BE} is the base-emitter voltage of the transistor and V_D is the forward drop across the two diodes in series. It is convenient to set the circuit parameters so that

$$V_{BE} = V_D \times \frac{R_1}{R_1 + R_2}.$$

Then Eq. (6-9) reduces to

$$I = V_{EE}\frac{R_2}{R_3(R_1 + R_2)} \qquad (6\text{-}10)$$

This shows that the current is independent of the properties of the transistor and the value of V_{CC}; it depends only on V_{EE} and the three resistors. The two diodes are needed in order to satisfy the conditions, since the voltage drop across each one is very nearly equal to V_{BE} and the voltage division between R_1 and R_2 requires that $V_D > V_{BE}$. The diodes and the transistor should be mounted close together to minimize temperature drift.

The FET circuit of Figure 6-11b is similar, but since no current can flow into the gate, the connections can be simpler. The action in

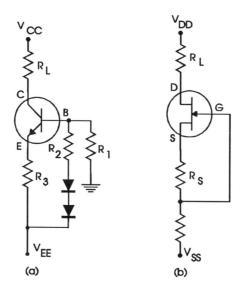

Figure 6-11　Transistors as current regulators.

this case is based upon the independence of the drain current from the source-drain voltage for a given gate potential (see Figure 6-7). Zero temperature coefficient can be attained only for a particular current, dependent on the transistor type, but it will be low for any reasonable current.

Cascaded Transistor Stages

The two transistors in a differential amplifier are not considered to be distinct stages, since the output of one does not feed into the other. If the gain of a single transistor is not sufficient, two or more stages can be connected in series. Figure 6-12 shows two ways in which this can be done. The circuit at (a) is the more straightforward of these, in that the two stages are identical. The capacitive coupling between stages restricts the signal to AC and each stage must have its own bias resistors. The circuit in Figure 6-12*b* shows a voltage follower stage providing a low-impedance output; this circuit can be coupled directly, avoiding the need for separate bias resistors. Op amps are better than multistage amplifiers for low frequencies.

　　Two bipolar transistors can be interconnected as shown in Figure 6-13 with the base current of the second stage fed directly by the emitter current of the first. This is called a **Darlington** circuit, and Darlington pairs are available as composite units that can be used as

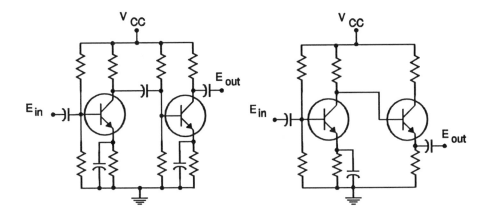

Figure 6-12 Two-stage transistor amplifiers: (a) capacitively coupled; (b) direct coupled.

though they were single transistors. The β of the transistor pair is equal to the product of the β's of the separate parts.

Transistors as Boosters

A frequent application of transistors is to enhance the current-carrying capacity of an operational amplifier. Consider an amplifier intended to drive a relay or a lamp that requires, say, 100 mA, whereas we know that the usual op amp limit is about 20 mA. Figure 6-14 gives a possible interfacing circuit using a pair of bipolar transistors as a

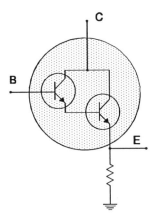

Figure 6-13 An *npn* Darlington amplifier.

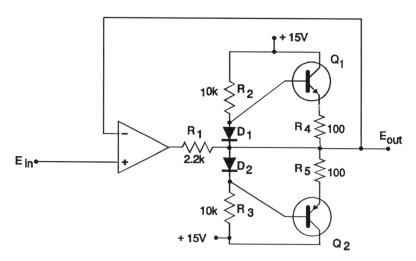

Figure 6-14 An op amp equipped with a booster that permits more current to be drawn than the amplifier could supply alone. If the signal is always to be of the same sign, a single transistor will suffice.

buffer stage, called a **booster**. The network formed by resistors R_2, R_3 and the diodes serves to set the proper operating points for the transistors. Resistors R_4 and R_5 provide short-circuit protection and improve transient response. The booster is always included within the feedback loop as shown.

A Darlington transistor is useful as a booster when its great current gain is needed. An example is given in Figure 6-15, a temperature control circuit for a small oven. A bridge is formed by the four resistors, of which R_4 is a negative temperature-coefficient sensor called a **thermistor**. The bridge output is sensed by a comparator that turns the current to the heater on if the temperature drops too low, or off if it is too high. (The voltage supplied to the heater is always positive in the circuit shown, therefore only a single transistor is sufficient as a booster.

The load for the boosted amplifier is a heater located in the same enclosure with the thermistor. We can assume that the system is initially in thermal equilibrium at the desired temperature. If the temperature should increase, the (+) input of the comparator would become more negative (i.e., closer to ground), and if this were sufficient to flip the comparator, the base of the transistor would become negative enough to cut off the current through the heater. A decrease in temperature would have the opposite effect. The amplifier is of a

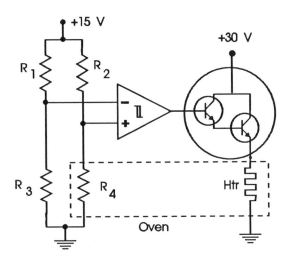

Figure 6-15   A temperature control for a small oven. The symbol inside the op amp triangle indicates the presence of hysteresis.

special type that provides **hysteresis** that stabilizes the comparator, against oscillating back and forth ("chattering").

Transistor Oscillators

Many types of bipolar transistors are applicable in high frequencies, up to hundreds of megahertz, where op amps are useless. An important application is to oscillators for the generation of sine waves. Several examples are shown in Figure 6-16. The first three of these circuits utilize a resonant LC "tank" as the frequency determining element. This parallel combination of capacitor and inductor resonates at the frequency that makes the impedances of the two arms equal:

$$\omega L = \frac{1}{\omega C} \qquad (6\text{-}14)$$

$$f = \frac{1}{2\pi\sqrt{LC}} \qquad (6\text{-}15)$$

In Figure 6-16*a*, the **Hartley** oscillator, the inductor is tapped to provide feedback to the emitter of the transistor, whereas in (b), the **Colpitts** circuit, feedback is taken between the two capacitors. The circuit in Figure 6-16*c* uses a small transformer to provide feedback.

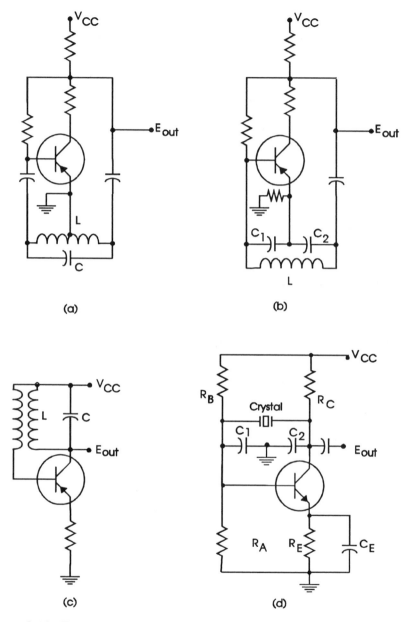

Figure 6-16 Four forms of transistor oscillators: (a) the Hartley circuit with a tapped inductor, and (b) the Colpitts circuit with a divided capacitor. In both, the frequency of oscillation is given by $f = 1/(2 \pi \sqrt{LC})$. (c) Oscillator with transformer coupling, and (d) the Pierce circuit with crystal control of the frequency.

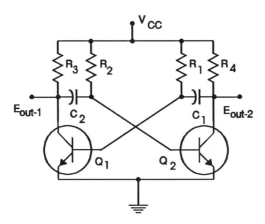

Figure 6-17   An astable multivibrator: the two outputs give signals that are 180°
out of phase with each other.

In all three circuits the output could be taken from an extra winding
on the inductor instead of from the collector as indicated.

The oscillator shown in Figure 6-16d, the **Pierce** circuit, is built
around a quartz crystal that resonates mechanically at a frequency
dictated by its physical dimensions. The several capacitors generate
appropriate phase relations to give the needed feedback. Crystal
oscillators are among the most precise devices available at moderate
cost. If the crystal is maintained at a constant temperature, frequency
stability within the sub-parts-per-million range is attainable.

Figure 6-17 gives a circuit for a square-wave generator, sometimes
called an **astable multivibrator**. The connections shown would lead
one to expect that both transistors would operate simultaneously in
a heavily conducting mode, since the bases of both are forward biased
through R_1 and R_2. However, when the power supply is first turned
on, inevitably one of the transistors, say, Q_1 will fire first as the result
of slight inequalities. During the turn-on of Q_1, a negative pulse will
be sent through C_2 to the base of Q_2, preventing it from turning on
immediately. Because of the capacitive coupling, this blocking is only
temporary. The flow of current through R_2 neutralizes the negative
charge on C_2, and brings the base of Q_2 to more and more positive
potentials, until it suddenly turns on. This generates a pulse that turns
Q_1 off, and the process repeats itself inedefinitely. The repetition
frequency, if the two transistors are identical, is given approximately
by the relation

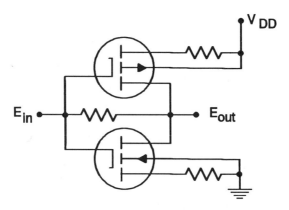

Figure 6-18　A two-MOSFET amplifier. The two transistors permit symmetrical amplification of a signal.

$$f = \frac{1.5}{C_1 R_1 + C_2 R_2} \qquad (6\text{-}16)$$

If $C_1 = C_2$ and $R_1 = R_2$, the square wave will be symmetrical.

MOSFET Circuits

Enhancement MOSFETS are particularly convenient for balanced applications, such as that shown in Figure 6-18. The two transistors, one n-channel, the other p-channel, are connected in tandem, so that a positive signal will be amplified by one, a negative signal by the other. This is not readily done with bipolar ICs, as it is difficult to form both npn- and pnp-structures on the same silicon substrate.

Transistors in Power Circuits

Another use of transistors is in power-handling circuits for which operational amplifiers are not usually applicable. All types of transistors (unipolar, JFET, and MOSFET) can be fabricated in forms that permit large currents and potentials to be amplified or controlled. They are usually packaged in metal cans provided with heat sinks for efficient heat dissipation. The actual circuits used are essentially those that have been described earlier, including booster service for op amps. The main difference is that all interconnections must be heavy enough to carry the expected currents, and insulating barriers must be able to withstand the voltages to be encountered.

High-power transistors usually cannot be mounted on the familiar printed circuit boards. Instead, they are attached directly to the metal chassis of the instrument, the chassis acting as heat sink as well as giving mechanical support. These transistors commonly have the metal case connected to the collector, and it is convenient, when possible, to design the circuit so that this element is at ground potential; otherwise a thin insulating layer of mica or plastic must be inserted to isolate the case from the supporting chassis.

Chapter 7

Transducers

Thus far we have discussed in detail many types of analog electronic circuits for modifying signals, but we have said little about where the signals are coming from. One of the most important applications of analog electronics is to that of converting physical properties of all sorts into electrical signals, often prefatory to entering the information into a computer. To do this we need, in addition to operational amplifiers, a variety of specialized devices, called **transducers**, that can make this conversion.[1]

Transducers can be classified according to the property that they measure. For example, we can speak of temperature transducers. They can also be classified according to the form in which the output appears. Thus we can distinguish between resistive, capacitive, or current transducers wherein the corresponding electrical quantity

---

[1] Actually the term "transducer" includes any device that converts one form of energy into another. In this general sense, a light bulb is a transducer in that it converts electrical to light energy. A more complete account of transducers can be found in G. W. Ewing, "Transducers," Chapter 4 of *Treatise on Analytical Chemistry*, 2d ed., (P. J. Elving, editor), John Wiley and Sons, New York, 1984, p 345.

varies as a function of the property measured. We will describe a number of transducers for measuring temperature, radiant energy (ultraviolet, visible, and infrared), pressure and displacement, as being representative of the field.

Temperature Transducers

Nearly every physical property has a temperature coefficient that can be expressed as a power series expanded about some point T_0:

$$X(T) = X_0[1 + \alpha(T - T_0) + \beta(T - T_0)^2 + ...] \qquad (7\text{-}1)$$

where X_0 is the value of the variable at temperature T_0 and $X(T)$ that at temperature T. In many cases the square and higher terms can be neglected. Differentiating with respect to temperature and solving for α gives

$$\alpha = \frac{1}{X_0}\frac{dX}{dT} \approx \frac{1}{X_0}\frac{\Delta X}{\Delta T} \qquad (7\text{-}2)$$

where Δ represents a small increment. In many systems α is nearly constant over small temperature excursions.

The most widely used temperature transducers are thermocouples, resistance thermometers, and semiconductor pn-junctions. We will discuss each of these in turn.

Thermocouples

The number of mobile electrons in a conductor varies with temperature, the effect being different in different metals. Hence any two unlike metals in contact will generate a difference of potential, called a **junction potential**, that is a function of temperature. Individual junction potentials cannot be measured directly, but the difference between two junctions at different temperatures *is* measurable. A pair of metals connected so as to permit measurement of this potential difference is termed a **thermocouple**. A typical temperature coefficient, that for the iron-constantan couple, is $\alpha = 0.6\ \dfrac{\mu V/\,V}{K}$ often expressed as 0.6 ppm/K. The response may deviate from linearity by a few tenths of one percent. A few representative thermocouples are listed in Table 7-1, together with their upper limits of usefulness.

Table 7-1

Some common thermocouples

Type[a]	Composition[b]	Sensitivity[c]	Upper limit °C
C	W (5% Re) / W (26% Re)	3.3	2200
E	Chromel / Constantan	59	1000
J	Iron / Constantan	51	760
K	Chromel / Alumel	41	1200
S	Pt / Pt (10% Rh)	5.5	1600
T	Copper / Constantan	40	300

[a]Symbols assigned by ANSI, the American National Standards Institute.

[b]Chromel is an alloy with the composition $Ni_{90}Cr_{10}$; Alumel is $Ni_{94}Mn_3Al_2Si$; constantan is $Cu_{57}Ni_{43}$. Chromel and Alumel are trade names of Hoskins Manufacturing Co.

[c] "Seebeck coefficient," microvolts/Kelvin.

Any circuit containing two metals necessarily has two junctions between these metals, either directly or with the interposition of a third metal. Thus thermocouples must be used in pairs, giving differential rather than absolute measurements. Traditionally, at least for laboratory use, the reference junction has been held at the ice point 0°C, but in many modern circuits the reference junction is simulated electronically.

There are two principal ways to measure the output of a thermocouple: with a low-impedance meter (Figure 7-1a) or with a high-impedance amplifier, (b) in the figure.

Figure 7-2 gives a practical circuit including cold-junction compensation. It is desired to determine the temperature of the enclosure marked T_x, while the reference junction is situated in a separate enclosure together with a temperature-sensitive current source such as the AD-590 (Analog Devices). The 590 is designed to pass a current of 1 μA per Kelvin over the range -55°C to $+150$°C. It develops across resistor R_T a voltage that is exactly the value required to compensate the voltage developed by the thermocouple junction. The resistor must

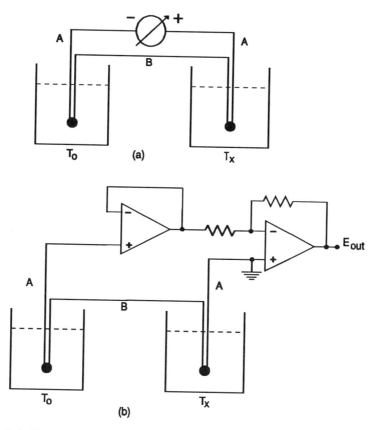

Figure 7-1 Two connections for a thermocouple. In each, T_0 is the temperature of a reference material (often a slurry of ice and water), and T_x is the temperature to be measured. The letters A and B represent two different metals. In (a) a current is measured; hence the resistances of the wires must be accounted for (or held constant). In (b) the voltage-follower prevents current from flowing in the thermocouple; hence the resistances of the wires becomes immaterial.

be carefully selected; it is different for the different types of thermocouples, as shown in the figure inset. The resistor string R_1, R_2, R_3 provides a a variable offset voltage to adjust the desired scale. The amplifier shown is an instrumentation amplifier provided with a suitable gain resistor.

Thermocouples are widely used in applications running from precision measurements of minute temperature differences in calorimetry, to rough monitoring of a furnace for heat-treatment of metallurgical samples.

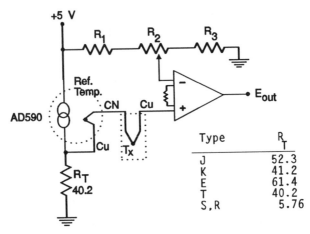

Figure 7-2 A thermometer using a thermocouple with cold-junction compensation. "CN" stands for constantan.

Thermosensitive Resistors

Any resistor made of a single metal has a positive temperature coefficient; that is, its resistance increases with rising temperature. The only metal that is much used as a resistance thermometer is platinum, which is advantageous in that it is very stable. A platinum resistor can be cycled repeatedly over a wide temperature range without permanent change in its properties. This convenience overcomes the disadvantages of a relatively low coefficient (0.3% K^{-1}). (The coefficient for nickel is twice as great, but nickel oxidizes badly and is much less reproducible.)

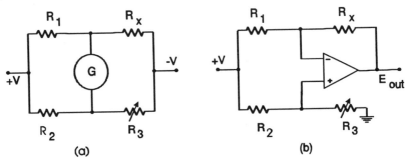

Figure 7-3 (a) A classical Wheatstone bridge; (b) an equivalent using an op-amp. In both circuits, at balance, $R_x = R_3(R_1/R_2)$. "G" in (a) designates a galvanometer, which is a highly sensitive microammeter used for null-balancing.

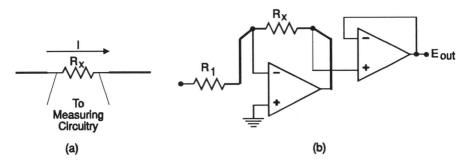

Figure 7-4 (a) A resistor or RTD provided with current leads (heavy wires) and sensing leads (fine wires); (b) an op amp implementation.

Precision platinum resistors, called **resistance temperature detectors** (RTDs) are widely used for highly precise temperature measurements. The resistance of an RTD is traditionally measured with a Wheatstone bridge (Figure 7-3*a*), but the bridge is often replaced with an op-amp circuit (Figure 7-3*b*). The classical bridge is always operated in a balanced mode: the variable arm of the bridge (R_3) is adjusted until a null is obtained, at which point the value of R_3 is read from calibrated dials. It is easily shown that at balance $R_x = R_3(R_1/R_2)$

It is often desirable to operate in an unbalanced mode so that the resistance of the RTD can be plotted automatically as a function of temperature, for example, when one wishes to follow the course of a chemical reaction by observing the accompanying temperature change. This could be done with the same circuit, but the output would not be linear. If, however, the RTD is connected in the feedback loop of an inverting op amp with constant input, the output will be a nearly linear function of temperature.

A potential source of error in these circuits is due to the resistance of the wires connecting the RTD to the measuring circuit. This is usually overcome when high precision is required, by adding sensing leads, as shown in Figure 7-4. The energizing current produces a potential drop in the connecting wires on both sides of the resistor being measured. To avoid error, particularly if the RTD is located at some distance from the measuring instrument, a second set of wires is provided, leading only to the measuring circuit and hence drawing negligible current. The follower now measures the voltage drop across the resistor alone.

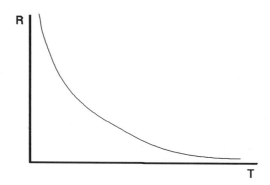

Figure 7-5 The resistance-temperature curve for a typical thermistor.

Semiconductor Temperature Sensors

The resistance of semiconductors follows an exponential relationship:

$$R \propto \exp\left[\frac{\Delta E}{2\rho kT}\right] \tag{7-3}$$

where ΔE has the dimension of energy,[2] k is the Boltzmann constant and T is the Kelvin temperature. The quantity ρ is a constant with the value 1 when the number of majority carriers (electrons or holes) is less than the number of impurity sites, and with the value 2 when this number is exceeded.

Over a range of temperatures, semiconductors show regions of positive and negative temperature coefficients. An example of a semiconductor device useful for temperature monitoring is the **thermistor,** made by sintering together the oxides of various transition metals. Thermistors can be made with room-temperature resistances from about 10 Ω to 10 MΩ. The behavior of a thermistor is not strictly represented by Eq. (7-3), since it is influenced by surface effects, particle size, and so on, but it follows a similar trend. A typical response curve is given in Figure 7-5. Thermistors are inherently nonlinear, but the response can be linearized by judicious use of two thermistors in a single package together with one or more resistors that have positive coefficients.

---

[2] ΔE is the forbidden energy gap between valence and conduction bands.

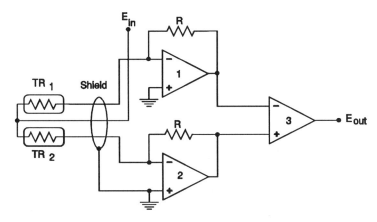

Figure 7-6 A differential circuit for measuring the difference between two temperatures. Thermistors TR_1 and TR_2 control the voltages at the outputs of amplifiers 1 and 2, which serve as inputs to difference amplifier 3. If the two thermistors are held at the same temperature, the output will be zero; if this is not true (because of manufacturing tolerances), then a small variable resistor can be inserted as required to compensate.

Most thermistors are used in a region in which the temperature coefficient is *negative* with values in the range of -3 to $-6 \% \, K^{-1}$. This gives the thermistor a much greater sensitivity than a platinum RTD, but with lower long-term stability and reproducibility. Thermistors are useful if utmost precision is not necessary. They are inexpensive and easily incorporated into a variety of electronic circuits; two of them are shown in Figures 7-6 and 7-7.

Figure 7-7 points to a factor that must not be overlooked in thermistor circuits: any appreciable current flowing through the thermistor will produce enough Joule heating to change the resistance. The self-heating is used in the anemometer, a device for measuring the flow rate of a gas, and in a thermal conductivity detector for gas chromatography. In general though, self heating is undesirable, and must be minimized or corrected for.

Junction Thermometers

Equation (7-3) showed that the resistance of a semiconductor is an exponential function of its temperature. For a *pn*-junction, the current that can flow is given as a function of temperature by

$$I = I_S \left[\exp(qV / kT) - 1 \right] \tag{7-4}$$

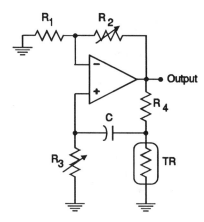

Figure 7-7 This circuit shows an op amp connected as a sine-wave oscillator stabilized with a thermistor. When the output increases, the current through the thermistor will also increase, thus raising its temperature and lowering its resistance. This reduces the positive feedback that controls the oscillation. The desired amplitude can be set by variable R_2, and the frequency by R_3.

where I_S is a constant, q is the charge on the electron, V is the applied voltage, k is the Boltzmann constant, and T the Kelvin temperature. The current I increases at about 7% per Kelvin. Since $(1.07)^{10} \approx 2$, it follows that I approximately doubles for every $10°$ rise in temperature.

This principle is used in a number of commercial temperature sensors, including, for instance, the Analog Devices AD590, a two-terminal device that produces an output current of exactly 1 μA/K, easily amplified to, say, 1 mV/K. Figure 7-8 gives a circuit for temperature control using the AD590.

Photosensitive Transducers

Transducers are available to cover any portion of the electromagnetic spectrum, but we will restrict our discussion to the ultraviolet, visible, and infrared regions as being the most significant for chemists, biologists, and many other scientists.

A widely used type is the **photomultiplier tube** (PMT). This consists of an evacuated envelope containing a number of electrodes. Radiation enters through a glass window and falls upon a metal plate (the cathode) which is usually coated with a semiconductor material. A fraction of the energy supplied by the radiation is converted into kinetic energy of electrons, some of which escape from the surface into the vacuum. (The efficiency of the cathode depends largely upon the

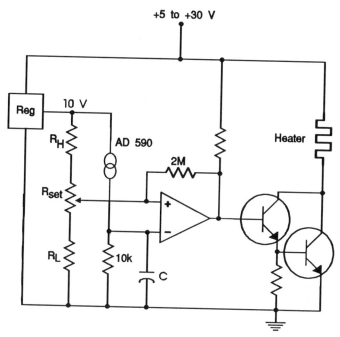

Figure 7-8 A simple temperature regulator using the AD590 sensor. The Darling-ton output must be able to carry the current needed by the heater; it may require a heat sink for cooling. The 2-MΩ resistor at the noninverting input of the comparator provides slight positive feedback and sufficient stabilization for efficient control. [Adapted from literature of Analog Devices, Inc.]

nature of the surface, a subject that space will not permit discussing here.)

A specially shaped electrode (called the first **dynode**) is placed near the photo-cathode and charged some 100 V more positive so that all the emitted electrons are accelerated toward it. Each electron strikes the dynode with enough energy to cause several secondary electrons to be emitted. These new electrons are in turn accelerated toward a second dynode, where each causes the emission of several more electrons, which hit a third dynode, and so on through perhaps 10 or 12 stages. The numerous electrons from the final dynode are collected on an anode which is shaped in such a way that no electrons can escape. Thus a current is generated that is delivered to the electronic measuring system. The multiplication factor in a PMT is dependent on the applied voltage. If each electron hitting a dynode produces n secondary electrons, and if there are m dynodes in the PMT, the multiplication factor is n^m. Thus for $n = 4$ and $m = 10$, the

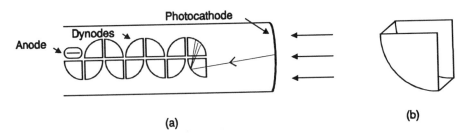

Figure 7-9 (a) A typical photomultiplier tube. Each dynode consists of a sector-shaped piece (b) acting as a secondary electron emitter. The radiation to be measured is incident from the right on a film of sensitive material deposited on the inner wall of the glass envelope. Each dynode in sequence (from right to left) is at a more positive potential.

factor would be $4^{10} \approx 10^6$, so that 1 nA of primary electron current would result in about 1 mA (which is the approximate upper limit for the output current in standard PMTs). There are many mechanical designs of PMTs, one of which is depicted in Figure 7-9.

The PMT acts as a high-gain current amplifier. Consequently the signal produced is in the form of a current rather than a voltage, so that the usual electronic circuitry includes an op amp acting as a current-to-voltage converter. Figure 7-10 shows a standard circuit, of which there are many variations in practice.

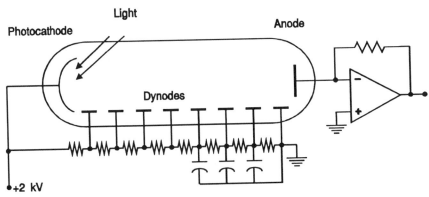

Figure 7-10 The basic schematic for a photomultiplier. The op amp functions as a current-to-voltage converter. Overall gain can be adjusted by varying the high voltage. The resistors form a voltage divider to provide the necessary potentials for the dynodes. The capacitors are needed if the signals have a high-frequency component, as with a pulsed light source.

Photosensitive Semiconductors

Semiconductors are inherently photosensitive in that the energy of incident radiation may be sufficient to cause separation of charges; that is to say, the incident energy can pull an electron away from a silicon–silicon bond producing an electron-hole pair. If this happens in homogeneous material, the result is decreased electrical resistance. If it takes place in the region of a pn-junction, it will induce a voltage across the junction that adds to the barrier potential. Both phenomena can be utilized, in the form of photoconductive or photovoltaic devices, respectively.

Photoconductive cells are fabricated from cadmium sulfide, lead sulfide, or a few other similar materials. They are widely used in commercial devices for automatic lighting control. The lead sulfide photoconductive cell is particularly useful as a detector for the near infrared region (about 0.7 to 3.0 μm).

It can be shown that the sensitivity of a photoconductor S is given by

$$|S| \approx \frac{1}{R_0}\left[\frac{\partial R}{\partial \varphi}\right] \tag{7-5}$$

where R_0 is the dark resistance, R is the observed resistance, and φ is the illuminance, expressed in lux or foot-candles. The parameter S is nearly constant but varies from one material to another. The sensitivity falls off with more intense ilumination and may be as high as 1000 reciprocal foot-candles for low light conditions.

A **photodiode** is an ordinary pn-diode optimized for sensitivity to light, and provided with a transparent window in the housing. Figure 7-11 gives the characteristic curves for a typical photodiode, showing the result of changing the illumination. A photodiode can be operated in either of two modes leading to an open-circuit voltage measurement or a short-circuit current measurement. For the first mode, the diode can be connected directly to the noninverting input of an op amp. No current can flow, so the diode is operating along the zero-current line of Figure 7-11, and the output voltage is a logarithmic function of illumination. In the second mode, the diode is connected to a low-impedance metering circuit that looks to the diode like a short circuit. The diode now operates along the vertical line

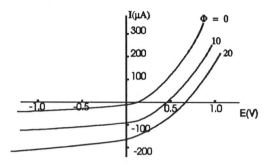

Figure 7-11 Current-voltage curves for a photodiode at three levels of illumination.

corresponding to zero voltage, and the output is linear with illumination. Choice between the two modes is a matter of convenience.

A *p-i-n* (or PIN) diode is a variety of photodiode that has a thin layer of intrinsic (i.e., undoped) silicon between the *n*- and *p*- segments. It is particularly useful when high-speed response is required.

Often, rather than using a simple diode, illumination is allowed to fall on the junction of a transistor. Such a unit is called a **phototransistor** (or a **photodarlington**). The advantage of course is the extra gain introduced by transistor action. The circuit symbols for these are shown in Figure 7-12. The relative spectral responses of the

Table 7-2

Spectral response of various photosensitive transducers (in μm)

	Low limit	Peak	Upper limit
PMT, multialkali cathode	0.29	0.60	0.95
CdS photoresistive cell	0.40	0.50	0.90
PbS photoresistive cell	0.40	2.00	3.00
Si photo-diode or -transistor	0.30	0.90	1.20

Note: The upper and lower limits stated are approximate; with an intense light source and high-gain amplifier, both limits could be extended somewhat. The limits are often decided by the transparency of the glass (or other) envelope.

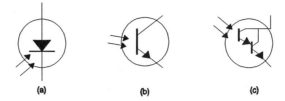

Figure 7-12 Symbols for (a) a photodiode, (b) an *npn* phototransistor, and (c) a photodarlington. The transistors in (b) and (c) are sometimes provided with external contacts to the bases, that can be used for bias adjustment or left unconnected.

various representative types of photosensitive devices discussed above is given in Table 7-2.

Figure 7-13 presents an example of a photometric amplifier for a spectrophotometer. Light from an incandescent lamp is broken down into its component wavelengths by a diffraction grating (not shown), and a small band allowed to fall on the photodiode. The output of the two-stage unit is fed into a digital panel meter. Notice the variable voltage source that can feed a very small current into the amplifier's summing junction. This is often a convenient arrangement for mak-

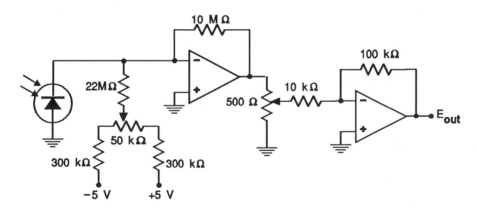

Figure 7-13 A photodetection circuit that can be used in a small spectrophotometer. The resistor network for injecting an adjustable voltage onto the summing junction is called a "dark-current" adjustment; it is used to compensate for any small current that flows when the photocell is in darkness. The 500-Ω variable is the gain control for the system. Thus one of the variables adjusts the photometer's zero point, the other its full-scale output. A precision dual op amp, such as the OP-215 or LF-412 would be well suited to this application.

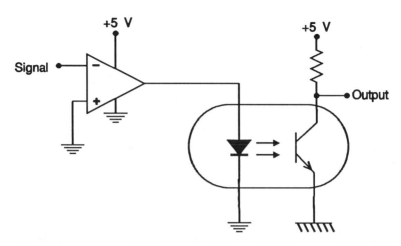

Figure 7-14 A comparator driving an optical coupler. The output of the comparator is either +5 V (which turns on the LED) or ground (which turns it off). The
phototransistor inverts the signal and is either turned on, grounding the output,
or it is turned off so that the output goes to +5 V. Note that "ground" is now the
digital ground, for which a different symbol is used.

ing fine adjustments in a circuit. The two 300-kΩ resistors at each end
of the 50-kΩ pot serve to narrow the voltage range available to less
than a volt, centered about ground. The selected voltage drives a
current through the 22 MΩ resistor, so that the added current is less
than 0.035 μA.

Optical Couplers

Sometimes, it is desirable to separate the grounds of two portions of
an electronic instrument, especially if analog and digital circuitry
coexist. If the two parts of the device use a common ground it will be
difficult to prevent the digital noise from interacting with the analog
circuitry. One way to avoid such interference is to substitute a light
beam for a conductive electrical connection. For this purpose a variety
of optical couplers are marketed, each containing a light source and
a photodetector in a common package. An illustration is given in
Figure 7-14. Such couplers are used mainly for digital signal transmission.

Infrared Detectors

The photons of infrared radiation contain less energy than those of
the visible and ultraviolet regions, insufficient to generate electron-

hole pairs in a room-temperature semiconductor. The infrared region (IR) of greatest interest to chemists and biologists extends from about 2 to 25 μm wavelength, so general purpose IR transducers cannot depend on photodiodes and phototransistors, nor on photomultipliers. Hence nearly all IR detectors make use of the less sensitive heating effect of the radiation. They often are mounted in pairs of duplicate sensors, one is exposed to the IR radiation, the other, a dummy, is shielded from it but subject to the same ambient temperature. Both thermocouples and thermistors have been adapted for this purpose.

One other infrared sensor that has become widely used is the **pyroelectric detector**. This typically consists of a crystal of triglycine sulfate (TGS), provided with metallic contacts on opposite faces. The heat produced by infrared radiation causes a realignment of electric charges within the crystal, producing a transient charge to appear between the electrodes. This high-impedance device must be mounted in close proximity to a FET preamplifier to act as an impedance transformer.

All the IR sensors described above are best used with a modulated (chopped) beam of radiation. The signal is then measured with an instrument that responds only at the modulation frequency, a **lock-in amplifier**. This amplifies the desired signal while at the same time eliminating the noise carried by other frequencies, thus greatly increasing the signal-to-noise ratio. A limitation of thermocouple and thermistor transducers for IR detection is their inherent slowness, restricting the rate of modulation to about 20 Hz. This arises because it is necessary to wait for the sensor to cool off before the next pulse of radiation. The pyroelectric sensor does not share this limitation, and can be used up to perhaps 100 kHz.

Strain Gauges

A force (stress) applied to a solid object will produce a deformation (strain). The strain is defined as the relative change in length, L, resulting from an applied stress $\Delta L/L$. As an example, if two lines are marked on a metal bar exactly 20 mm apart, and under stress the distance changes to 20.01 mm, the strain is $0.01/20 = 0.0005$. This type of strain can be conveniently measured with a **strain gauge**.

A strain gauge may consists of an array of fine wires mounted on a suitable frame, or an equivalent array made from a thin metallic foil (Figure 7-15). In either case, the array is mounted on the object under study, often by cementing it to the surface. The applied force

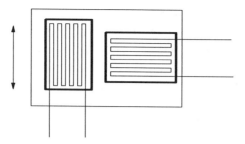

Figure 7-15 One form of dual strain gauge. The entire array within the rectangu-lar border is cemented to the surface of a mechanical member that may be subject to strain along the dimension indicated by the arrow. The resulting minute motion will tend to stretch the left-hand gauge, increasing its resistance, whereas it will have negligible effect on the right-hand unit (dummy).

always results in at least a very minute motion or stretching of the body and hence an equal motion of the strain gauge. The latter changes its resistance when its linear dimension is changed, and this resistance change can be measured by a Wheatstone bridge arrangement or other comparable circuit. Often a second strain gauge is mounted along with the principal one but oriented at right angles, to serve as a reference arm of the bridge. Both gauges are effected equally by any temperature change.

Figure 7-16 shows a practical circuit for a strain gauge bridge. Most gauges are low resistance, about 350 Ω, so they cannot be given more than about 4-V excitation without overheating. The bridge is usually

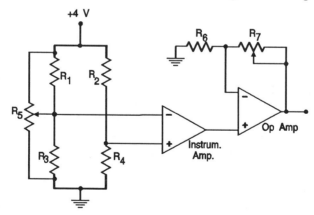

Figure 7-16 Connections for a strain gauge in the Wheatstone bridge configu-ration. The bridge itself is denoted by heavy lines. Pot R_5 permits zero adjustment, while R_7 controls the sensitivity.

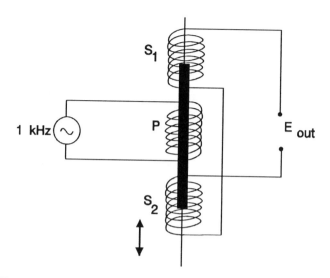

Figure 7-17 A linear differential transformer. As the iron rod is moved up or down within the windings, a variable 1-kHz voltage is induced in the secondaries.

powered from an isolated supply, such as a battery, since its ground level cannot be the same as that for the amplifiers. One pair of diagonally opposite bridge arms, such as R_1 and R_4, constitute the active members of a quadruple strain gauge, while the others, R_2, and R_3, form a dummy gauge to compensate for temperature changes.

The electronic circuitry for strain gauges is available as integrated units (e.g., Analog Devices types 1B31 and 1B32). These include an excitation power source, an instrumentation amplifier and an output filter, everything needed for the operation. Other strain gauges are used in some types of pressure transducers (mentioned in the next section), and also as **load cells**, the heart of many electronic balances.

A related transducer is the **linear variable differential transformer** (LVDT), shown in Figure 7-17. This device consists of three aligned coils with a hollow core. An iron slug can be moved within the core in accordance with the motion to be observed. The energy from the oscillator is transferred from the primary coil P to the two secondary coils S_1 and S_2 via the iron slug. If the slug is exactly centered between the windings, no voltage will appear at the output because the two secondaries are in phase opposition. If the slug moves in either direction a voltage will appear at the frequency of the oscillator. The output is of one phase if the slug moves up, but 180° different if it moves down. Detection is by a phase-sensitive amplifier.

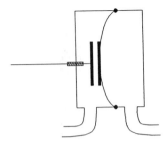

Figure 7-18 A diaphragm pressure gauge. The curvature is exaggerated for clarity.

The LVDT is used in some types of balances to detect the null point, and in general is appropriate wherever a small linear motion is to be monitored.

Pressure Transducers

One of the simplest pressure gauges, at least conceptually, is the mechanical gauge illustrated in Figure 7-18. A change in the gas pressure on one side relative to that on the other will cause a deflection of the flexible diaphragm detectable by a corresponding change of capacitance between the diaphragm and a fixed plate. This capacitive element can be made part of a sine-wave oscillator, such that the frequency of oscillation will be a measure of pressure, albeit nonlinear. The determination of a frequency by means of digital counting circuitry can have high resolution, hence this pressure gauge is

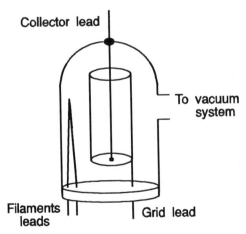

Figure 7-19 The Bayard-Alpert ionization vacuum gauge.

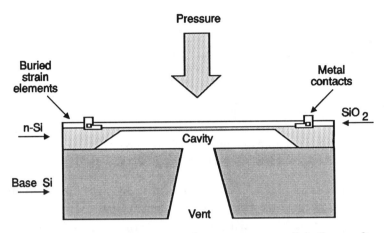

Figure 7-20 Cross section of a semiconductor pressure cell. In the configuration shown, the difference in pressure between the region above the diaphragm and the free atmosphere is sensed. For absolute pressure measurements, the vent is sealed off, with a vacuum in the cavity. (Adapted from literature of SenSyn, Inc.)

capable of highly precise measurements. It can be useful down to the order of 10^{-6} torr.

For lower pressures (higher vacua) the Bayard-Alpert gauge of Figure 7-19 can be used to advantage. This consists of a heated filament and a positively charged mesh called the grid, surrounding an axial collector wire that is given a negative charge. Electrons thermally emitted from the hot filament are accelerated by the electrostatic field between it and the grid. Most of the electrons pass through the grid and are then decelerated by the reverse field and sent back again toward the grid. They tend to oscillate around the grid wires several times before finally being caught. This lengthened trajectory gives the electrons a good opportunity of striking gas molecules and ionizing them. The positive ions so formed are attracted toward the central collector where they produce a current that can be measured with a high-gain amplifier. The operating range is about 10^{-5} to 10^{-10} torr.

A number of manufacturers now offer pressure transducers fabricated directly on a silicon substrate using techniques developed for integrated circuits. Such a device can be mounted on a printed circuit board if desired, or incorporated into a probe for such applications as blood-pressure measurement. The design is possible because of the unusually high elasticity and strength of silicon. A thin diaphragm is etched into the silicon wafer that changes its dimensions slightly with

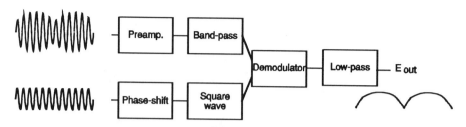

Figure 7-21 Block diagram of a typical lock-in amplifier. The upper path carries the signal, the lower, the reference sine wave.

pressure changes (Figure 7-20). A strain gauge is fabricated directly on the same chip. These transducers are useful from a few torr to perhaps 10 atmospheres, with an accuracy of about $\pm 0.1\%$

Instrumentation

Lock-in Amplifiers

The output from the majority of transducers is in the form of analog signals always accompanied by noise. For signals that are repetitive (AC), one of the widely used detection devices is the **lock-in amplifier**, often called merely a "lock-in." It is especially useful at recovering faint AC signals buried in noise. Two inputs are required, one the data-carrying signal itself, and the other a reference signal of the same frequency and adjustable phase. The reference signal can usually be derived from the same source as the excitation for the transducer. The reference can be quite strong, hence immune from noise interference. Figure 7-21 is the block diagram for a typical lock-in.

The signal is passed through a band-pass filter tuned to the appropriate frequency, thus cutting out some of the noise and harmonics. It then enters a phase-sensitive rectifier (demodulator) where it encounters an electronic switch controlled by the reference signal. If the signal and reference are exactly in phase, the signal will be passed without attenuation when the reference is high and inverted when it is low. All those components of noise that are of other frequencies or are random in nature will be blocked.

One type of demodulator consists of an analog multiplier, a unit that accepts two inputs and calculates their product. The reference waveform is first converted to a square wave which is phased in such a way that it has a value of $+1$ for one-half cycle and -1 for the next. Multiplication of this square wave by the signal, gives a full wave

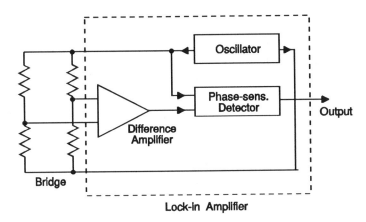

Figure 7-22 A Wheatstone bridge both energized and measured by means of a lock-in amplifier.

rectified output. The output may be then passed through a low-pass filter, if a DC output is desired.

Commercial lock-in amplifiers have a built-in oscillator that can provide excitation to the transducer as well as the reference signal for the demodulator. Figure 7-22 gives as an example a strain-gauge Wheatstone bridge that is both energized and measured by a lock-in.

Signal Averaging

In many measuring systems a process can be triggered repetitively by a suitable excitation. For example, in a luminescence measurement a series of identical flashes of light may be applied to the system, resulting in similar responses time after time. The measuring instrument can be triggered by the same initial flash, and the responses superimposed in such a way as to give their sum over a large number of sequential events. This result, divided by the number of flashes, gives the average value of the responses. The effect of noise is greatly reduced, since noise pulses are as likely to be negative as positive. It can be shown on statistical grounds that the S/N ratio is improved by a factor equal to the square-root of the number of repetitions.

The **multichannel analyzer** is a widely used form of signal averager that consists of a series of channels–for example, 500–each containing a switch and a memory device (see Figure 7-23). A timing circuit turns on the numbered switches sequentially one at a time (with switch S_{in} closed, and S_{out} open), following the trigger. The result can be displayed as a curve of the signal against time, but with

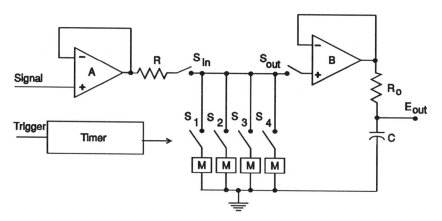

Figure 7-23 Schematic representation of a multichannel signal analyzer. A practical instrument would have many hundreds of switches, each feeding its own memory unit, similar to the four shown here.

insufficient S/N ratio. To increase the signal, we must take responses from a large number of triggers, each one of which gives rise to a similar curve. These are averaged by taking readings at succesive time intervals. Each memory cell is fed information repeatedly at a precisely the same point in the signal. For instance, channel 85 might be turned on at exactly 85 ms (or 85 μs) after each triggering. The information in each memory unit is cumulative, increasing with each repetition.

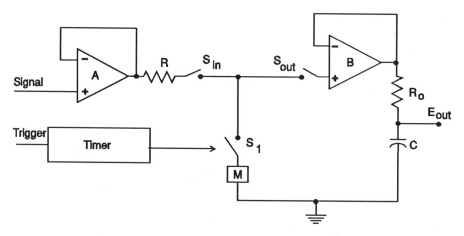

Figure 7-24 A boxcar integrator. This is similar to one channel of the analyzer in Figure 7-23, but the action of the timer is different.

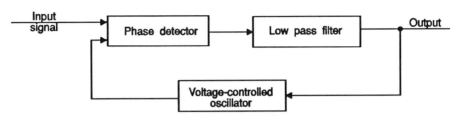

Figure 7-25 Basic circuit of a phase-locked loop.

To retreive the data, switch S_{in} is opened and switch S_{out} is closed. Then the timer closes each memory switch in turn, passing the data on to a read-out device such as an oscilloscope, or to a computer.

The **boxcar integrator** is a simpler integrating device that has only a single data channel, as shown in Figure 7-24. The timer now selects a particular time delay after each repetitive excitation event. Thus the response at a specified time, such as 100 ms after the initial trigger, can be observed for a large number of events, again building up the S/N ratio.

Phase-locked Loops

The **phase-locked loop** (PLL) is a versatile component for manipulating AC signals. It is made up of three functional parts: a phase-detector that measures the difference of phase between two AC voltages, a low-pass filter, and a voltage-controlled oscillator (VCO). These are connected to permit the VCO to form a feedback loop around the other two blocks, as shown in Figure 7-25.

The operation of the PLL can be explained as follows: With no input signal, the VCO runs freely, generating a square wave at some frequency f_0 determined by its internal time constant. When an AC signal is presented to the input at a frequency f_S within about 10 % of f_0, the phase detector will generate a DC signal that is proportional to the difference between f_S and f_0. This DC voltage, called the **error voltage**, is isolated by the filter and impressed on the VCO, changing its frequency so as to reduce the $|f_S - f_0|$ difference nearly to zero. This brings the error voltage close to zero and the VCO frequency remains "locked" onto the signal frequency, differing only by a small phase angle. The frequency span centered around f_0 within which frequency locking can take place is called the **capture range**. Once locked, the signal can move further away from f_0 in either direction without losing lock, over an extent called the **lock range**.

There are many application for the PLL. We will mention a few of these that are especially pertinent to laboratory instrumentation. In signal conditioning, the PLL can be used as a lock-in amplifier to extract a repetitive signal from a noisy background. The VCO will duplicate the incoming frequency but greatly attenuate the noise. Since the output of the VCO follows the signal frequency, the PLL can be considered to act as a **tracking filter**.

The PLL makes a convenient demodulator for amplitude-modulated signals, and thus is often used in AM radio receivers. For this application, an analog multiplier is needed, with the signal connected to one input and the VCO to the other. Integrated circuit PLLs are available that have a multiplier on the same chip for this service.

Another application is in the precision generation of harmonics. A digital binary counter (described in a later chapter) is connected between the VCO and the phase detector. This effectively forces the VCO to run at a higher rate. If for example the counter produces one pulse for every 32 pulses it receives the VCO will have to run 32 times faster to keep up with it, and the output will be a square wave with frequency 32 times the input, its 32nd harmonic.

Grounding and Shielding

Most circuit schematics have numerous ground connections as indicated by the conventional symbol, and the matter is often given very little thought. Yet inadequate grounding is a major source of improper instrument operation, especially in low-level systems.

By definition, all analog ground terminations must be connected together. This guarantees that all currents can find a way to return to their respective sources and ensures a common voltage reference, a zero level. These connections, however, cannot be made in a random way without the likelihood of impairing the precision of measurement.

An example of improper grounding is shown in Figure 7-26, where the signal ground has simply been connected to the output return line on its way back to the power supply. This demonstrates a significant error resulting from the current through the load feeding back into the input. The actual error may well be greater than this because the common to the power supply will also be carrying return current from other parts of the circuit.

To avoid grounding problems, two rules should be followed: (1) the grounding wires should be as short as possible and of heavy gauge; (2) each ground wire should be connected to a single master ground

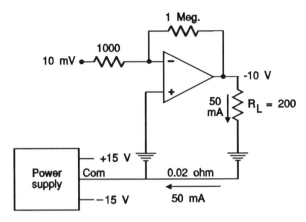

Figure 7-26   The effect of improper grounding. An error of 1 mV (10%) is produced at the noninverting input by 50 mA flowing through wires estimated at 0.02 Ω. "Com" denotes the common terminal of the power supply.

point in a star configuration. If this is not practicable, several satellite stars can be established, each connected by heavy wires to the principal ground point. For instance, a plug-in circuit board can include a star for connections to the components on the board.

Even for properly grounded systems, interference from external fields may still be strong. If the impedances involved are large, electromagnetic fields can be especially troublesome since they consist of high voltages originating in very high impedance sources. Fields of hundreds of volts per millimeter may easily be present, either of atmospheric origin or simply from rubbing one's shoes on a carpet. Electromagnetic fields generated by transformers, fluorescent lamps, radio stations, and the like, induce currents and resulting voltage drops in conductors. As a rule of thumb, one can assume that stray fields are apt to become important if impedances greater than about 1 MΩ are involved.

Interference from these sources can be minimized by observing the following precautions:

1. The AC power section of an instrument should be physically remote from the signal section.

2. Any line carrying high current, especially AC, should have the two insulated conductors twisted together so that electromagnetic fields largely cancel out.

3. High-input impedance amplifiers should be connected to low-impedance sources, if possible.

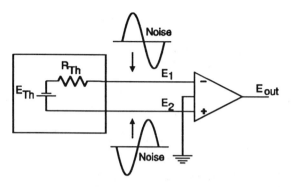

Figure 7-27   A differential system. Noise, assumed to be AC, enters both inputs with the same phase and is largely rejected.

4. Differential input devices should be employed to minimize common-mode noise.

5. Shielding should be incorporated where appropriate, as described in the following paragraphs.

The principle of differential operation is shown in Figure 7-27. The amplifier depicted responds to the difference $E_2 - E_1$. If the noise is impressed equally on both inputs (common mode), it is largely subtracted out. The ability of an amplifier to eliminate such interfer-

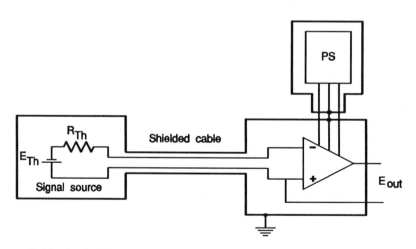

Figure 7-28   A shielded system. The shield arround the source and amplifier, connected through a shielded cable, only connects to the power supply shield through the power-supply common.

ence is measured by its CMRR. Instrumentation amplifiers are specially designed to enhance this quantity.

Shielding consists of surrounding either the source of interference or the sensitive stages of an instrument with a grounded barrier of copper foil or screen. This enclosure provides a low-impedance path to ground for the various induced stray currents. The power supply should have its own shield (see Figure 7-28).

In conductors carrying very small currents, leakage to the shield may produce significant errors. In this case the **guard** method can be used effectively. The guard is a second shield connected to a voltage close to the signal level rather than to ground. The leakage then will be from the guard to ground, not involving the signal-carrying wires. Many commercial measuring instruments are provided with a separate panel connection for the guard; if no guard is used, this should be connected to ground.

Chapter 8

Switching Circuits

This chapter will present a brief summary of the various means by which connections between circuits can be turned on or off. Transistor switches have been described earlier.

Manual Switches

There are a great many different kinds of manual switches with various voltage and current ratings, some intended for breaking or making a single circuit and some with multiple contacts for more complex operations. A commonly used system for describing switches is in terms of **poles** and **throws**, as illustrated with typical examples in Figure 8-1. Rotary switches, like that in (c) of the figure, can be designed so that the wiper breaks contact completely with one point before touching the next ("break-before-make"), whereas in others the space is momentarily bridged while shifting from one point to the next ("make-before-break"). In some applications this difference is important. For example, one might wish to change feedback resistors in an op amp circuit by means of a switch. If the switch were open-circuited between contact points, the amplifier would be left momentarily without any feedback path, causing undesirable saturation.

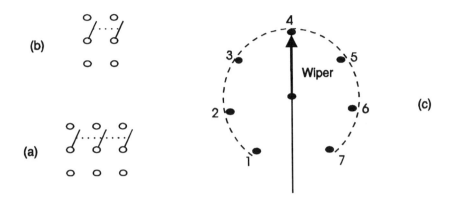

Figure 8-1 Three types of manual switches: (a) double-pole, double-throw (DPDT), (b) triple-pole, double throw (TPDT), (c) single-pole, 7-throw, or 7-position. The dashed lines in (a) and (b) indicate that the two (or three) movable arms operate in synchronism. Style (c) may have multiple segments, sometimes called **wafers**, connected to the same shaft.

Relays

A relay is a switch, such as any of those discussed above, that is made to operate electrically by responding to a controlling current flowing through an electromagnet or solenoid. Some relay circuits utilize one pair of contacts electrically in series with the magnet winding so that once the relay is closed it will remain so (it will **latch** closed) until the current is interrupted.

Relays can be designed to be actuated by either AC or DC at various voltages. In addition, they are classified by the type of construction. One type is the **reed-relay** which consists of a pair of thin electrical conductors made of spring metal, one of which is of magnetic material; it is surrounded by a coil of wire to produce a magnetic field and energize the switch. The response is sudden ("snap action").

The reed relay may also be energized when a magnet is in close proximity; this could be used, for instance, to limit motion in a machine tool or as a sensor in an intrusion alarm system.

Electromechanical relays are not as widely used in low-power circuitry as formerly, since a variety of semiconductor devices have been developed to replace them. They continue to be used extensively in applications involving high currents or voltages.

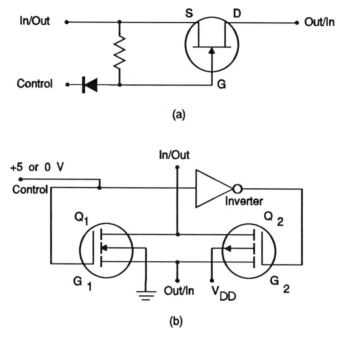

Figure 8-2 Transmission gates: (a) using a single JFET, and (b) made from a complementary pair of MOSFETs; the function of the inverterr is to give +5 V when the control is at ground and zero when the control is at +5 V.

Analog Transmission Gates

A field-effect transistor, as we have seen, can be used as an on/off switch by giving it appropriate signals that will cause it to go either into saturation or into cutoff. The current can pass in either direction and still be interrupted by the switching action. A FET used this way is called a **transmission gate** or **analog gate**, contrasting with the digital gates[1] to be discussed in detail later.

    Figure 8-2a shows the basic transmission gate using an n-channel JFET. If a voltage more negative than either source or drain is applied to the control input, the switch will be turned off, whereas if the gate is left open or grounded, the switch will be turned on. The gate shown in Figure 8-2b, made up from two enhancement-mode MOSFETs, one

---

[1] It is unfortunate that the word "gate" is used in so many senses, including the control element of a FET. One must guard against ambiguity.

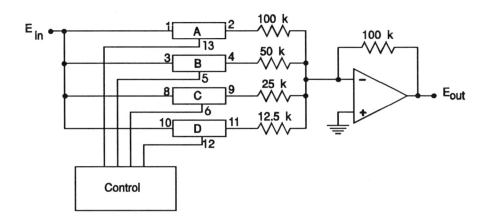

Figure 8-3 A variable-gain amplifier using a CD4066 to select the input resistor. The numbers designate the pins of the CD4066 (pin 14 is V_{DD}, usually +5 V, and pin 7 is ground).

n-channel, the other p-channel, is widely used in integrated circuits. The output of the inverter is zero when the control input is positive, and positive when the control is zero. Hence gate G_1 and gate G_2 are opposite in their voltage but identical in action. Both transistors, Q_1 and Q_2, are on for positive control voltage and off for negative. This design gives greater speed of operation than a single MOSFET would, because the time constant of each member is lowered by the presence of the other.

Transmission gates are available as quads, that is four gates combined into a single IC, packaged in a 14-pin unit. An example is the CD4066 quad switch. Figure 8-3 shows a typical application wherein the four segments of the CD4066 are used to connect any one of four input resistors to an op amp, thereby giving control of the gain. It can readily be seen that when switch A is on, the gain is 1, when B is on, the gain is 2, etc. Closing both A and B at the same time gives a gain of 100/33 = 3. The reader should be able to calculate the gain for each possible combination of closed switch elements. The box marked "control" contains circuitry that energizes the proper control lines.

Transmission gates are characterized by their on-resistance, which is ideally zero but in practice may be 100 Ω or so, and by the leakage current that can still flow when the switch is off. The leakage should

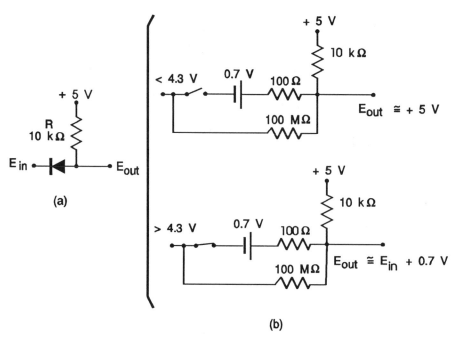

Figure 8-4   A diode switch (a), and its equivalent circuits (b). By applying the principle of the voltage divider, one can see that the output must be either +5V or (Ein + 0.7V).

not be greater than about 10^{-7}A. These limitations must always be kept in mind in designing applications such as that discussed above.

Diode Switches

Diodes can act as automatic switches activated by the signal itself. When forward biased, the diode acts as a closed switch (low resistance), and when reverse biased, as an open switch (high resistance). This analogy to a mechanical switch is demonstrated in Figure 8-4. As long as E_{in} is less than one diode drop below +5 V (i.e., 4.3 V), the diode will be in its low-resistance state, current will flow through R and the diode. The whole potential drop (except for the forward voltage drop, 0.7 V) will appear across the resistor, so that $E_{out} = E_{in} + 0.7$. When E_{in} is greater than about +4.3 V, the diode becomes nonconducting, no current flows through R, and $E_{out} = +5$ V. In our model this corresponds to the switch being turned off. The plot in Figure 8-5 shows this relationship. Note that the switching action resides in the diode alone; E_{out} is either connected directly to E_{in} or allowed to

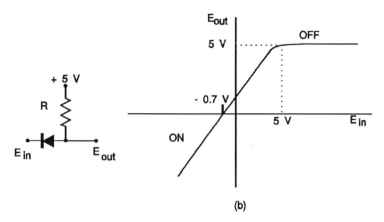

(b)

Figure 8-5 The diode switch of Figure 8-4 and its response plot. Note that zero input gives nearly zero output and +5 V in gives +5 V out. The circuit acts as a leveler, since any higher voltage gives the same +5 V output. It is assumed that the impedance of the load is large compared to R, so that the current drawn from the output is negligible.

assume 5 V. Another form of diode switch with its plot is shown in Figure 8-6. Other configurations of diode and resistor can also be utilized in switching circuits.

Thyristors

The thyristor is a power-handling device that can remain in a stable condition in either of two modes, conducting or nonconducting. A

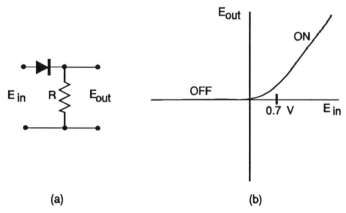

(a)   (b)

Figure 8-6 A diode switch that acts as a clamp for "0", since any negative input results in zero output. This is a half-wave rectifier.

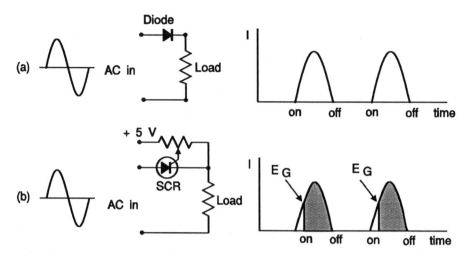

Figure 8-7 Comparison of (a) a diode, and (b) an SCR, as rectifiers. In (b), the point in each cycle where the switch turns ON (E$_G$) is determined by the setting of the variable resistor.

pulse applied to a gate terminal will cause the thyristor to break into conduction and remain latched in that condition until the voltage across the diode drops to zero or the current is interrupted externally. Thus the action is similar to that of a diode, except that it latches in the off state until it is triggered, at which time it latches in the on condition.

The basic type of thyristor is called a **silicon controlled rectifier** (SCR). To understand its operation, refer first to the half-wave rectifier using a standard diode, as in Figure 8-7*a*. The diode acts as a switch

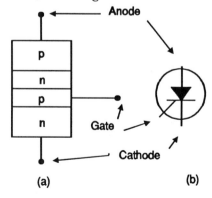

Figure 8-8 An SCR: (a) construction and (b) the symbolic representation.

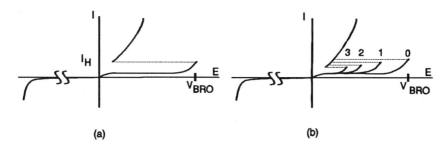

(a)   (b)

Figure 8-9 Characteristics of an SCR: (a) with zero gate current, where I_H designates the holding current below which the device turns off; (b) the effect of making the gate voltage successively more positive.

that is turned repeatedly on and off as the AC voltage crosses zero. The SCR represents an extension of the same mode of operation (Figure 8-7b). Here the switching can be made to occur anywhere in the positive half-period, but the switching off occurs only on crossing zero.

Physically, the SCR is a combination of four semiconductor layers in the sequence *pnpn* (Figure 8-8). The current-voltage characteristic of such a device is shown in Figure 8-9a. The general behavior is that of a conventional silicon diode, except for the region corresponding to small forward currents. If the anode-cathode potential be increased from zero, with no connection made to the gate electrode, the current will assume a very small value (a few milliamperes in a device capable of carrying tens of amperes), and remain almost constant for an appreciable voltage range, finally turning upward. At this point, labeled V_{BRO}, a forward "breakover" occurs (dotted line), and the

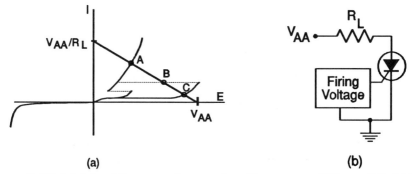

(a)   (b)

Figure 8-10 (a) A load line superimposed on the curves of Figure 8-9; (b) the basic SCR control circuit.

device reverts to normal diode operation, with current limited only by the resistance of the external circuit. At the same time the potential drop across the device falls to a low value. This abrupt decrease in impedance is called "turning on," or "firing" of the thyristor. Once fired, it can only be turned off by interruption of the anode current. The breakover voltage can be decreased by supplying a positive voltage to the gate, relative to the cathode, as seen in Figure 8-9*b*.

The turn-on process can be better understood by reference to Figure 8-10*a*. The unusual feature of this graph is that there are *three* intersections of the load line with the characteristic curve. The curve passing through point A is no different from the response of any conducting diode. The fact that the voltage drop in the conducting mode is small keeps the power (heat) dissipation from being excessive. Point C corresponds to an off condition, with the voltage across the diode nearly equal to V_{AA}. In this case the anode current is very low so that, again, the heat dissipation is minimal. Point B is unique in that it does not correspond to any stable state.

When first connected, the SCR normally goes to the off state[2]. Firing can be brought about by applying a positive pulse to the gate electrode. This changes the characteristic curve, which now follows the inner segment of the curve of Figure 8-10a. Points B and C now disappear and only A remains, so that the SCR is forced to assume the current and voltage of this point, and the device is successfully fired.

The SCR will continue to conduct until the cathode-anode potential is brought close to zero (or negative). When the SCR is used as a rectifier, the current automatically becomes zero once in every AC cycle.

There are several types of thyristors in addition to the SCR. One of these is the **triac,** which is a bidirectional modification. Its characteristic curve is the same as that of the SCR in the first quadrant, and repeats that curve symmetrically in the third quadrant (Figure 8-11).

In principle a thyristor can be triggered by a DC signal, such as the output of an op amp, but generally only pulses are used. One way by which a train of appropriate pulses can be generated is with the **unijunction transistor** (UJT), a three-terminal device shown in a typical circuit in Figure 8-12. The important characteristic of the UJT

---

[2] If the voltage V_{AA} is applied too suddenly, the thyristor may self-trigger, and go directly into conduction.

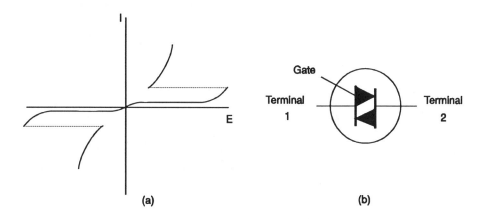

(a)   (b)

Figure 8-11 The characteristic curve of the TRIAC (a), and its symbol (b). The curve in quadrant I applies when terminal 2 is positive, quadrant III when negative.

is that the resistance between the emitter E and the second base B_2 suddenly decreases from its normal level of several kilohms to nearly zero when the potential of the emitter is raised above some critical value, thus permitting a sizable current to pass. When the power supply is first connected, the capacitor begins to charge through R_T. When its potential reaches the critical value, the resistance between E and B_2 drops abruptly and the capacitor discharges through the UJT and R_2. When it reaches some low level, depending on the sizes

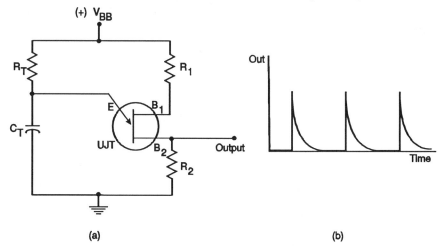

(a)   (b)

Figure 8-12 (a) A pulse generator using a unijunction transistor. (b) The waveform at the output. The pulse repetition rate is proportional to $1/R_T C_T$.

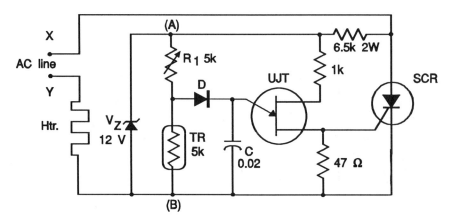

Figure 8-13 A temperature-control circuit using a UJT to fire an SCR. The component marked TR is a thermistor.

of the resistors, the base-emitter resistance returns to its high level, the capacitor starts to recharge, and the cycle repeats. The output exhibits a sharp pulse suitable for triggering an SCR.

Figure 8-13 shows an application of both UJT and SCR to a temperature regulator. Current can flow to the heater through the SCR, which is controlled by the pulses from the UJT. The latter is in turn controlled by the resistance of the thermistor. Since this circuit contains neither a transformer nor a relay, it can be constructed in a conveniently compact form. The circuit must be analyzed separately for successive half-cycles of the AC power. When the line connection marked Y is positive, the SCR cannot conduct, regardless of what

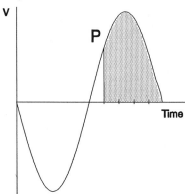

Figure 8-14 Operating principles of the circuit of Figure 8-13. The heater is on during the shaded portions of the cycle. The tick marks correspond to successive firings of the UJT.

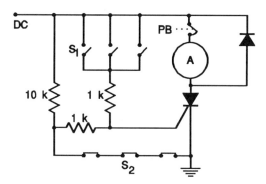

Figure 8-15 An SCR applied to a burglar alarm system. The alarm is designated by A. The alarm is turned off by depressing the normally closed pushbutton PB.

happens to its gate, whereas the Zener diode conducts freely in its forward direction. Hence points A and B will be at essentially the same potential, and the UJT cannot operate. On the next half-cycle power lead X is positive, while the Zener maintains point A at 12 V more positive than B. The capacitor proceeds to charge through R_1 at a rate determined by the resistances of R_1 and the thermistor which is in contact with the medium whose temperature is to be controlled. The lower the temperature, the larger will be the resistance of the thermistor, and the greater will be the fraction of the current through R_1 available to charge C. This increases the frequency of production of pulses by the UJT. The direct connection between the base of the UJT and the control gate of the SCR ensures that the SCR is turned ON at that time within each cycle when the first pulse appears (see Figure 8-14). The SCR turns OFF automatically when the voltage crosses

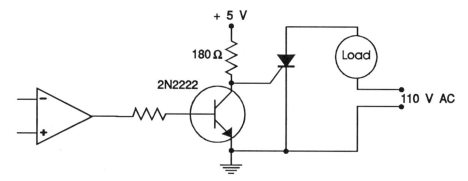

Figure 8-16 An SCR-controlled load triggered by the output of a comparator. The load receives only the positive half-cycles.

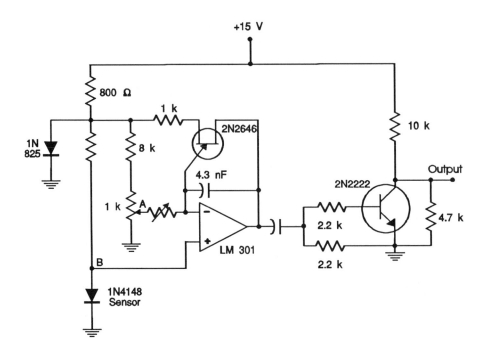

Figure 8-17 A temperature-to-frequency converter using a 1N4148 diode as sensor. The 1k variable pot permits trimming the zero point, while the 10 k variable determines the span.

zero. The time in which energy is supplied to the heater, indicated by the shaded portion of the diagram, is greater when the pulse rate is faster, corresponding to a temperature lower than desired. The temperature level can be set by adjustment of the variable resistor R_1.

Figure 8-15 shows a simple but effective burglar alarm circuit built around an SCR. The gate voltage is determined by a network of resistors that includes several switches. If any one of the switches in the bank marked S_1 is *closed*, or any one of bank S_2 is *opened*, the SCR will be triggered, thus energizing an alarm. It can only be turned OFF by depressing the normally closed pushbutton.

It is also possible to trigger a thyristor from the output of an op amp or comparator. An example of this interface, shown in Figure 8-16, makes use of a comparator boosted by an *npn* transistor to turn on an SCR. In this case, no switch is needed for turn-off, because the AC power goes through zero on each half-cycle, turning the load off automatically as soon as the comparator changes states.

Figure 8-17 shows a circuit that will develop a frequency proportional to temperature sensed by a diode. It makes use of a unijunction transistor to discharge the capacitor of an integrator. The rate of repeatedly charging the capacitor, and hence the frequency, is determined by the difference of potential between the points marked A and B, with the latter determined by the forward voltage drop across the sensing diode. The readout requires only a rather simple frequency counter.

Chapter 9

Power Supplies

Batteries

Every active electronic device must have a source of DC power. The simplest, conceptually, depends on the use of batteries. The main advantages of batteries are that they give pure DC directly without additional circuitry and that they release the user from dependency on the AC power mains. Battery power is essential for units to be operated in the field, for small portable items such as pocket radios and calculators, and for automotive devices. Their principal disadvantages are that they have a limited lifetime, even when they can be recharged, and that they cannot be relied upon to give a fixed voltage independent of the amount of current drawn and of the temperature.

Batteries can be classified in various ways, either according to the materials of which they are made or by their electrical properties. Table 9-1 lists some examples. The low internal impedance guarantees that the voltage will be nearly independent of load and likely to continue at the original level until the battery is practically exhausted. Mercury and lithium cells are widely used in applications where constancy of output outweighs the expense. Small lead-acid batteries are employed if large current drain for short periods is expected; chemically they are identical to automotive batteries. "Gel cells" are a form of lead-acid cells in which the electrolyte is formed into a gel to prevent spillage.

Table 9-1

Representative small batteries

Type	Recharge?	Volts	Comments
Alk. Ni/Cd	yes	1.35	Low impedance, long shelf life
Zn/MnO2/C	no	1.5	Inexpensive, high impedance
Alk. Zn/MnO2/C	no	1.5	Lower impedance, longer life
Alk. Zn/HgO	yes	1.35	Very low impedance
Alk. Zn/AgO	yes	1.85	Very low impedance
Lithium/MnO2/C	no	3.0	High energy density, long life
Lithium/iodine	no	3.0	High energy density, long life
Lead/acid	yes	2.0	High energy density, long life,

As indicated in the table, many types of batteries can be recharged by passing current through them in the reverse direction. For most of these types, the charging current must be limited carefully in accordance with the manufacturer's recommendations, to avoid overheating or production of gases that might damage the cell[1].

Rectifying Power Supplies

The term "power supply" almost universally refers to a system for converting AC (50 or 60 Hz) to DC with suitable voltage and current ratings. There are two classes called, respectively, **linear** and **switching supplies**. We will present an overview of power-supply designs. Be aware that it is economically wise to purchase ready-made power supplies rather than to construct them from discrete components.

Linear power supplies consist of three, possibly four, segments, as illustrated in the block diagram of Figure 9-1. The transformer serves to change the AC voltage from 110 or 220 V to a lower value, such as 25 V. The rectifier unit that changes AC to DC is most commonly a bridge of four diodes, as shown in Figure 9-2a, though small supplies

---

[1] Recharging cells requires a number of precautions that cannot be detailed here. See J. Williams, *EDN*, May 24, 1990, p.147.

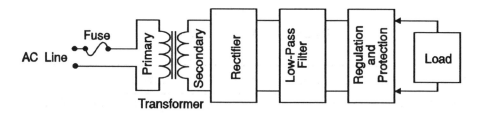

Figure 9-1 Block diagram of a typical linear power supply.

in which economy is more important than efficiency use only one or two diodes. The output of the diodes is effectively a pulsating DC, which must be processed in the filtering unit to produce a steady voltage. The filter consists of two capacitors and a resistor, as in part Figure 9-2*b* of the figure. The function of stage (*c*) is to maintain the output voltage at a constant (regulated) level.

There are many circuits that can be used as regulators, but those available as integrated circuits will fill almost any needs. A popular example, shown in Figure 9-2*c*, is the 78xx series, where the "xx" is replaced with 05, 08, 12, or 15, indicating the nominal voltage produced. This protects the output voltage from variations due to

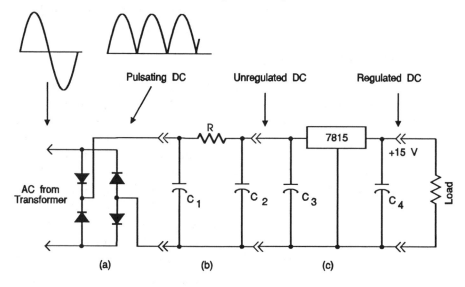

Figure 9-2 The rectifier (a), filter (b), and regulator (c) portions of a typical 15-volt power supply. Capacitor C_3 can be omitted if the regulator and filter are physically close to each other.

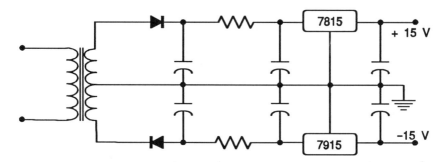

Figure 9-3 A dual regulated power supply using a split-secondary transformer.

either a fluctuation in AC line voltage or a change in the load that is powered by the supply.

A power supply that is to give both positive and negative voltages can be built in similar manner, using a regulator of the 79xx series that can handle negative outputs. Figure 9-3 shows such an arrangement.

In designing a power supply, attention must be paid to the power dissipation of the various components. The rectifying diodes can be had in various power levels from 1/4 watt to 10 watts or more. The regulators are generally rated as to their current-carrying capability rather than wattage dissipation. The largest to be met with in laboratory instruments is rated at 2 A. Both rectifiers and regulators can be fitted with heat sinks if necessary.

Linear power supplies are inherently inefficient since a considerable fraction of the power taken from the line ends up as heat rather than as useful power delivered to the load. Much more efficient in the use of power are **switching power supplies** (often called "switchers"). These are more complex and difficult to design, and hence are not worthwhile unless considerable current is to be drawn. There are

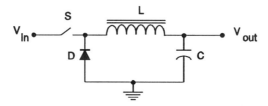

Figure 9-4 Essential parts of a step-down switching power supply.

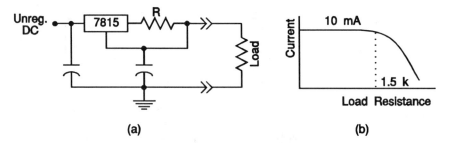

(a)   (b)

Figure 9-5 A constant current supply, with its loading characteristic.

several arrangements that can be used in a switcher, one of which is shown in Figure 9-4.

This circuit operates as follows: The input V_{in} originates in an unregulated power suppy. Switch S is alternately closed for a time T_{on} then opened for a period T_{off}. The total, $T_{on} + T_{off} = T$, is called the switching time. The switch S can be an analog gate energized by an oscillator operating at a few tens of kilohertz. Thus the voltage at the input to the inductor L is $+V_{in}$ during time T_{on} and zero during time T_{off}. The current that flows through L cannot increase instantaneously when S is closed, but rather builds up linearly at the rate $L(dI/dt)$, storing energy in the inductor. While the switch is open, the stored energy is transferred to the load and the current path is completed through the diode and ground. The output voltage is $V_{out} = V_{in}(T_{on}/T)$. Regulation is effected through a voltage feedback from the output that controls the duty cycle of the oscillator which operates the switch. The major advantage over a linear power supply is that very little power is dissipated, thus leading to high efficiency[2].

Current Supplies

The regulated power supplies described in the preceeding section are **voltage sources**. They maintain a constant output voltage over a wide range of currents. They can readily be converted to **current sources** which, in contrast, provide the correct variable voltage needed to maintain the output current at a constant value. Figure 9-5 shows a

---

[2] This is only a very brief overview of switching power supplies. For more information, the reader should consult: (a) S. Ciarcia, *Circuit Cellar Ink*, Issue 8, April/May 1989, p.10; (b) F. Goodenough, *Electronic Design* July 27, 1989, p.33: (c) J. Williams, *EDN*, Jan 18, 1990, p. 151.

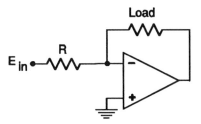

Figure 9-6 A constant current regulator using an op amp.

commonly used connection for a constant current supply based on a three-terminal regulator. The reference connection for the 7815 is removed from the ground, and applied to control the voltage across resistor R. The current sent through the load is $15/R$ amperes. The maximum current depends on the rating of the regulator and the DC voltage source.

To see how it works, suppose $R = 1.5$ kΩ so that the current is controlled at 10 mA. As long as the load is less than 1.5 k, the current will be exactly 10 mA, as desired. For larger load resistances more than 15 V will be necessary to push that much current through, and the characteristic curve will drop off. The maximum voltage that is available (15 V in this case) is called the **voltage compliance**. A larger current than 10 mA would be available if the resistor R were decreased in magnitude, but the compliance would still be 15 V.

Another way to regulate current is to place the load in the feedback loop of an operational amplifier that is supplied with a constant voltage, as in Figure 9-6. The constant voltage E_{in}, together with the resistor R, enforces a current E_{in}/R to flow through the load. The compliance is simply E_{in}. The application of this circuit is limited to situations were the load has no direct connection to ground.

Part II

Digital Electronics

Chapter 10

Binary Numeration

We have now completed our basic treatment of analog electronics and will turn to the digital field. Later we will need to return to the analog world again with a discussion of techniques of transferring information from one domain to the other. Only then will we be ready to discuss computers.

Digital circuits can only operate on signals in the form of a succession of states that are either on (usually +5 volts) or off (zero volts). Such signals are best treated in terms of a system of numeration based on only two digits, 0 and 1. This is called the **binary** system, analogous to the familiar **decimal** system based on ten digits. These are not the only possibilities, and we shall be concerned also with the **octal** and **hexadecimal** systems, based on eight and sixteen, respectively.

To count in any of these systems, one starts with 0 and increments by one repeatedly until there are no more digits. Then a 1 is placed in the next column to the left and the entire sequence (zero to the maximum) is repeated in the right-hand column. This is so familiar in decimal counting that it seems obvious to us, but in the other systems it takes a bit of getting used to.

In hexadecimal notation, more then ten digits are needed, and usage has established these extra digits as the letters A through F. Table 10-1 compares the binary, octal, decimal, and hexadecimal

Table 10-1

Comparison of numeration systems

Decimal	Binary	Octal	Hexadecimal	BCD
0	0000 0000	0	0	0000 0000
1	0000 0001	1	1	0000 0001
2	0000 0010	2	2	0000 0010
3	0000 0011	3	3	0000 0011
4	0000 0100	4	4	0000 0100
5	0000 0101	5	5	0000 0101
6	0000 0110	6	6	0000 0110
7	0000 0111	7	7	0000 0111
8	0000 1000	10	8	0000 1000
9	0000 1001	11	9	0000 1001
10	0000 1010	12	A	0001 0000
11	0000 1011	13	B	0001 0001
12	0000 1100	14	C	0001 0010
13	0000 1101	15	D	0001 0011
14	0000 1110	16	E	0001 0100
15	0000 1111	17	F	0001 0101
16	0001 0000	20	10	0001 0110
17	0001 0001	21	11	0001 0111
31	0001 1111	37	1F	0011 0001
32	0010 0000	40	20	0011 0010
33	0010 0001	41	21	0011 0011

systems of numeration. (The column headed "BCD" will be discussed later.)

One might ask: What exactly does binary "1001" mean? Fundamentally, it is merely a number, just as 1234 is a number in the decimal system. As the table tells us, binary 1001 is the equivalent of decimal 9, and so it could refer to nine apples or nine minutes or nine millimeters. On the other hand, a 4-digit binary number might be used to indicate the positions of four on/off switches; specifically, 1001 would tell us that switches numbered 1 and 4 are on while 2 and 3 are off.

We have said that a computer can only handle binary numbers. This is simply because the computer is designed so that any connecting wire must have only one of two possible voltages, usually zero and +5 V. Hence a switch could be used to change the state of any of these conductors, either connecting it to a 5-V source or to ground.

If this is the case, one might wonder why we are also interested in octal and hexadecimal numeration. The principal use of octal numbers is to organize binary numbers in more manageable groups. For example, the binary number 110100111010 is difficult to identify and memorize, but arrangement in groups of 3 bits each, 110 100 111 010, is more acceptable. As Table 10-1 shows, every possible grouping of three binary digits, from 000 to 111, corresponds to one specific number in the octal system. This means that we can replace the groups in the example above with the corresponding octal digits, making it easier yet to recognize and remember:

$$\begin{array}{ccccc} 010 & 001 & 011 & 101 & 111 \\ 2 & 1 & 3 & 5 & 7 \end{array} = 21357_8$$

where the subscript 8 indicates that the number is in octal.

Another way in which binary numbers can be organized is in groups of four, representing the values between 0 and $1111_2 (= 15_{10})$. This classification parallels the division of computer memory in **bits** (single digits), **nibbles** (groups of four bits) and **bytes** (groups of eight bits). Any group of one or more bytes acting as a unit form a **word**. Thus a computer word consisting of two bytes might be grouped as

$$0010 \quad 0010 \quad 1110 \quad 1111$$

The 16 possible values of each group are represented by a single digit in the hexadecimal system. According to this notation the above word becomes $22EF_{16}$. Hexadecimal numbers are often designated by the abbreviation "hex" or simply the letter "h." Thus $22EF_{16}$ can be written as $22EF$hex , $22EF_h$, or $22EF$h.[1] It is well to keep in mind that some of the familiar decimal terms are not appropriate in other systems. For example, 100hex should *not* be pronounced "one hundred hex," but rather "one, zero, zero, hex." The hexadecimal system is widely used in connection with computers, particularly in specifying locations in memory.

From time to time we find it necessary to convert a quantity from one system to another. To understand how to do this, we must introduce the concept of "place." Consider the decimal number 2459. We can say that the "2" is in the "thousands place," the "4" in the "hundreds place," and so on, each successive "place" corresponding to a power of 10. In a binary number, such as 11010, the places take successive powers of 2. In octal, they correspond to powers of 8, and in hex to powers of 16.

This can be clarified by writing the decimal identity:

$$2459_{10} = (2 \times 10^3) + (4 \times 10^2) + (5 \times 10^1) + (9 \times 10^0)$$

Similarly,

$$4347_8 = (4 \times 8^3) + (3 \times 8^2) + (4 \times 8^1) + (7 \times 8^0)$$
$$= 2048 + 192 + 32 + 7 = 2279_{10}$$

and in binary,

$$1101_2 = (1 \times 2^3) + (1 \times 2^2) + (0 \times 2^1) + (1 \times 2^0)$$
$$= 8 + 4 + 0 + 1 = 13_{10}$$

Thus we have demonstrated a method for conversion from other systems into decimal.

The most convenient way to convert from decimal into other systems, is to divide successively the decimal number by the base of

---

[1] Examination of Table 8-1 will show that OCT 31 = DEC 25, This has been taken by numerologists to prove the identity of Halloween and Christmas.

the desired system, the remainders forming the new number, beginning with the least significant bit. Thus conversion of 44_{10} to binary follows the scheme given below, where "rem" designates the remainders:

$$44/2 = 22 \quad \text{rem } 0$$
$$22/2 = 11 \quad \text{rem } 0$$
$$11/2 = 5 \quad \text{rem } 1$$
$$5/2 = 2 \quad \text{rem } 1$$
$$2/2 = 1 \quad \text{rem } 0$$
$$1/2 = 0 \quad \text{rem } 1$$

The resulting binary number is read in the remainder column, *bottom to top*, as 101 100$_2$. Let us convert 2279_{10} into octal. Using a similar format,

$$2279/8 = 284 \quad \text{rem } 7$$
$$284/8 = 35 \quad \text{rem } 4$$
$$35/8 = 4 \quad \text{rem } 3$$
$$4/8 = 0 \quad \text{rem } 4$$

Hence $2279_{10} = 4347_8$, consistent with the reverse example given previously. The reader can try checking out the hex-to-decimal example. To convert from one base to another, other than decimal, it is often easiest to convert first to decimal, then to the second base. The program below written in QuickBASIC provides an easy method to convert between the different bases.

BCD Notation

It is often necessary to transfer data from binary to the decimal equivalent in a form that is convenient for operating a digital display. For this purpose simple binary-to-decimal conversion will not serve, because the circuitry that energizes the display must act on individual digits, yet still be in binary. A separate system must be devised: the BCD or **binary-coded decimal**. This code follows the usual binary counting from 0000 to 1001, but instead of proceeding to 1010, it skips directly to 0001 0000, then continuing to 0001 1001 followed by 0010 0000 and so on. This is shown in the last column of Table 10-1.

Program to change number bases *

```
DEFINT A-Z
10: INPUT "Convert number ";A$
20: INPUT "From base ";R1
IF R1>36 THEN PRINT "Base too large, 1 to 36 only":GOTO 20
IF R1<2 THEN PRINT "Base must be 2 or larger": GOTO 20
30: INPUT "to base ";R2
IF R2<2 OR R2>36 THEN PRINT "Base must be 1 to 36 only":
GOTO 30
    T# = 0
    FOR C = 1 to LEN(A$)
        V = ASC(MID$(A$,C))
        IF V>47 AND V<58 THEN  V2 = V − 48
        IF V>64 AND V<91 THEN  V2 = V− 55
        IF V>96 AND V<123 THEN V2 = V − 85
        IF V2>=R1 THEN PRINT "Numeral too large for base ";R1
        PRINT:GOTO 10
        T# = T#*R1 + V2
    NEXT C
    B$ = " "
    WHILE T#<> 0
        V2 = T# − INT(T#/R2)*R2:  REM: V2 − T# MOD R2
        T# = (T# − V2)/R2
        IF V2<10 THEN V = V2 + 48
        IF V2>9 THEN V = V2 + 55
        B$ = CHR$(V) + B$
    WEND
    PRINT "The answer is: ";B$:PRINT
    GOTO 10
    END
```

* (Adapted by G. W. Ewing from David L. Kahn, BYTE, March 1985, p. 153)

Chapter 11

Binary Logic Gates

The field of digital electronics makes extensive use of logic gates. These building blocks are generally composed of a few transistors with associated resistors and sometimes diodes. They are manufactured in several families, characterized by particular circuit design. We will examine a number of these logic gates in some detail, but first it is necessary to introduce some basic mathematical concepts.

Boolean Algebra

Boolean algebra is a mathematical discipline which permits the handling of variables that can take only one of two values, in other words, **binary** variables. Boolean algebra is based upon a series of four operators:

> 1. The AND operator ($\bullet$):
>   > A $\bullet$ B $\bullet$ C means A AND B AND C
> 2. The OR operator ($+$):[1]
>   > A $+$ B $+$ C means A OR B OR C
> 3. The EXCLUSIVE-OR operator ($\oplus$):
>   > A $\oplus$ B $\oplus$ C means any *one* of A OR B OR C
> 4. The NOT operator: (overbar)

---

[1] The $+$ sign in this context is not to be confused with the familiar addition sign.

$\overline{A}$ means NOT-A or "complement of A"

Electronic devices have been invented that implement each of these four operators by accepting binary inputs and generating from them the appropriate output voltages. The AND, OR, and XOR operations are performed by gates with corresponding names: AND-gates, OR-gates, and XOR-gates. (XOR is read exclusive-OR). The NOT operator is handled by a special type of gate called an inverter. In addition, gates are available to perform two operations simultaneously:

NAND   NOT $(A \bullet B \bullet C)$ = $\overline{(A \bullet B \bullet C)}$

NOR   NOT $(A + B + C)$ = $\overline{(A + B + C)}$

These functions can be combined in various ways, much as can the functions in ordinary algebra. Thus $A + B + C$ is the same as $B + C + A$, but $A \bullet (B + C)$ is not the same as $(A \bullet B) + C$. The operation within the parentheses must be carried our prior to combination with other variables. A few examples may help to make these relations clear:

A house door has three locks, designated X, Y, and Z. The door is locked if *any one* of the locks is on. If 1 refers to a locked condition and 0 to an open one, the door is locked if X OR Y OR Z = 1, and we can write in symbolic terms, $X + Y + Z = 1$. On the other hand, for the door to be open, *all* locks must be open, that is, if X AND Y AND Z = 0, or $X \bullet Y \bullet Z = 0$.

A house has two doors, each with one lock. For the house to be secure, *both* locks must be closed, which means X AND Y = 1, $(X \bullet Y = 1)$. For entry to be possible, only *one* lock need be open: $A + B = 0$.

If one of the two doors on the house has two locks (X AND Y) while the other door has only one (Z), the house will be secure if $(X + Y) \bullet Z = 1$, and so on.

A well-established principle in symbolic logic is **De Morgan's theorem,** which states that [NOT-A AND NOT-B] is equivalent to [NOT-(A OR B)], and that [NOT-A OR NOT-B] equals [NOT-(A AND B)], or in Boolean notation,

$$\overline{A} \bullet \overline{B} = \overline{A + B} \qquad \overline{A} + \overline{B} = \overline{A \bullet B} \qquad (11\text{-}1)$$

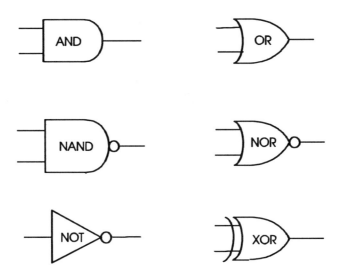

Figure 11-1 The standard circuit symbols for logic gates. All except the NOT gate can have any number of input lines. (The XOR gate is seldom used with more than two inputs.) The small circle indicates a negation wherever it appears.

The preceeding examples of doors and locks can be used to verify DeMorgan's theorem; the reader is urged to explore this further. There is much more to Boolean algebra that can be of use in electronic logic, but that outlined above is all we need for present purposes; more detail is presented in Chapter 20.

Implementation of Logic Functions

As mentioned earlier, each of the Boolean operators can be simulated by electronic gates shown in Figure 11-1. We will follow the notation that letters such as A, B, and C will designate input variables. Each gate can be described operationally by means of a **truth table**, a listing of all possible combinations of inputs and the corresponding outputs. Truth tables for the basic three-input gates are shown in Table 11-1. The possible values are listed in the usual sequence of binary numeration. (The NOT gate is absent from Table 11-1, since it cannot have more than one input.)

The De Morgan theorem makes possible a duality in logic gates. For example, a NOR-function can be implemented either by a simple NOR-gate shown in Figure 11-1 or by an AND-gate with two inverters (NOT-gates). This is shown in Figure 11-2a, along with a similar duality

Table 11-1

Truth tables for the basic logic gates

Inputs	Outputs				
ABC	AND	NAND	OR	NOR	X-OR
0 0 0	0	1	0	1	0
0 0 1	0	1	1	0	1
0 1 0	0	1	1	0	1
0 1 1	0	1	1	0	0
1 0 0	0	1	1	0	1
1 0 1	0	1	1	0	0
1 1 0	0	1	1	0	0
1 1 1	1	0	1	0	0

applied to the NAND-gate in (b). A small circle appended to any gate symbol, either at an input or at the output, indicates negation.

Synthesis of Gate Systems

Logic gates can be combined to make more elaborate systems. A good example to demonstrate this process is the combination of two AND-gates, two inverters, and an OR-gate to implement the XOR-function, as shown in Figure 11-3a. To demonstrate the correctness of this assemblage, one can construct a truth table, shown at Figure 11-3b, by including columns giving the states of the circuit at points P and Q, and then combining these intermediate values to give the output C. It will be seen that the result is consistent with the prior definition of the XOR function. The same approach can be applied to complex combinations of logic gates, by exhausting all possible input combinations.

In addition, it is often necessary to use the reverse approach and to devise a logic system that will obey a specified truth table. Suppose, as an example, that it is required that the output be HIGH (i.e., equal to "one") for the binary inputs 0110, 1010, and 1110, but not for any others. It is often possible to design a circuit with only a few gates to implement the desired behavior. However, if ingenuity fails, one can always follow the "brute force" method illustrated in Figure 11-4.

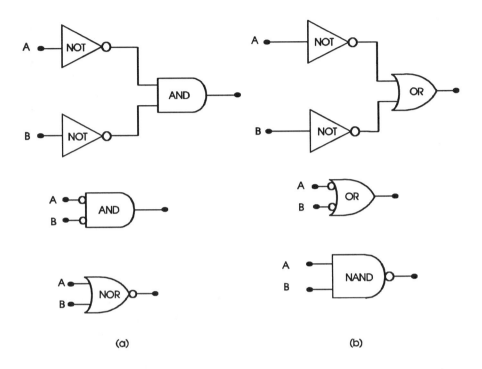

(a)   (b)

Figure 11-2 Combinations of gates to illustrate the equivalence of functions. (a) Three symbols for an NOR-gate; (b) Three symbols for an NAND-gate.

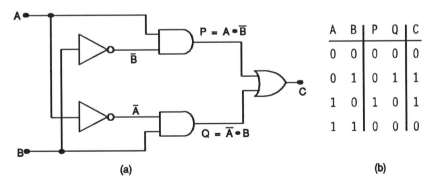

A	B	P	Q	C
0	0	0	0	0
0	1	0	1	1
1	0	1	0	1
1	1	0	0	0

(a)   (b)

Figure 11-3 An EXCLUSIVE-OR gate composed of discrete gates (a), and its truth table (b). Note that here (and henceforth) the names of the individual gates are not printed, but can be identified by their shapes.

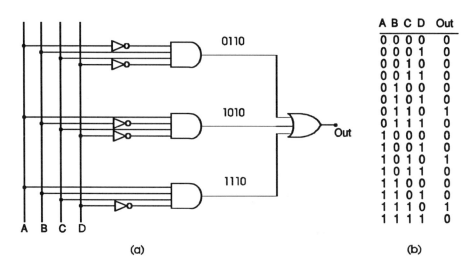

A	B	C	D	Out
0	0	0	0	0
0	0	0	1	0
0	0	1	0	0
0	0	1	1	0
0	1	0	0	0
0	1	0	1	0
0	1	1	0	1
0	1	1	1	0
1	0	0	0	0
1	0	0	1	0
1	0	1	0	1
1	0	1	1	0
1	1	0	0	0
1	1	0	1	0
1	1	1	0	1
1	1	1	1	0

(a)   (b)

Figure 11-4   A "brute-force" implementation (a) of the truth table given in (b).

All four inputs, marked A, B, C, and D, are fed into each of the three AND-gates, with inverters as needed. Each AND-gate responds to one of the three specified cases, and their outputs are combined into a three-input OR-gate to give the correct response. A more efficient solution, is shown in Figure 11-5 where a system with fewer gates gives the same truth table.

Transistor Implementations: TTL Logic

One of the most widely used logic systems is a family known as **transistor-transistor logic** (TTL). To understand the TTL operation,

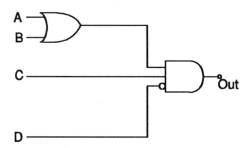

Figure 11-5 A simplified logic system implementing the same function as in Figure 11-4. The reader should construct a truth table for this circuit to verify that it is identical in function to the circuit of the preceding figure.

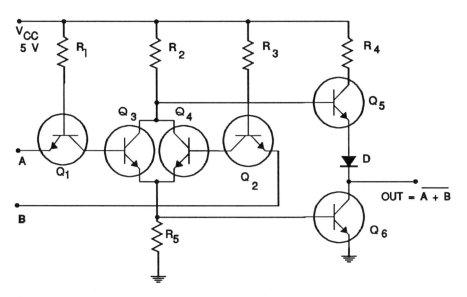

Figure 11-6 A TTL NOR-gate, one of four identical circuits in the type 7400 integrated device.

consider Figure 11-6, which shows a circuit that performs the NAND function. It is necessary to follow closely the interaction of the several transistors. If either transistor Q_3 or Q_4 (or both of them) are turned on, current will be able to pass through the resistors R_2 and R_5, which will make the base of Q_5 less positive and that of Q_6 more positive than would be the case if both Q_3 and Q_4 were off. This will turn Q_5 off and Q_6 on, causing the output to go to ground potential (logical 0); if Q_3 and Q_4 are turned off, the two output transistors will change state, and the output will become equal to V_{CC} (logical 1).

The two central transistors are controlled by the inputs acting through Q_1 and Q_2. If A is at ground potential, current will flow to it from V_{CC}, producing a voltage drop in resistor R_1, turning Q_3 off, whereas if A is +5 V, no current will pass through the resistor, and Q_3 will be turned on. Similarly with B and Q_2. Consequently when either or both inputs are HIGH, the output is LOW, and vice versa. The last stage, consisting of Q_5, Q_6, and the diode, is called a "totem-pole" circuit. It provides a low impedance output, which is the condition needed for the gate to drive other gates. The diode ensures that Q_5 is OFF when Q_6 is saturated.

Figure 11-7 shows the schematic of a modified TTL gate. It is more economical to fabricate, because it uses fewer parts. Note that it requires a transistor with two (or more) emitters, which, although not

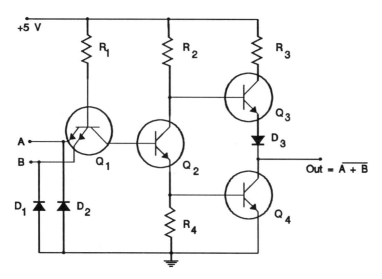

Figure 11-7   A more economical version of the TTL NOR-gate, the 7400. Transistor Q_1 is provided with dual emitters, which enables it to replace several conventional transistors. The diodes D_1 and D_2 protect the transistors from negative inputs.

available in discrete form, can be produced on the silicon chip just as easily as the single-emitter types.

A modification of this appears in Figure 11-8. The upper portion of the totem pole is omitted, namely transistor Q_5, resistor R_4, and the

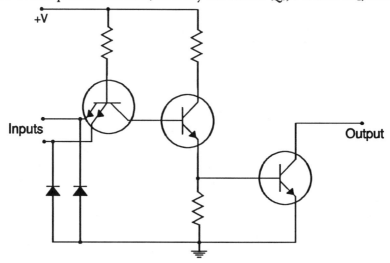

Figure 11-8 An open-collector TTL NOR-gate, the 7403.

diode, leaving the collector of Q_6 hanging, the so-called "open-collector" configuration. This is useful in situations where the gate must drive some component that requires a higher voltage (e.g., a relay). A "pull-up" resistor must be supplied externally from the collector pin to the positive voltage.

A long series of logic gates have been developed in the TTL sequence, designated by numbers starting with 74 (54 for the corresponding parts made to military specifications). These gates cover a multitude of logic functions, from simple inverters to flip-flops, to special-purpose units used mostly in computer interfacing. They are all similar in the general plan of using bipolar transistors together with resistors and only very few other circuit components. Inputs lead directly to emitters, usually with diodes to ground as shown in Figures 11-7 and 11-8. The diodes are protective devices, shunting to ground any negative voltage that might accidentally be applied to the input. The normal output consists of two transistors end-to-end in the totem-pole configuration.

One drawback of the standard TTL gates is that they are comparatively slow operating because inherently either one or the other of the output transistors goes into saturation, depending on whether a 1 or a 0 is present. In saturation, a bipolar transistor stores up a considerable amount of charge across its junction, and it takes a period of time for that charge to drain away when the state of the transistor changes. In a device with only a small number of gates, this may be negligible, but large-scale integrated circuits contain thousands of separate transistors, and a delay of even a few nanoseconds per gate would result in significant delay.

Another TTL series has been developed to increase the speed. It is distinguished by the insertion of letters LS after the 74 (or 54) in the type designation. The "S" indicates that the device is designed with Schottky diodes and transistors, and the "L" refers to the lower power required. Saturation is avoided by connecting a Schottky diode across the transistor paralleling the base-collector junction. The junction voltage of the Schottky is less than that of the pn-junction within the transistor, so the voltage can never build up to the point of saturation. In almost every case, the electrical functions of 74LSxx gates (aside from the speed) are identical with those of the parent device. Since the power consumed is less, heat generation is not as great a problem. An example is presented in Figure 11-9. Even the 74LS series may be on the way to becoming obsolescent and replaced

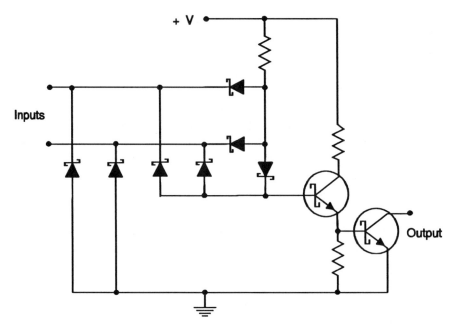

Figure 11-9 The 74LS03 open-collector Schottky NAND gate.

with logic series using newer technology, such as the 74ALS (advanced schottky TTL) series.

Wired-OR

In general, the outputs of two gates cannot be connected together. One gate might command the output to be HIGH while the other requires it to stay at ground. The resulting battle is inadmissable and would be likely to end in the destruction of one or both gates. The only exception

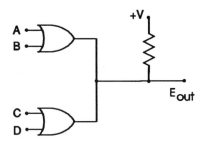

Figure 11-10 A wired-OR circuit. The output is A + B + C + D, which could be obtained just as well with a single 4-input OR-gate.

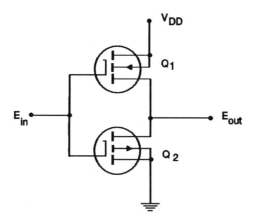

Figure 11-11 The inverter circuit in CMOS, consisting of two transistors, one *p*-channel, the other *n*-channel.

to this rule occurs with open-collector gates, rather than the usual TTL totem pole as in Figure 11-9. It is possible to connect the outputs of two or more of such gates, which are effectively OR-ed together, as in the circuit of Figure 11-10 which results in $E_{out} = A + B + C + D$.

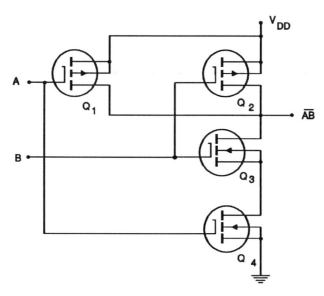

Figure 11-12 A 2-input NAND-gate in CMOS.

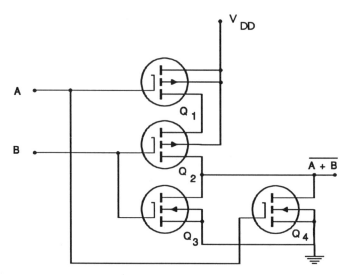

Figure 11-13 A 2-input CMOS NOR gate.

CMOS Logic

A distinct series of logic gates is available in complementary metal oxide semiconductor (CMOS) technology, which employs MOSFETs, both n- and p-channel types, rather than bipolar transistors. CMOS logic is used extensively in large computer chips. There are considerable practical differences between CMOS and bipolar technologies as applied to logic systems, so we must treat this subject in some detail.

Figure 11-11 shows the basic inverter in the CMOS system. Note its simplicity. When the input is high, Q_1 is turned off and Q_2 is on, so the output is essentially grounded (i.e., LOW). Grounding the input, in turn, gives a HIGH output. Figure 11-12 gives the schematic for a 2-input NAND-gate, the CMOS equivalent of the 7400 TTL gate previously described. A comparable NOR-gate is given in Figure 11-13. The reader should have no trouble in deducing how these circuits work.

It is to be noted that the basic gates in both TTL and CMOS are the NAND and NOR. To obtain the corresponding AND / OR functions, an inverter must be added, either internally or externally.

The supply voltage V_{DD} for CMOS can have any positive value from 5 to 18 V. CMOS gates exhibit extremely low power dissipation and high noise immunity. The speed of operation is comparable to

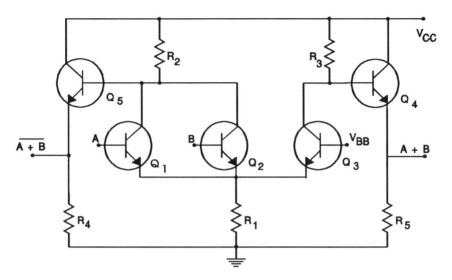

Figure 11-14 An ECL OR / NOR gate.

TTL. The CMOS devices are prone to injury from static electric charges such as those that may be present in ordinary handling. Most CMOS devices are provided with built-in protective diodes intended to shunt any extraneous charges harmlessly to ground. However, suitable precautions should be observed in their storage and manipulation. It is good practice for a person handling CMOS parts to wear a metallic wristband connected to ground to avoid the build up of static charges.

A number of series of CMOS gates are available. One of these makes use of the same numerical designations as TTL, with the letter C, as 74Cxx. Another widely used series has an entirely different set of numbers, 4xxx, with no letter code.

ECL Logic

Yet another logic system makes use of bipolar transistors with coupling through the emitter connections. This is known as **emitter-coupled logic** (ECL). The basic gate structure in this series, an OR/NOR gate, is shown in Figure 11-14. The heart of the circuit is the Q_2-Q_3 pair of transistors with their emitters connected together. Q_1 parallels Q_2 to provide two inputs. Two additional transistors, Q_4 and Q_5, supply two outputs, that are out-of-phase with each other, a characteristic of ECL not shared with TTL or CMOS. The voltage V_{BB} is generated internally by a bias circuit, not shown. This bias voltage establishes a quiescent current through transistor Q_3 that effectively cuts off Q_4, so the right-hand output is zero. If a signal is impressed upon Q_1 (or Q_2) so that it is turned on, the current through R_1 is immediately shifted from Q_3 to Q_1, cutting off Q_3. This means that the base of Q_4 is raised so that this transistor conducts, giving a HIGH output, while at the same time, Q_5 is cut off and its output becomes LOW. ECL logic can operate at a higher speed than TTL or CMOS, because none of the transistors can become saturated. Another advantage is that, because the total current remains nearly constant when the gate changes state, no voltage spike (**glitch**) is created. The logic levels for ECL are different from those we have discussed before. A HIGH is usually 0.9 V, whereas a LOW is 1.75 V.

Comparison of Logic Types

Each of the logic families described above has its particular advantages and disadvantages, all of which must be weighed before deciding on which system to use in an instrument design. Table 11-2 gives a comparison of nine different logic classes, including three each for TTL, CMOS, and ECL.[2] We will discuss briefly each of the listed parameters.

Speed

High speed is inherently advantageous, but it is often accompanied by increased noise, power consumption, and cost. Speed is here described in relation to two measurable quantities: the input-to-output propagation delay through a typical gate (D flip-flop), and the maximum frequency at which a specific device can cycle. (The first of these should ideally be small and the second large.) It can be seen that the ECL gates excel in all these speed criteria. Flip-flops are discussed in the next chapter.

---

[2] Table 11-2 is taken, with permission, from G. Tharalson, *Electronic Products*, May 1989, p. 53. The accompanying discussion is a condensation of the same article.

Table 11-2
Characteristics of Logic Families

PARAMETER	TTL			CMOS			ECL		
	LS	ALS	FAST	MG	HC	FACT	10KH	100K	ECLinPS
Speed									
OR-gate propagation delay, ns	9	7	3	25	8	5	1	0.75	0.33
D-FF toggle rate (MHz)	33	45	125	4	45	160	330	400	1000
Power consumption per gate									
Quiescent (mW)	5	1.2	12.5	0.0006	0.002	0.003	25	50	25
Operating at 1MHz (mW)	5	1.2	12.5	0.04	0.6	0.8	25	50	25
Supply voltage (V)	+4.5 to	+4.5 to	+4.5 to	+3 to 1	+2 to 6	+2 to 6	-4.9 to -5	-4.2 to -4	-4.9 to -5.
Output drive (mA) or ohms load	8	8	20	1	4	24	50 ohm	50 ohm	50 ohms
DC noise margin (%)									
High input	22	22	22	30	30	30	28	41	28
Low input	10	10	10	30	30	30	31	31	31
Device types	190	210	110	125	103	80	64	44	30
Relative cost per gate	0.9	1	1	0.9	0.9	1.5	2	10	28

LS: Motorola Low-power Shottky TTL
FAST: Motorola Advanced Shottky TTL
FACT: Motorola Advanced CMOS
100K: National 100K Series ECL

ALS: Texas Instruments Advanced Low-power Shottky
MG: Motorola High Speed Silicon Gate CMOS
10KH: Motorola 10KHG Series ECL
ECLinPS: Motorola Advanced ECL

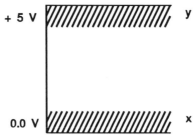

Figure 11-15 Illustrating the noise margin concept for 5-V logic. If the voltage at an input lies within the region marked "x," the gate accepts it as zero. Similarly a voltage within area "y" is taken as logical 1. The range between the shaded areas is known as the noise margin.

Power Consumption

If many gates are to be mounted close together, the amount of power comsumption determines if forced cooling is required. Low power requirements are of course critical in battery-operated equipment. Table 11-2 shows that power consumption is far less with CMOS gates than with the others. The power requirement is quite sensitive to the frequency of operation, particularly in the CMOS series. The power drawn by TTL logic increases markedly above about 10 MHz. ECL may be the most favorable at high frequencies, though it is the most power-hungry at low frequency.

Supply Voltage

The operating voltage for TTL and especially ECL must be closely regulated. This requires the power supply to be designed with very low output impedance. If many gates are involved, rather large currents will be flowing through the power-supply leads, complicating the design of printed-circuit boards and other components, to avoid local overheating and glitch problems. CMOS has the great advantage of permitting wide latitude in supply voltage.

Output Drive

This is a measure of **fan-out**, the number of other gates that can be driven simultaneously by a given gate. It is quoted in milliamps rather than number of gates, as a more general parameter. Clearly ALS is superior to the older TTL and CMOS families. ECL is capable of driving directly a 50-Ω load often encountered in high-frequency data transfer.

Noise Margin

As pointed out previously, the greater the voltage gap between high- and low-logic voltages, the less likely is it that a noise pulse will produce false states. In the table, the noise margin is specified as a percentage of the total swing between high and low voltages that cannot be exceeded without false triggering (see Figure 11-15). Both CMOS and ECL are better than TTL in this regard.

Device Types

The figures are given for the combined offerings of National Semiconductor and Motorola, two of the largest producers. The list of available gates in ECL is shorter, since ECL devices are only manufactured where their special properties are required. It is quite permissible to mix units from different series in a single instrument.

Relative Cost

Except for some of the super-fast ECL units, the differences in cost of the various logic series is usually negligible for individual users.

Three-State Logic

Occasions can arise that call for outputs of several gates to be individually connected to the same point as was the case with the wired-OR configuration. This can also be handled by the use of **three-state** output circuitry. Each logic element is provided with an extra input called the **enable** line. If the enable is activated, the expected logic appears at the output, but otherwise the output changes to a high-impedance condition that effectively cuts it off from all other points to which it is connected permitting some other gate to take control.

Figure 11-16 shows the circuit for the enable system of a typical TTL gate. In standard TTL gates with a totem-pole output one of the two output transistors is on while the other is off. In the three-state gate, a HIGH on the enable line causes *both* Q_4 and the Q_2-Q_3 darlington to go off together, which puts the output in a high-impedance state, taking no commands from the prior logic.

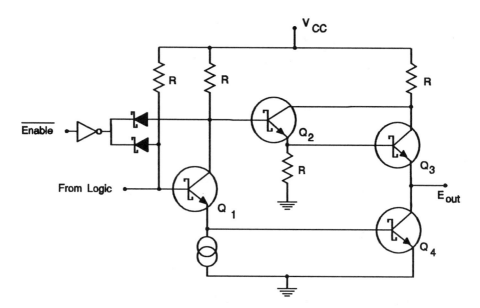

Figure 11-16 A circuit for adding three-state logic to a typical TTL Schottky logic gate. The bar over "enable" signifies that the circuit is enabled (i.e., the signal from logic appears at the output) when the "enable" input is low.

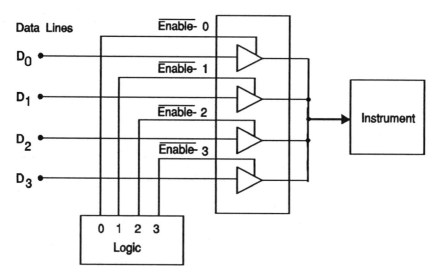

Figure 11-17 A quad three-state buffer connecting four data lines to an instrument. The logic turns on the desired enable line so that only one of the data inputs has access to the instrument at any moment.

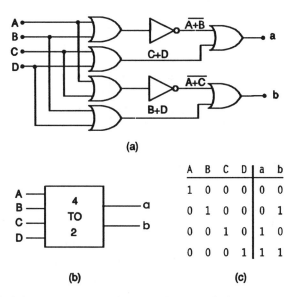

(a)

A	B	C	D	a	b
1	0	0	0	0	0
0	1	0	0	0	1
0	0	1	0	1	0
0	0	0	1	1	1

(b)   (c)

Figure 11-18 A four-to-two encoder (a) with its symbolic representation (b) and truth table (c). (Note that only one input is permitted to be high at any given moment.)

Three-state gates[3] are widely used in computer technology, allowing several devices to have access to a bus line one at a time. They can also be used in many other applications, such as that shown in Figure 11-17, which shows an example where an instrument can be allocated a number of input data lines, but with only one being enabled at any given time. The figure shows a discrete three-state buffer. Many other logic gates are provided with three-state controls.

Encoders and Decoders

Sometimes one needs to condense the information transmitted along several parallel lines so that the flow can be reduced to only a few lines. This can be done by a process known as **encoding.** Encoders can translate the information from four lines to two, from eight lines to three, from sixteen to four, and so on. To make this possible, the input lines must be restricted so that only a single line can be high at any given time. Figure 11-18 gives an example of a 4-to-2 encoder. The

[3] Also called "Tri-state," a registered trademark of National Semiconductor Corporation.

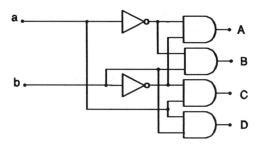

Figure 11-19 A circuit for a 2-to-4 decoder.

data originally come on the four wires A, B, C, and D, and are condensed to lines **a** and **b**. It might seem that some information content would be lost in this process, but this is actually not true. The inputs are restricted to only four cases, as seen in the truth table. In general an encoder with 2^n inputs will have n outputs.

The inverse process, called **decoding,** permits using signals on a few lines to control the flow of information over a larger number of lines. For a 2-to-4 decoder, for instance, the truth table would be the same as that in Figure 11-18, except that the incoming information is carried by **a** and **b**. Figure 11-19 shows the corresponding circuit. The reader should verify the decoding action by building a truth table showing the four outputs corresponding to the four possible input

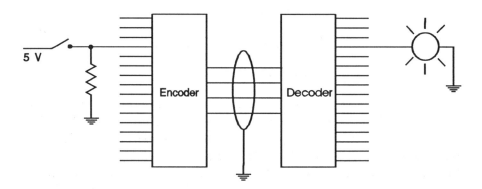

Figure 11-20 A four-wire cable connecting one of 16 lines to light a lamp corresponding to a particular switch position. The grounded pull-down resistor on the input line ensures logic zero when the switch is open. Sixteen such resistors are needed.

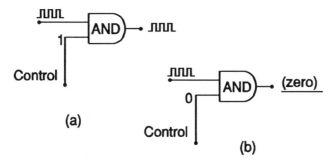

Figure 11-21 A two-input AND-gate acting as a switch to control the flow of digital data.

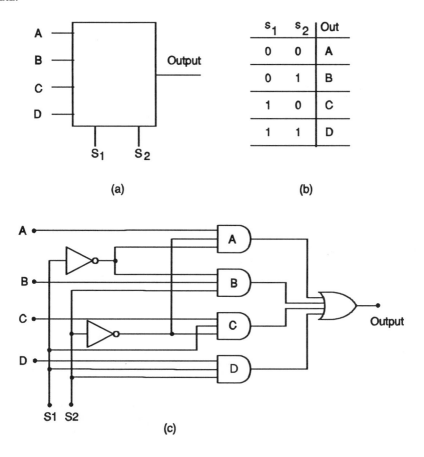

Figure 11-22 A four-line multiplexer (MUX): (a) component symbol; (b) truth table; (c) logic diagram. S_1 and S_2 are the address ports; note that any given address disables three out of the four data paths.

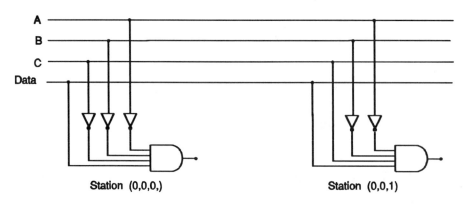

Figure 11-23 An example of remote control by means of a bus. The gates reproduce the signal on the data line only if the proper address appears on the A, B, and C lines.

combinations (00, 01, 10, and 11), with the restriction that each output line is either on or off, like the truth table of Figure 11-18 in reverse.

An example of coding and decoding may be seen in Figure 11-20. The state of a 16-position switch must be known at a distance. One way to achieve this would be to install a 16-wire cable to select one of 16 signal lamps. But the on-off information about each switch position could be fed into a 16-input encoder. This would give all the necessary information on only four output lines. Thus a four-wire cable (plus ground) would serve in place of 16 wires. At the receiving end, the signals from the cable would be decoded by a four-to-sixteen decoder, which in turn would energize the appropriate lamp.

Gating and Multiplexing

Implementation of certain on/off switching functionalities dealing with the flow of digital information can make use of logic gates. Consider the gate in Figure 11-21. When the control line is at logic 1, the AND gate has no effect on the signal flow. Aside from a few nanoseconds of propagation delay time, the output reproduces the input. On the other hand, a zero on the control line (b in the figure) forces the output to remain at zero regardless of what happens to the input data line. Thus the signal is stopped or allowed to flow in a manner nearly like that of a simple mechanical on/off switch,

A similar result can be had from an XOR-gate in place of the AND, but the sign of the control is reversed. The reader can easily discover what would happen if any other gates OR, NOR, NAND were substituted for the AND in Figure 11-21.

A special type of encoder that permits individual selection of channels is known as a digital **multiplexer**; an example is shown in Figure 11-22. The address lines serve to select which of the inputs will be steered to the output. Thus if the address code stands at 00, the AND-gate marked "A" will be enabled so that the high or low at the output will correspond to the HIGH or LOW at the "A" input. At the same time the other three inputs are disabled so that their signals cannot affect the output. The data path could be selected successively by connecting the S inputs to a binary counter (described in the next chapter) or they could be selected by other digital logic.

Figure 11-23 shows a portion of a **bus**, a type of structure widely used in computer systems. This configuration represents another use of address lines to exercise control over the transmission of data. The three lines marked A, B, and C can carry any three-digit binary number (000 to 111), giving a total of eight possibilities. Each "station" connected to the bus (two are shown) will respond to its own address and to no other. The device that is addressed can then receive the information destined for it from the data line, one bit at a time. The "bus" consists of the three address lines and the data line. One more conductor would permit 16 separately addressed stations to be attached to the bus. In a computer, as we shall see, much more extensive bus arrangements are required.

Chapter 12

Flip-flops and Registers

Up to this point, we have considered only those logic systems in which there is an instantaneous correlation between input and output. Any change in the states of input voltages is immediately reflected in a corresponding change in the output. This is called **combinatorial logic**. There are, however, many situations in which it is necessary to retain or remember the state of logic present at some previous moment. To accomplish this, logical elements with *memory* are required.

Memory in an electronic context requires holding a series of 1's and 0's for as long a time as they may be needed. This is not to be confused with simple time delay, which may allow values to be held passively until they die away of their own accord. Logical memory devices can be grouped in two major classes, called **nonvolatile** and **volatile**, according to whether they retain or lose their content upon loss of power. Nonvolatile memory includes such magnetic devices as recording tape and both floppy and hard computer disks. In the present chapter we will be entirely concerned with volatile memory.

Digital computers require very extensive active memory capability, almost always volatile. This memory is usually measured in terms of bytes, kilobytes (kbytes), and megabytes (Mbytes). To give perspective, consider the memory requirements of small computers. For example, the computer used in writing the third edition of this book,

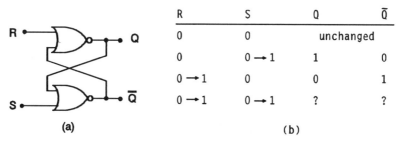

R	S	Q	$\overline{Q}$
0	0	unchanged	
0	0 → 1	1	0
0 → 1	0	0	1
0 → 1	0 → 1	?	?

(a)   (b)

Figure 12-1 (a) A NOR-based flip-flop, and (b) the corresponding state table. The symbol "0 → 1" indicates a transition from LOW to HIGH. Note that the output depends on the *order* in which one of the inputs goes HIGH. The question mark indicates that the states of the outputs are undetermined upon return LOW, whereas "unchanged" means that there is no change from the previous state.

in 1984, had 64 kbytes of memory, whereas the computer now in use has 8 Mbytes.

The basic memory unit is a **flip-flop** (FF), a circuit with two complementary outputs the values of which can be interchanged by pulses given to the proper input. This is readily visualized by means of Figure 12-1 which shows a simple flip-flop, called an RS flip-flop, constructed from two NOR gates.[1]

It will be seen that the feedback connections restrict the output in the following way. (Asume that initially both inputs are low.):

1. If one input goes HIGH and the other remains LOW, then one of the outputs will go LOW, the other HIGH.

2. If both inputs return to LOW, there will be no change from the preceding state.

3. If both inputs go HIGH and return to LOW simultaneously, there are two possible states with equal probability, namely $Q = 1$ and $\overline{Q} = 0$, or $\overline{Q} = 1$ and $Q = 0$, the result being indeterminate (a situation to be avoided).

The results of changes at the two inputs are summarized in Figure 12-1*b*. (Note that a truth table, as used with simple or compound gates, is not sufficient for defining the operation of a flip-flop, since we must now specify the *change* to HIGH and LOW conditions as well. This is the

---

[1] The flip-flop is a modernized version of the "multivibrator," invented by Eccles and Jordan in 1919. Another version of a multivibrator has been seen in Figure 6-17.

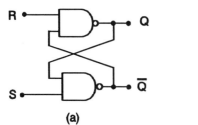

R	S	Q	$\overline{Q}$
1	1	unchanged	
$1 \to 0$	1	1	0
1	$1 \to 0$	0	1
$1 \to 0$	$1 \to 0$	?	?

(a)   (b)

Figure 12-2 (a) A NAND-based RS flip-flop. The inputs are normally HIGH. (b) The corresponding state table showing the result of momentarily grounding one or both inputs. The notation is as in Figure 12-1.

reason for specifying that an input *goes* HIGH or LOW, rather than *is* HIGH or LOW.)

The state in which Q is HIGH is called the **set** condition; whereas if Q is LOW, the circuit **resets**. Momentarily pulsing the S input *sets* the circuit to give $Q = 1$, whereas pulsing the R *resets* it to $Q = 0$. Whenever the flip-flop is operated under the specified restriction, the two outputs are complementary, hence the notation, Q and $\overline{Q}$.

These facts indicate the suitability of flip-flops as memory devices provided that the inputs are maintained normally LOW and are not allowed to go HIGH simultaneously. The circuit can remember whether a pulse to logical 1 has or has not occurred at the S input. Let us assume that the circuit is initially in its normal condition, with both inputs LOW. The state of the Q output answers the question "Has S been pulsed more recently than R?" If $Q = 0$, this may be interpreted as the answer, "No, it has not." A pulse at S changes Q to 1, meaning, "Yes, it has." The answer is still "Yes" after repeated pulses at S, but a pulse at R resets the circuit to $Q = 0$ so that the answer becomes "No." Alternatively, one can think of the circuit as indicating *which* input has most recently been pulsed.

This circuit has a major flaw. If both inputs are pulsed HIGH simultaneously, there is no way to predict which of the two outputs will go HIGH. This situation may well occur in practice, for as we shall see, logic systems are usually controlled by a series of clock pulses generated for the purpose, so that all logic changes throughout the system occur together. The circuit designer has the responsibility of making sure that such undefined states are avoided.

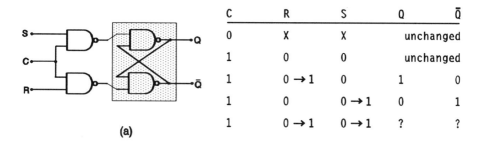

C	R	S	Q	Q̄
0	X	X	unchanged	
1	0	0	unchanged	
1	0 → 1	0	1	0
1	0	0 → 1	0	1
1	0 → 1	0 → 1	?	?

(a)

Figure 12-3 The gated RS flip-flop. (a) The schematic. The portion within the shaded box is the basic RS unit; input NAND gates and a clock ihave been added. (b) The state table. The clock is assumed to be initially in its HIGH state. When the clock changes to 0, the preceding state is retained, in all but for the last case, which is undefined.

A very similar circuit can be assembled from NAND gates, as shown in Figure 12-2. The action is analogous to that of the NOR-gated flip-flop, but here the outputs in their normal condition are both HIGH. Note that the letters R and S are in reversed positions relative to Figure 12-1. Because RS flip-flops are so easily implemented from NOR or NAND gates, they are not commonly manufactured as discrete ICs. If several inputs are available, as when the circuit is constructed from three-input gates, a pulse at any input is effective in triggering the response.

Figure 12-3 shows a more complicated circuit, the **gated,** or **clocked** RS flip-flop, in which two additional gates are inserted at the inputs. A series of synchronizing pulses, called **clock** pulses, may be applied to the common input (marked C) in order to synchronize or "strobe" the flip-flop with other parts of the system. When the clock pulse is high, the NAND gates act simply as inverters and the overall action is the same as for a simple NOR flip-flop. When the clock signal drops to ZERO, the mode of operation changes because both input NAND gates are now forced HIGH, regardless of the states of R and S. The output is thus frozen in whatever state it held just before the change in clock level. The two modes of operation, **enable** and **disable,** are analogous to the sample and hold that we encountered in op amp circuitry.

If at any point during the sampling period (clock HIGH) S is 1 while R is 0, the output Q will become 1 and will remain so after the clock goes to 0. If, however, R is 1 and S is 0, Q will be 0. If both inputs were

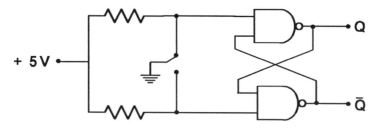

Figure 12-4 An RS flip-flop acting as a bounceless switch. The reader should be able to predict which position of the switch should make Q = 1, and $\overline{Q}$ = 0 and vice versa.

at 0 during the sampling period, the states of the outputs would remain the same as before the clocking cycle.

One use of an RS flip-flop is in interfacing mechanical and electronic logic devices. An example of such use is shown in Figure 12-4. A mechanical switch is liable to suffer from "bounce," a series of fast on/off transitions that result from almost imperceptible vibrations occurring as the switch is closed. These repeated brief contacts can adversely affect logic systems, as each momentary contact is interpreted as a separate logic change. If a switch is connected to the input of a flip-flop, this situation does not occur, since both R and S, once activated, are no longer affected by changes in their respective lines. In other words, the gate *latches* onto its state, and the device becomes a **bounceless switch.**

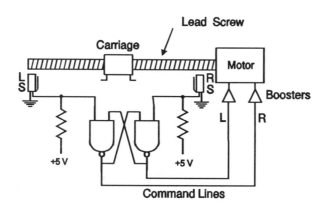

Figure 12-5 A limit control system for an automatic scanning mechanism. The carriage, when driven to the left, eventually closes the left limit switch (LS), which triggers the flip-flop, reversing the motor. When the carriage reaches the right-hand limit switch, the motor reverses again, and so on.

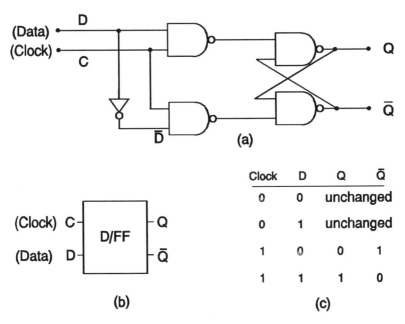

Figure 12-6 A D-type flip-flop: (a) the schematic block diagram; (b) the abbreviated symbol, in which D stands for "data," and C refers to the "clock," also known as "enable" in this type of circuit; (c) the corresponding state table.

Another example of a latching device is shown in Figure 12-5. This shows the mechanism of an optical scanner in which a sensor carriage (i.e., a photocell mount) is moved across a picture repeatedly from one side to the other. As the carriage reaches one limit, it activates a switch that causes a flip-flop to change state, and this in turn reverses the drive motor, sending the carriage to the opposite limit, where the action repeats.

The D Flip-Flop

A characteristic of the clocked RS flip-flop is the need for three input lines R, S, and C. A convenient modification is the D flip-flop (Figure 12-6). This requires only two lines, one of which determines the logic state, while the other (the clock line) determines the exact moment when the input state is to be loaded into memory. The D input replaces the former R and S inputs. The state table corresponds to that in Figure 12-3, except that only the second and third entries are possible, and these are active only if the clock is HIGH. When the clock is LOW, the output is unaffected by changes in the D input. In short, the D flip-flop

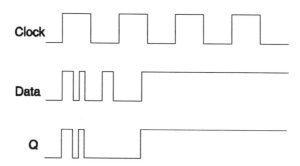

Figure12-7 Behavior of a D flip-flop. Successive data pulses are shown as irregular in order to illustrate the various situations that may arise. When the clock is LOW, changes in the data line are *not* transferred to the output.

accepts and memorizes the value of D when the clock is HIGH and retains it in memory when the clock drops to logic 0. The relationships between the various states can better be visualized from Figure 12-7.

Edge-Triggered D Flip-Flops

As we have seen, the output of a D flip-flop follows the input continuously as long as the clock signal remains high. Thus it acts as a **track-and-hold** device. More useful is the digital **sample-and-hold** operation, comparable to the analog device with the same name. This differs from the track-and-hold device in that the loading of information is done only during a very short period. An example of such a device is the **edge-triggered D flip-flop**, such as the TTL model 74174 which has six similar units in one IC package. As the name suggests, the sampling operation takes place on the 0 to 1 transition of the clock pulse, the **positive-going edge**.

Flip-flops of this type are widely used for loading data into a computer. Even if it takes some time (many microseconds) to obtain the data, this can then be transferred into the computer at a precisely known instant, controlled by the clock.

The JK Flip-Flop

An even more versatile component is the **JK flip-flop** shown in Figure 12-8. In its fully extended form it has no less than seven connections to external circuitry. The presence of the clock input is essential to ensure lack of interference between an incoming signal and the information previously stored. This results from the fact that the unit contains two cascaded elementary flip-flops, the master (composed of

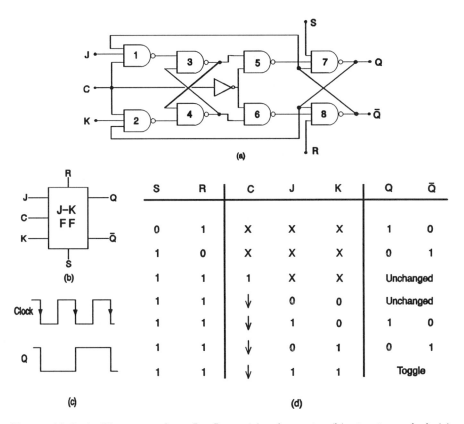

S	R	C	J	K	Q	Q̄
0	1	X	X	X	1	0
1	0	X	X	X	0	1
1	1	1	X	X	Unchanged	
1	1	↓	0	0	Unchanged	
1	1	↓	1	0	1	0
1	1	↓	0	1	0	1
1	1	↓	1	1	Toggle	

Figure 12-8 A JK master-slave flip-flop: (a) schematic; (b) circuit symbol; (c) waveforms at C and Q when J, K, R, and S are all at logic 1 corresponding to the last line of the table; (d) the state table. Other versions of JK flip-flops may vary in details of state table and waveforms. The symbol "X" means "don't care," either 0 or 1. The down-arrow (↓) denotes a step transition from 1 to 0.

NAND gates 3 and 4) and the slave (gates 7 and 8). Information is fed into the master only when the clock signal goes HIGH; in this state there is no communication between master and slave. The data is passed on to the slave only as the clock goes LOW. The result of this sequence is that data from both the present and previous cycles are stored simultaneously. Output changes occur only on the downward clock transitions. The operation is best described by means of a set of four statements:

- If both J and K are LOW before and during a clock pulse, there will be no change in the outputs following the pulse.

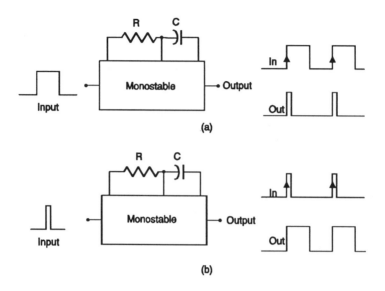

Figure 12-9 Examples of (a) pulse shortening, and (b) pulse lengthening by means of a monostable that responds to a positive-going logic transition. The duration of the output pulse is determined by the RC product in each case.

- If $J = 0$ and $K = 1$ before and during a clock pulse, the outputs will go to $Q = 0$, $\overline{Q} = 1$

- If $J = 1$ and $K = 0$ before and during a clock pulse, the outputs will go to $Q = 1$, $\overline{Q} = 0$

- If both J and K were HIGH before and during a clock pulse, the outputs will toggle; that is to say, if Q was 1 before the pulse, it will change to 0, or if it was 0 it will go to 1, with $\overline{Q}$ making the opposite change.

Any of these changes will take place only on the *falling* edge of the clock pulse. The circuit is independently set to $Q = 1$ by grounding the S input or reset by grounding R. These latter commands override any signals present at the J, K, or C inputs. The table of Figure 12-8*d* shows all these relations; the symbol ↓ designates the falling edge of the clock pulse.

Monostable and Astable Flip-Flops

The flip-flops we have been describing are inherently **bistable de-vices**; they will remain in either of two states indefinitely. Similar

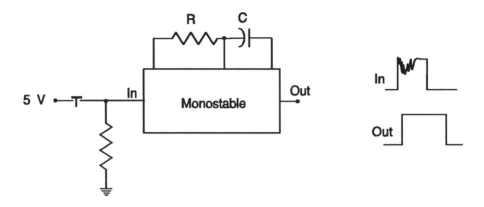

Figure 12-10 A momentary contact pushbutton switch rendered bounceless with a monostable.

circuits with other modes of operation are also possible. For example, some types have internal feedback connections that cause the device to oscillate continuously between its two output states, without the need for any input.. This is called an **astable** flip-flop, or a **multivibrator**. On the other hand, if the feedback action is only one-sided, a **monostable** circuit results, in which a change of input generates a single output pulse then resets to the original state. The monostable is sometimes called a **one-shot,** or a **univibrator**. Any of these flip-flops can be assembled from discrete components but are also available as ICs. A widely used monostable is the TTL model 74121, which can give pulses varying from about 40 ns to 40 s, following an input transition. The adjustment of the pulse width τ is established by an external resistor and capacitor, according to the formula $\tau = 0.7\ RC$.

A monostable can be made to alter the width of pulses, as shown in Figure 12-9, making it possible to change the duration of whatever operation is controlled by each pulse, without altering the repetition rate. A monostable can be used to eliminate the bounce effect in momentary contact pushbutton switches (Figure 12-10). The stretched pulse time must be longer than the duration of the bounces from any one closure.

For applications that require the timing of a series of events with moderate precision, the chain of monostables shown in Figure 12-11 presents a convenient and inexpensive solution. In this system each monostable after the first is triggered by the trailing edge (1 → 0 transition) of the pulse from the preceding unit. Applications to

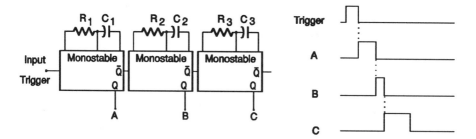

Figure 12-11   A sequential controller using a string of three monostables. The action can be initiated by a trigger from a manual switch or by a logic signal. Many monostables are provided with complementary outputs. The duration of each pulse is determined by the corresponding *RC* network. Some monostables are triggered by the 0 → 1 transition and others by 1 → 0.

automation are limitless, especially as one can form branched chains of any desired complexity. It should be noted, however, that since the precision of any one step is limited to perhaps 1%, the overall timing error will increase with increased chain length. Higher timing precision can be obtained with shift registers and with counters, to be considered later in this chapter.

The 555 Timer

A very useful and versatile device that can be employed as a monostable or an astable oscillator is the model 555. This is shown, configured as a square-wave generator, in Figure 12-12. The heart of this IC consists of a tandem pair of comparators, A and B, that actuate an RS flip-flop. Both comparators sense the voltage present on the capacitor C. The two comparators use as reference voltages, respectively, 1/3 and 2/3 of the power-supply voltage. Initially, the capacitor charges through R_A and R_B until it reaches 2/3 V_{CC}. At this point comparator A sets the flip-flop which causes the transistor T to turn on. The capacitor then discharges through R_B and T until it reaches 1/3 V_{CC}, at which point it triggers comparator B, resetting the flip-flop and completing the cycle. The capacitor then begins to recharge, and the process repeats itself *ad infinitum*, with the capacitor alternating between 1/3 and 2/3 of V_{CC}. The result is a square wave of frequency

$$f = \frac{1.44}{(R_A + R_B)\, C}$$

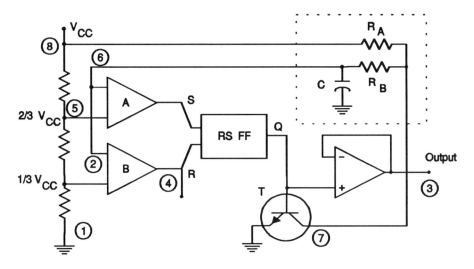

Figure 12-12   The model 555 timer connected as an astable square-wave generator. All components shown, except R_A, R_B, and C are contained within the IC. The encircled numbers designate the eight pins of the IC device.

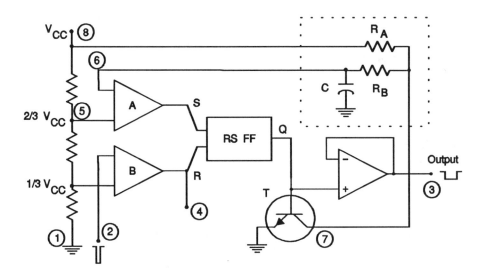

Figure 12-13 The 555 configured as a monostable. The pin numbers are the same as in Figure 12-12.

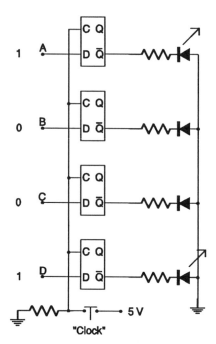

Figure 12-14 Four D flip-flops connected as a four-bit register used as a data latch. The values of the binary inputs 1001 are shown at the moment when the pushbutton is closed, by lighting the corresponding LEDs; the resistors are to protect the LEDs from excess current. Note that each LED lights when its $\overline{Q}$ is HIGH. (A transistor booster may be needed at each output to supply sufficient drive current for the LEDs. An MPQ2222 quad *npn* would be convenient.)

The duty cycle, which is the ratio of the time spent in the high state to the time in the low state, is given by $R_B/(R_A + R_B)$. (Note that both of these expressions are independent of the value of V_{CC}.) The output is taken through a buffer amplifier to give a low output impedance that permits significant current to be drawn without affecting the frequency.

The 555 has provision for overriding the reset command by a pulse applied to pin 4, and for changing the reference potential for the A comparator to some voltage other than 2/3 V_{CC}, by supplying it at pin 5. For operation as a monostable, the 555 is connected as shown in Figure 12-13. A momentary voltage of less than 2/3 V_{CC}, applied to pin 2 triggers the device, the output goes LOW, and the transistor turns off. This starts the capacitor charging (through R_A and R_B), its voltage increasing with the time constant $t = RC$. The voltage reaches 2/3 V_{CC}

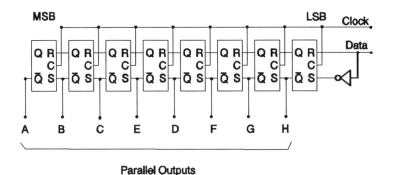

Parallel Outputs

Figure 12-15 A shift register. The logical state at the serial input is shifted through the system on consecutive transitions of the clock.This is a slightly simplified 74LS164 register. The data enter on the right, so that the MSB of the output appears on the left.

in 1.1 time constants. At this point the A comparator changes state, and the transistor is turned on. The capacitor discharges rapidly; the timer has completed its cycle and awaits another trigger pulse. A great many ingenious circuits have been devised using the 555 or its dual (the 556). The reader is referred to the manufacturer's literature for details.

Registers

A number of flip-flops can be connected so that they operate as a unit called a **register**. Figure 12-14 shows a simple register composed of a series of four D flip-flops sharing a common trigger or clock, so that they can serve as a **data latch** for a four-bit input word. Whenever the clock goes HIGH (or a manual pushbutton gives a HIGH input at the clock pin), the outputs, either Q or $\overline{Q}$, will adopt the corresponding input logic and will hold it until the next clock transition.

Shift Registers

The flip-flops in the register of Figure 12-14 act individually in that no single unit has any effect on any of the others. A particularly useful arrangement can be obtained by connecting a series of flip-flops head-to-tail so that an output (either Q or $\overline{Q}$) of each unit drives the data input of the next, as shown in Figure 12-15. The clock inputs of all units are tied together so that they operate simultaneously. The signal at the data input of the first flip-flop is passed on to the second at the next clock transition, then to the third, the fourth, and so on,

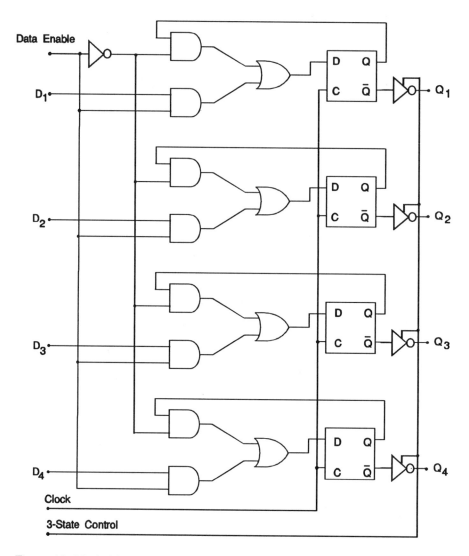

Figure 12-16 A 4-bit register made up of D-flip-flops. This is a simplified form of the 74LS173, allowing parallel input and parallel three-state output.

hence the name **shift register**. The register shown can accept inputs only in serial form, though other models may also accept a parallel input. Information can be read from it in parallel form; again, other models may permit both serial and parallel output.

Another type of register, built up from D-flip-flops, is shown in Figure 12-16. The Q output of each flip-flop is connected through AND-OR logic to the D input of the same stage, serving as a latch as long

Parallel Inputs

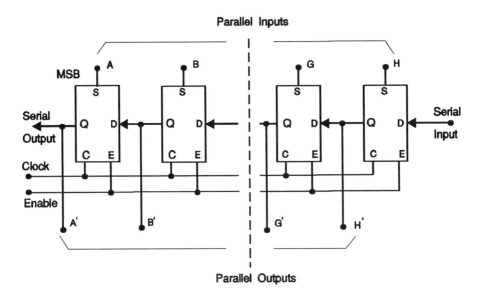

Figure 12-17   A shift register permitting either serial or parallel operation at either inputs or outputs.

as the "data enable" line is HIGH. A LOW on this line disables the latch and permits data to be entered through the other side of the AND-OR gate system. Since serial input and output is not provided in this register, there is no direct connection between the flip-flops other than a common clock. The three-state outputs permit control over the time at which data is passed on to a four-bit bus or other circuitry.

For many applications it is desired to **preload** the register with a binary word, entered in parallel fashion, then to shift it one or more places to right or left before reading it out (see Figure 12-17). This would be true, for example, in multiplying a binary number by two (10_2), which is equivalent to moving the digits one place to the left. Suppose that it is desired to multiply 00101011_2 by 100_2. We could do this with an 8-bit shift register (with eight flip-flops) by entering the number 00101011 through the inputs A, B, . . ., H. The serial input line would then have to remain unchanged for the duration of two successive clock pulses. The first clock pulse would move the H-digit to the G place, the G-digit to the F position, and so on. The next pulse would shift the digits one further place to the right, thus effecting the desired multiplication by 100_2 or 4_{10}. The answer, 10101100_2, can then be read out through the outputs marked A', B', . . ., H'

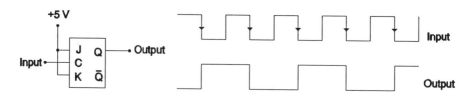

Figure 12-18 A JK flip-flop acting as a divide-by-two unit.

Counters

Consider a JK flip-flop operated in the toggling mode as in Figure 12-18. The output Q will change state for every *falling* edge of the clock transitions. It becomes evident, by comparison of the input and output, that the frequency of the square wave has been divided by two. Thus a toggling JK flip-flop acts as a divide-by-two unit. Cascading similar units permits frequency division by successive powers of 2 (i.e., 4, 8, 16, 32). Other types of flip-flops than JK can be toggled and thus used as counters. Some of these change states on the rising rather than falling edge of the clock pulse.

To better understand how a counter works, refer to Figure 12-19. Four flip-flops are cascaded to permit division by sixteen. Clearly each successive stage undergoes half as many transitions as its predecessor. Thus, if the input is energized by a 32-kHz square wave, the output will have a frequency of 32 kHz /16 = 2 kHz. The circuit can be used for counting the number of pulses elapsing between two events and treat them as a measure of time. Hence counters are sometimes called **timers.**

The left-to-right labeling of outputs in the drawing has the advantage that the outputs A, B, C, and D, have a direct numerical significance. By examining the logical values for a few completed input cycles, we can observe that the outputs represent at any moment the binary equivalent of the cycle number. Thus the circuit has two functions: (1) it divides the frequency by powers of two, and (2) it counts the number of input pulses.

It is often desirable to obtain an output reading directly in the familiar decimal system. The underlying difficulty is that 10 is not an integral power of 2. Three flip-flops are not enough to count to 10, and four are more than enough. Therefore some means must be devised for interrupting the counting process at 10. One way to do this is to arrange logic circuits to detect when binary 1001 has been

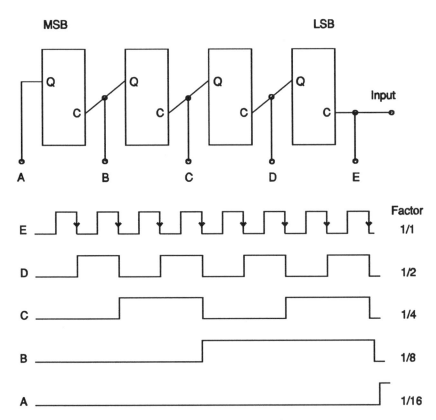

Figure 12-19 A divide-by-sixteen register, and successive waveforms.

reached, since this corresponds to decimal 9, and to reset all flip-flops at this point so that the next count is 0000 rather than 1010 (Figure 12-20). The circuit forces the register to skip to 0000 following the 1001, rather than to continue binary counting. The BCD counter, also called a **decade scaler** consists of several such modules, cascaded.

Sequential Timing Counters

A counting register is useful in quite another way, in that it provides a flexible and convenient way to program events with greater precision than afforded by a string of monostables. Every binary number can be uniquely identified by means of an AND gate with its inputs taken from successive outputs of a binary counter. Figure 12-21 shows circuitry for the general implementation of such a decoding procedure. Sixteen AND gates would be required to exhaust all the possibilities with this

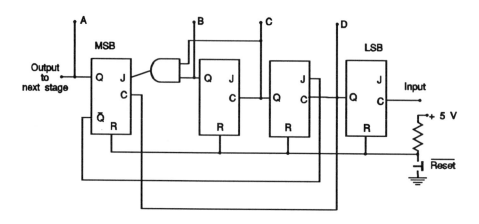

Figure 12-20 A single-stage decimal counter (74LS90), based on interrupting the count after the ninth pulse. The output is in the BCD code.

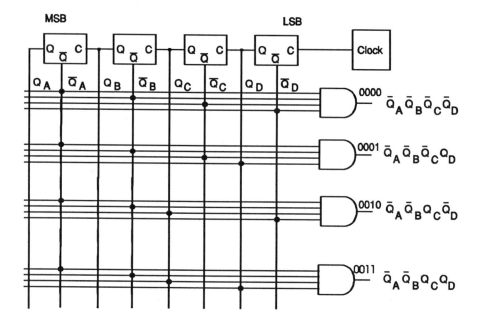

Figure 12-21 Timing circuitry, showing the output gates corresponding to the first four logical combinations. Each AND gate will deliver a high corresponding to the indicated binary number produced by the shift register at the top.

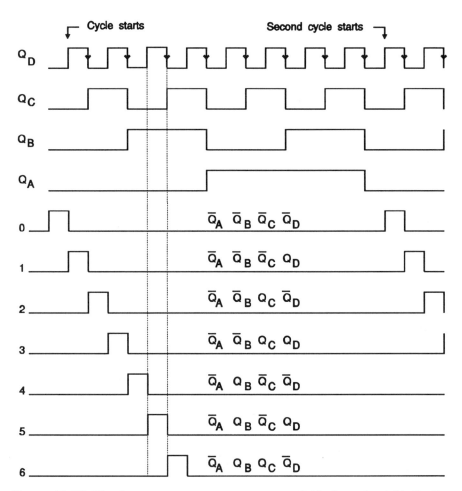

Figure 12-22 The first seven timing sequences available from the 4-bit flip-flop of the previous figure. At number 16, the sequence will return to all ZEROS.

four-bit counter. The actual timing diagram thus obtained is shown in Figure 12-22.

Let us suppose that the clock gives a square wave of 1-s period. The output of the first flip-flop (Q_A) is then a square wave with a period of 2 s. Each state ($Q_A = 1$ and $\overline{Q} = 0$) lasts for exactly 1 s, which is the duration of the pulses at each of the output AND gates. The experiment to be sequenced is made to start at the same instant that the counter starts counting clock pulses. Then if the event is to occur exactly 3 s later and last for 5 s, it can be turned on and off respectively by the signals from the AND gates marked 0011 and 0101. Unless inhibited by a logic command, it will repeat every succeeding 16 s.

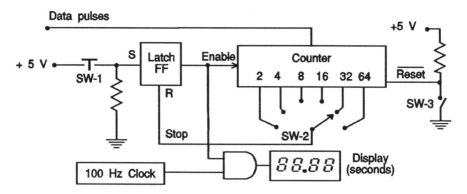

Figure 12-23 A preset counter. The display shows the number of seconds elapsed during the time interval required for the binary counter to amass 32 pulses from the data source, if switch SW-2 is set at 32 as shown.

Counters can also be used in a **preset** mode, meaning that they will indicate the time when a preselected number of counts will have arrived. Figure 12-23 shows an example. The 100-Hz clock drives a display counter provided that the AND gate is open. The experiment is started by closing the pushbutton switch SW-1. This enables a latch that remains on, even when the switch is released. The counter and the AND gate are energized and the display starts incrementing and counting seconds. When the data reaches 32, the "stop" line goes HIGH and resets the latch, thus terminating the display count. The indicated time shows the duration of the experiment.

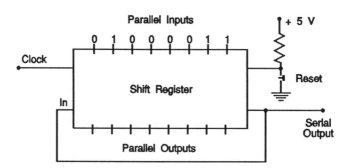

Figure 12-24 A shift register used to generate a repetitive pattern of ONES and ZEROS. Once loaded by means of the parallel inputs, the pattern circulates through the register, appearing also serially at the output. Note that the contents of the register exit in the right-to-left order.

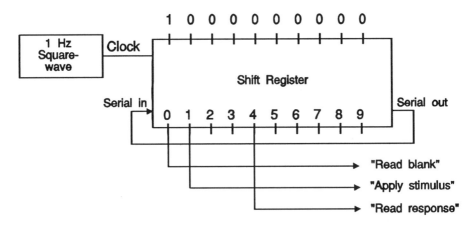

Figure 12-25 A shift register used to control a stimulus-response cycle. The complete cycle is 10 s in duration. Cycling through the unused outputs provides time for the "relaxation" of the system to its original state.

Ring Counters

It is possible to feed the serial output of a shift register back into the initial serial input, as in Figure 12-24. This arrangement, called a **ring counter**, can be used to generate a repeating pattern of 1's and 0's. In the example shown, the binary word 1100 0010 has been entered via the parallel inputs. It then circulates through the register. The output will give a continuous train : 1100 0010 1100 0010 1100 0010. . . unless provision is made for disabling the circuit, for instance, after 8 clock pulses.

Ring counters have practically no inherent time limitation, so that cycle durations from microseconds to days are equally possible. Ring counters, however, must be initialized whenever the power is turned on; this can be done by a combination of a *clear* command to eliminate all previous highs and a *preset* command to load the desired pattern.

Ring-counters can be used to control a sequence of events. Typical of many experiments, one might want to execute the following operations: (1) take a background measurement before the experiment, (2) apply a stimulus, (3) wait for the response, and (4) measure the response. Such a scheme can be implemented in elegant style by a single shift register, as shown in Figure 12-25. The initial pattern, 1 000 000 000 (read blank), changes after a 1s delay to 0 100 000 000 (apply stimulus). After an additional 3 s it arrives at 0 000 100 000 (read response), and after a further 6 s the cycle repeats. This same

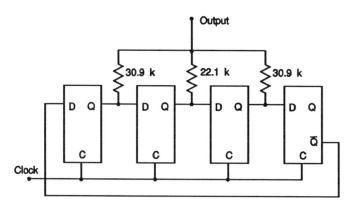

Figure 12-26 An eight-step digital sine-wave generator. (After Lancaster <u>Radio Electronics</u>, January 1989, p. 37).

basic system can be applied to most so-called pulse methods that are of the stimulus/response type.

An interesting variant of the ring counter is obtained by feeding back the $\overline{Q}$ output to the serial input, a configuration sometimes known as the **Möbius ring counter**. Figure 12-26 shows an application of this type of counter to a digital sine-wave generator. The circuit shown, with four D-flip-flops, will produce the wave form of Figure 12-27a, an 8-step approximation to a sine wave. A similar shift register with eight flip-flops will give a better approximation, as in Figure 12-27b.[2]

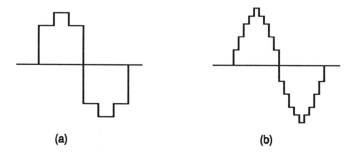

(a)   (b)

Figure 12-27 Stepwise sine waves obtained from a Möbius ring counter: (a) with four flip-flops; as in figure 12-26; (b) with eight flip-flops. (After Lancaster, op.cit.).

[2] The resistors should be 22.1, 41.2, 53.6, 57.6, 53.6, 41.2, and 22.1 kΩ, respectively

Providing the resistors are selected with care, these stepped wave forms can easily be converted to excellent sine waves by means of a low-pass filter. The lowest harmonic in the first circuit is the seventh, and this is only 1/7 of the fundamental in amplitude. For the second circuit, the lowest harmonic is the fifteenth at only 1/15 the amplitude. The fundamental frequency produced for the two circuits is 1/8 and 1/16 of the clock frequency, respectively.

Chapter 13

Interdomain Converters

In this chapter we discuss methods by which information can be transferred between analog and digital domains. Such conversion is essential in several applications for example: (1) To transform data from analog transducers into the digital equivalent for presentation to a computer. (2) To exert control over external devices such as those used in physiological measurements.

Encoding Considerations

The interpretation of **resolution** differs in the analog and digital domains. In digital measurements, an inherent limitation lies in the essential discontinuity of the process, whereas there is no similar limitation in the realm of analog signals. Another factor to be considered is that the frequencies present in the analog signal must be compatible with the digital clock rate, since this determines the highest frequency that can be present in digital systems.

To gain insight into the type of considerations encountered in adapting analog information to digital instruments, let us consider the measurement of an optical spectrum. The experiment consists in scanning the wavelength over a range of 500 nm and recording the corresponding light intensity. Both the wavelength and the intensity data must be made available in digital form. Figure 13-1 indicates the kind of information needed.

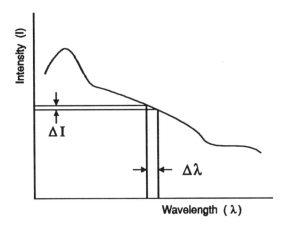

Figure 13-1 A hypothetical spectrum indicating the step increments required in digitizing the analog curve.

The wavelength drive mechanism can conveniently be scanned by means of a stepping motor that advances the drive by successive increments. If the device is programmed to give one pulse per step, the determination of the wavelength reduces to counting the number of pulses from the start of the scan. It might seem appropriate to make each step correspond to exactly one nanometer of wavelength, so that the desired range of 500 nm would provide 500 pulses. The step, however, should be smaller than the expected error in the analog electronics, thus ensuring that the stepwise nature of the count does not add to the overall instrument error. Suppose that statistical considerations require that the wavelength step be no greater than 0.5 nm; the drive mechanism should then be designed to give at least $500/0.5 = 1000$ steps.

The situation is different with respect to the measurement of the light intensity. In the first place, we do not know in advance the range of values to be expected, though the maximum output of the instrument at full scale is known. Once the desired resolution is established at, say, 0.5% of full scale, we can determine the parameters that the digital system must possess in order to attain this resolution. In this example, the intensity scale must be divided into at least 200 steps, each corresponding to 0.5% of full scale. In practice this would require the signal to be described by an 8-bit word, since this corresponds to a maximum of $(2^8 - 1)$ or 255 steps.

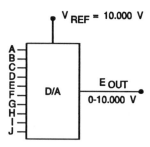

Figure 13-2 Block representation of a typical D/A converter. The input has 10 lines (lettered A through J), giving $2^{10} = 1024$ possible combinations. The device can be adjusted to give other output ranges, usually by changing the reference voltage.

Glitches

As mentioned briefly in the discussion of logic gates, unwanted voltage spikes, called **glitches**, can occur in any circuit that contains many digital devices. These result from the fact that typically the gates and flip-flops are required to change state simultaneously, subject to a common clock. In general the time required to turn a gate off is different from the time required to turn it on. Hence it may be that many gates will turn on some nano- (or micro-) seconds before others turn off, thus producing a momentary false combination of states. At the same time, a relatively large momentary change occurs in the power required for the circuit. The resulting sudden change in voltage may have the effect of transferring the glitch to gates that are not supposed to change state at this time. Glitches can also cause interference (noise) in analog circuits, especially if the digital and analog portions are referred to a common ground.

Sometimes a sample-and-hold circuit can help to "deglitch" a system. The signal is sampled just prior to the state change and holds this value until the change has been completed and all transients have died away.

Digital-to-Analog Converters

A component designed to perform this transformation is called a **digital-to-analog converter,** a **D/A converter,** or simply a **DAC.** As suggested in Figure 13-2, it accepts binary numbers through its multiple input lines, and produces an equivalent analog voltage at its single output. If there are n input lines, the output voltage can assume

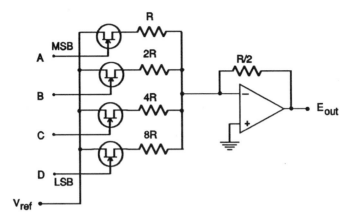

Figure 13-3 Schematic diagram of a 4-bit DAC. If a variable signal is substituted for the reference, the device functions as a programmable amplifier with digital controls.

any of 2^n discrete values from zero to the maximum determined by the reference voltage, V_{ref}. Thus for n = 10 and V_{ref} = 10 V, the output can have any of 1024 evenly spaced voltages, including the limits of 0 and +10 V. Any voltage between these limits can be specified to within 9.8 mV (10/1024 = 9.8 mV), giving a resolution of slightly better than 0.1%. Compare this precision with that of most analog panel meters, which are rated at perhaps 2% of full-scale.

If still better resolution is needed, it can be obtained either by selection of a DAC with more inputs, or by decreasing the reference voltage. For instance, if V_{ref} = 5.00 V, the smallest output increment will be 4.9 mV, doubling the resolution but sacrificing the range.

One type of DAC takes the form of a programmable amplifier, an example of which is shown in Figure 13-3. A series of transmission gates (FETs) are controlled by an equal number of digital input lines, being turned on when the signal is HIGH and off when it is LOW. The output voltage can readily be calculated by simple algebra, using the rules for parallel combinations of resistances to determine R_{in} (the input resistance to the op amp) for each input code.

It becomes impractical to use this simple circuit for more than 4-bit resolution because of the need of very high-valued precision resistors, which lead to instability and slow speed.

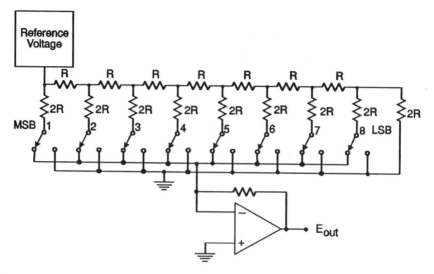

Figure 13-4 An 8-bit DAC with an *R/2R* ladder network. Of the numbered switches, No. 1 is the MSB and No. 8 the LSB. The switches are usually transistors or FETs.

The Ladder-Network DAC

Another type of DAC is shown in Figure 13-4. This differs from the preceding chiefly in the configuration of the resistors, commonly called a **ladder network**. A repeating network of this kind has the same impedance as seen by the reference voltage, no matter how many rungs there are in the ladder. This circuit has the practical advantage from the design standpoint of requiring only two resistance values, some of *R*, others of 2*R*. Note that the input closest to the reference voltage has the greatest effect on the result, and so is the most

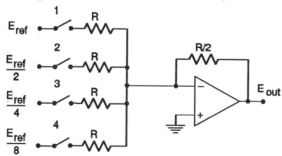

Figure 13-5 A circuit analogous in operation the DAC of the previous figure. All resistances are equal.

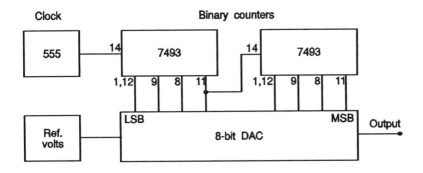

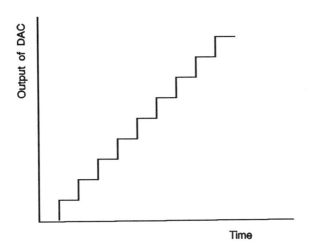

Figure 13-6 (a) A staircase generator using an 8-bit DAC. (b) A staircase; the circuit above would give 256 steps. The voltage interval per step depends on the reference voltage. The numbers in (a) are pin designations for the binary counters (type 7493).

significant bit (the MSB), whereas the farthest is the LSB, the least significant bit. When the nth switch is closed the output of the op amp is increased by $1/2^n$ of the reference voltage.

The action of this DAC can perhaps be visualized more readily by means of Figure 13-5. The DAC ladder serves to divide the reference voltage by successive powers of 2, and each term is used or not according to which switches are closed. For example if the reference is 10 V and the digital value is 1001 0000 (decimal 144), only two switches will be closed correponding to the two ONES. The output will

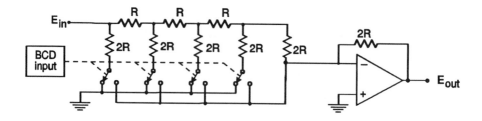

Figure 13-7 A BCD-controlled DAC as a voltage divider. Compare this to the DAC shown in Figure 13-4.

be $(10 + 10/8)/2 = 5.63$ V. This agrees with the expected value of $10 \times 144/256$.

Either of the DACs described above can be operated with a fixed reference voltage, as shown, or with an analog signal applied at the same connection. In the latter case the output consists of this analog input (E_{in}), multiplied by the digital input N:

$$E_{out} = N \times \frac{E_{in}}{N_{max} + 1} \qquad (13\text{-}3)$$

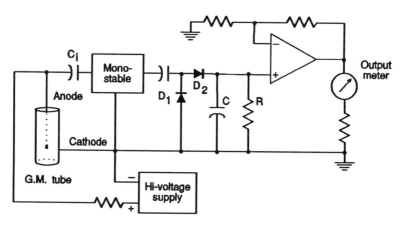

Figure 13-8 A ratemeter, as used with a Geiger-Müller counter. The op amp is configured as a follower-with-gain. The monostable acts as a pulse shaper, that converts a sharp peak into a square pulse of fixed duration. Capacitor C_i must be rated at least 1000 V.

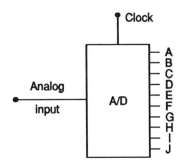

Figure 13-9 Block symbol for an A/D converter. The output is shown as having 10 digits, but others are available with fewer or more digits.

Here N_{max} is the largest digital number available (all 1's). So used, the unit constitutes a **multiplying DAC,** since it multiplies the analog input by a digital gain (of less than 1).

DACs can be used in various applications besides converting signals to a computer-compatible format. For example, a DAC driven by a binary counter will produce a staircase, as in Figure 13-6, useful in a variety of experiments. A multiplying DAC, connected as shown in Figure 13-7 provides a digitally controlled voltage divider. The model AD7525 (Analog Devices) is designed for this service; it can give an output anywhere between 0 and 1.998 times E_{in}, with a resolution of 0.001, as controlled by a BCD input.

An example of an instrument with a built-in D/A converter is the ratemeter shown in Figure 13-8, which is used in the measurement of X-ray and nuclear radiations. The photons or particles produce individual pulses through the action of the Geiger-Müller tube. Each pulse contributes an increment of charge to capacitor C via the monostable, and the resulting voltage is indicated on the meter. At the same time the resistor R discharges the capacitor at a rate proportional to its voltage, so that a steady state is attained determined by the number of pulses per unit time. The RC-network and associated diodes constitute a special-purpose DAC, converting a pulse train to an analog voltage.

Analog-to-Digital Converters

As the name implies, the action of an **analog to digital converter** ADC is the reverse of the DAC. The principle is shown in the block symbol of Figure 13-9, which might be called the complement of Figure 13-2. The essence of the ADC is a digitizer that expresses the analog signal as one of a large set of numerical values. The block shown in Figure

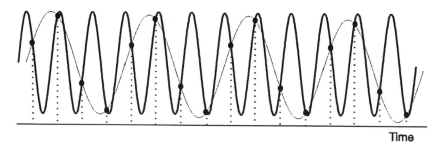

Time

Figure 13-10 Aliasing. The higher-frequency sine wave, if sampled at the equal time intervals, marked with dots, would give the false impression that the wave had a frequency much lower than is actually the case.

13-9 divides the signal into 2^{10} or 1024 segments, so that if the full-scale value of the analog input is 5.000 V, a change of 1 in the least significant output digit (the LSB) would correspond to $(5.000)/(1024) = 4.883$ mV change in the input. This would be fully accurate enough for many purposes.

In general the analog signal to be converted is made up of a number of superimposed AC frequencies, and the number of digital readings per second (the clock rate) must be carefully considered. Let us assume first that a pure sine wave of frequency f is to be converted into its digital equivalent. If the signal were sampled at frequency f (i.e., once per cycle), the samplings would all be identical and would not show that the signal is actually AC (this might be useful as a means of synchronous rectification, but not as A/D conversion). Sampling *less* frequently than once per cycle could be interpreted as a sine wave with a frequency smaller than f that was not originally present, called an **alias**. Figure 13-10 should make this clear. As a rule, the sampling

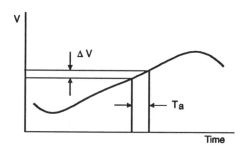

Figure 13-11 The relationship between the aperture time of a DAC and the rate of change of the input signal.

frequency must be at least twice that of the highest frequency component of the signal, if such distortions are to be avoided.

Another limitation relates to the finite time required for the DAC to accomplish its conversion, called the **aperture time**, $T_{A,}$. The analog signal will vary during the time T_A by the amount ΔV, which is the maximum voltage uncertainty, depending on the slope of the signal curve at that point, as can be seen in Figure 13-11 and is given by

$$\Delta V = \frac{dV}{dT} T_A \qquad (13\text{-}4)$$

We want to establish the maximum allowable aperture time that will still give the required resolution $\Delta V/V$.

Let us consider a signal with an information bandwidth extending to some given frequency f_{max}. The largest permissible value of T_A is related to this maximum frequency

$$T_A = \frac{1}{f_{max}} \frac{\Delta V}{V} \qquad (13\text{-}5)$$

This means that the aperture time cannot be greater than the period of the greatest frequency component of the analog signal, multiplied by the relative resolution.

Suppose, for example, that we need to find the appropriate aperture time for sampling a 400-Hz sine wave (or a non-sinusoid in which the highest frequency component is 400 Hz) with a resolution of 1%. Equation (13-5) gives

$$T_A = \frac{\Delta V/V}{f_{max}} = \frac{0.01}{400} = 25 \ \mu s \qquad (13\text{-}6)$$

Thus the aperture time must not be greater than 25 μs. If the DAC is not fast enough to make a conversion in this short time, we can introduce a sample-and-hold (S/H) amplifier. The S/H must be programmed to sample the signal for 25-μs intervals or less, with a suitable repetition rate, holding its value for the time required for the DAC to make its conversion. Remember that both this and the previous restrictions must be obeyed.

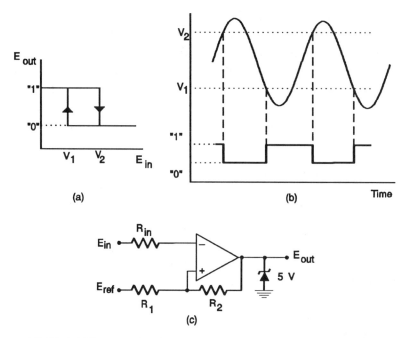

Figure 13-12 (a) The transfer characteristic of a Schmitt trigger. (b) The forma-
tion of a square wave from a sine wave by means of a Schmitt trigger. (c) An op
amp connected as a Schmitt trigger; for minimum offset error the input resistor
R_{in} should equal the parallel resistance of R_1 and R_2. Reasonable values are R_1
= 2 k, R_2 = 2 M. For faster digital applications, special ICs are to be preferred.

The Schmitt Trigger

Digital signals generated by mechanical switches or op amps may not
have suitable wave shapes for connection to a D/A converter or other
logical devices. The difficulty may be due to switch *bounce*, described
earlier, or it may be that a varying signal does not change its state
abruptly enough to be interpreted unambiguously. In either case, the
signal must be further processed, and this can be done by a device
known as a **Schmitt trigger**. The chief feature of this module is its
hysteresis. As shown in Figure 13-12a, if E_{in} = 0, then E_{out} = +5 V
(logical 1), and when E_{in} = +5, E_{out} = 0, the behavior of an inverter.
As the input is increased from zero, the output stays at logical 1 until
voltage V_2 is reached, when it suddenly changes to 0; but when the
applied potential is lowered again, no change takes place until the
lower threshold voltage V_1 is reached.

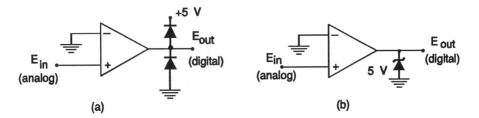

Figure 13-13 A comparator serving as a 1-bit digitizer. The two diodes in (a) or the single 5-V Zener in (b) limit the output voltage excursions to the two extremes of 0 and +5 V, as needed for TTL logic.

Figure 13-12*b* shows how a Schmitt trigger can be used to convert a sine wave (or any other periodic signal) into a square wave of the same frequency. When the analog signal gets to level V_2 the Schmitt changes state from 1 to 0, and stays that way until the signal decreases to level V_1, at which point it goes HIGH again. This sequence repeats indefinitely. The resulting square wave is suitable for connection to TTL or other logic systems. It is possible to achieve the same effect with an op amp connected as in Figure 13-12*c*, but the transition speed is smaller.

The Comparator as a Digitizer

The comparator, either with or without hysteresis, can be considered to be a single-bit digitizer. When the output of a fast comparator is bounded by diodes as shown in Figure 13-13, it gives TTL-compatible digital levels according to whether the input is higher or lower than ground. Some special comparators, such as the high-speed LM160, are powered from +5 V and ground (rather than plus and minus 15 V), and thus give TTL levels directly without need for the bounding diodes.

Types of ADCs

The simplest multibit type of analog-to-digital converter consists of a binary counter that drives a DAC, as shown in Figure 13-14. The analog input is compared with the signal from the DAC. The output of the comparator, via the AND gate, controls the flow of clock pulses reaching the counter.

Initially both the counter and the DAC are at zero, $E_{in} > E_{DAC}$, the comparator output is HIGH, and the clock pulses are allowed to reach

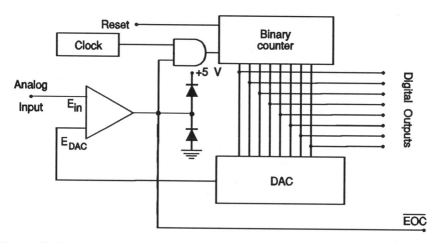

Figure 13-14 A simple form of ADC. The bar over the "EOC" is the Boolean NOT operator, signifying that the active signal is LOW, and goes HIGH when the conversion is completed.

the counter. This causes the DAC to increase its output step by step. When the DAC output becomes equal to the analog input, the AND gate causes the counting to terminate. A signal called the **end-of-conversion flag** (EOC) is simultaneously produced at an auxiliary output, indicating that the conversion is complete. This flag is essential if the digital output is fed into a computer. At this point the counter contains a binary number that is a measure of the analog input. The counter must be reset to zero before starting another measurement.

Figure 13-15 shows a modification of this circuit to permit continuous measurements. An up/down counter is substituted for the up-

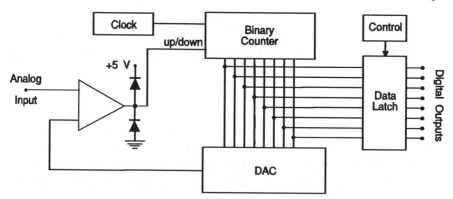

Figure 13-15 An ADC using a bidirectional counter.

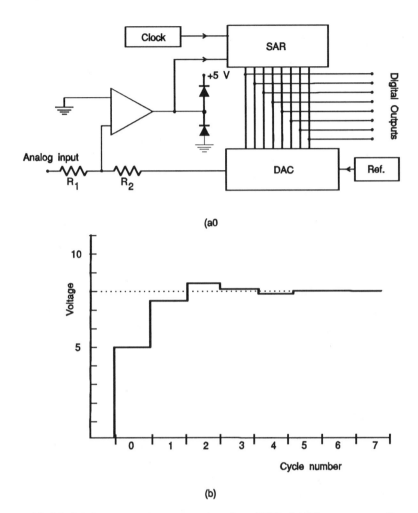

Figure 13-16 (a) A successive approximation ADC. (b) The output voltage for successive bits. The dashed line at 8.00 V represents the true value.

counter of the previous figure. Now the AND-gate is no longer needed; the counter counts up if the comparator output is LOW, and counts down if the output is HIGH. Also the reset switch is not needed, for one conversion can immediately follow another. No matter whether the new input calls for a higher or lower count, the system can respond. Since conversion is continuous, no EOC flag is needed, but the data latch is necessary to hold the output long enough to accomplish its purpose, whatever that may be.

Successive Approximation Converters

One of the most widely used methods of A/D conversion utilizes a **successive approximation register** (SAR). The example shown in Figure 13-16*b* assumes that an input signal of 8.00 V is to be converted to its digital counterpart. The operation starts with the SAR set to zero. When conversion begins, an initial clock pulse creates a 1 in the MSB position, causing the register to take a value of 1000 0000 = 5 V, corresponding to half-scale, as a first approximation to the input. If, as it happens in this example, the input is greater than half-scale, the 1 is allowed to remain as the MSB, and a second approximation is made by giving a tentative 1 to the next bit. Since the converted value of 1100 0000 is 7.50 V (which is less than 8.00 V), the 1 is retained as the second digit. A third approximation gives a tentative 1 to the third bit, but this makes the contents of the register greater than the input, so the 1 is replaced by a 0. This trial-and-error method is repeated until the register is filled. Each successive approximation creates a step that is just half the magnitude of the previous one, converging to the final value. This type of ADC, though inexpensive, is too slow for many applications. It is frequently encountered in digital voltmeters.

Dual-Slope ADCs

The dual-slope ADC (Figure 13-17) uses still another approach. In this circuit a comparator is fed by the output of an analog integrator. Initially an analog switch steers the incoming signal to the integrator for a predetermined period, $t_1 - t_0$, determined by the time needed for the counter to reach 1111 1111. The next pulse resets the counter to 0000 0000. At this point a new pulse is generated, called **overflow**. The charge Q accumulated on the capacitor during this time interval bears a direct proportion to the input E_{in}. Thus

$$Q = \frac{E_{in}}{R} \int_0^{t_1} dt = \frac{E_{in}}{R} t_1$$

The overflow signal throws the switch connecting the (negative) reference voltage to the integrator. This starts a downward ramp of slope E_{ref}/RC volts per second, which continues until the comparator

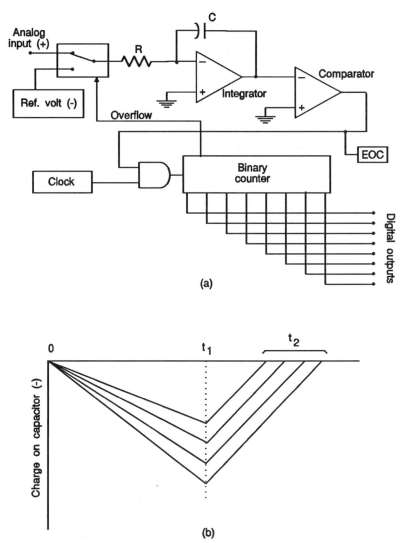

(a)

(b)

Figure 13-17 A dual-slope ADC. (a) Schematic. Note that no internal DAC is required. The overflow pulse, produced on crossing zero, reverses the position of the analog switch. (b) Timing diagram for several input values. Observe that the first period (t_0 to t_1) is characterized by a fixed duration and variable slope, whereas the second (t_1 to t_2) has a fixed slope and variable duration.

senses zero volts (at time t_2) and stops the counter by means of the AND gate. The duration of the downward integration ($t_2 - t_1$) is a measure of the accumulated capacitor charge and thus of the input. The counter output at this point measures $t_2 - t_1$ and consequently the analog input

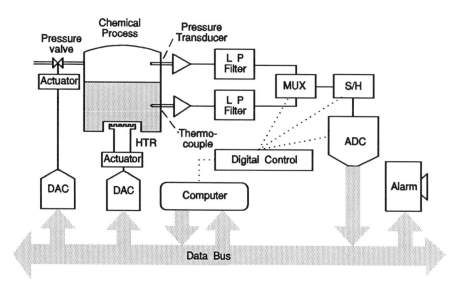

Figure 13-18 An example of an industrial process controlled by a computer. Note the essential use of both ADCs and DACs. The unit marked "digital control" might contain one or more shift registers to coordinate the various stages of operation, all under the control of the computer. LP: low-pass filter. MUX: multiplexer. S/H: sample-and-hold.

voltage. The dual-slope ADC is highly accurate and inexpensive, but fairly slow. It is often used in applications not requiring high speed.

An Example

Figure 13-18 is included as an example of computer control of a process in chemical engineering, illustrating applications of analog-to-digital and digital-to-analog converters.

Voltage-to-Frequency Converters

The V/F converter is a special type of ADC that, instead of generating a digital word corresponding to the numerical value of the incoming analog voltage, produces a square wave or string of pulses with the **frequency** proportional to this voltage. Figure 13-19 shows a circuit using a commercial model, the LM-131.

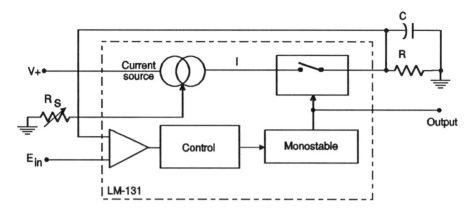

Figure 13-19 Equivalent circuit of the model LM-131 voltage-to-frequency converter.

The principle of operation is based on a forced equality between the voltage on capacitor C and the input voltage E_{in}. Because of this equality, the discharge of the capacitor through R occurs at a rate proportional to E_{in}/R. Feedback through the system compensates for this discharge current by providing a train of constant amplitude current pulses. This in turn requires the frequency of the pulse train to be proportional to E_{in}/R, which is the desired effect.

When starting the operation, E_{in} is compared with the voltage V appearing on the capacitor. If E_{in} is higher, it will cause the monstable to fire, thus closing the analog switch and turning on the current, I. The pulse from the monostable stays on for time t, during which an amount of charge $Q = I \times t$ will be injected onto the capacitor. This charge will increase the voltage by a small increment. If V is still less than E_{in}, the monostable will be fired again. This process repeats until V becomes equal to E_{in}, at which point the system will be in equilibrium and the monostable will continue sending pulses of charge to the capacitor at exactly the rate required to compensate for the charge that is disappearing through R to ground.

The frequency of the pulse train so produced is proportional to the input analog voltage to within about 1%, and the time required to respond to a large change of input is of the order of 100 μs. The circuit can be linearized (to about 0.05%) and the response time decreased, with the aid of an op amp that forces the charge and discharge of the capacitor to be linear with time.

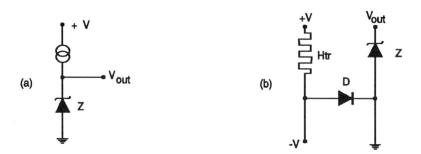

Figure 13-20 (a) The equivalent circuit of the AD580 voltage reference. The double-circle represents a constant-current device. The output is close to 2.50 volts. (b) The equivalent schematic of the LM-199 voltage reference. The diode D is an inherent part of the silicon substrate, and is always reverse-biased. The heater actually consists of an amplifier with its input shorted, a configuration that dissipates a constant amount of power, so the unit quickly settles to an equilibrium temperature and holds it indefinitely. Both devices have temperature coefficients measured in μV/K, compared to about 5 mV/K for a simple Zener.

Many commercial ICs, including the 555 timer, the 8038 signal generator, and many varieties of phase-locked loops can also be used to convert a voltage to a corresponding frequency. The advantage of the LM-131 and similar units lies in their being optimized for this service.

Voltage References

Nearly every type of interdomain converter needs a precision source of known voltage as an internal reference. It may be satisfactory to utilize the regulated power supply for this purpose, but often greater precision is needed. The design of a reference supply is simplified since negligible current is to be drawn.

The simplest reference voltage would be that established by a Zener diode, but this is beset by two major drawbacks: the Zener voltage varies (albeit slightly) with the current passed, and it is restricted to 5 or 6 V if a significant temperature coefficient is to be avoided. Both of these drawbacks can be circumvented. A current source can be connected so as to maintain constant current through the diode (the AD580 is an example, Figure 13-20a. The type LM-199, (b in the figure) is a successful reference source incorporating a small heater embedded in the same silicon chip with the Zener, effectively thermostating the diode.

Part III

Computer Electronics

Chapter 14

The Computer as an Electronic Device

The influence of the computer on everyday life is so pervasive that essentially all aspects of human activities are affected by it. For this reason we shall present only a special view of small computers, a discussion that emphasize their electronic roots. In this treatment, we shall concentrate on the personal computers introduced by IBM, usually called "PCs."

Historical Note

Calculating devices have been known for a long time. The first one appears to have been the ancient Chinese abacus which is still in use today. It consists of a set of markers that slide on wires: its operation implies memory (in terms of the markers position), and data processing (by shifting the markers.)

In the West, we must wait until the beginning of the seventeenth century for a true computing instrument, the slide rule, a device based on logarithmic transformations. This invention turned out to be very long-lived, and has dominated engineering calculations until very recently. Actually, the slide rule may be more correctly considered as the grandfather of the *analog* computer, since it uses lengths as analogs of numbers. The electronic analog computer uses operational amplifiers to do mathematical operations (which is the reason for the qualifier *operational*). It was successfully employed for solving complicated linear differential equations, at a time when the digital computer was in its infancy. Later, as the digital computer progressed, the analog computer was gradually abandoned, on account of its limited precision.

In contrast, the limits of precision for digital (meaning numerical) devices are much more favorable. The first such numerical computing devices were also developed in the seventeenth century, with the work of Blaise Pascal. He created the first mechanical calculating machine, operating on digital principles, the "Pascaline." It was very successful, but only a few copies were made. Subsequent mechanical calculating machines have enjoyed a very long development life, and an occasional unit may still be seen in operation.

It is interesting to speculate why, if solid bases for general purpose analog and digital computers were laid three hundred years ago, so little happened in the intervening centuries. Why did computers fail to evidence themselves to the world until few years ago, when suddenly an explosive development took place?

The reason is that several additional technological developments had to converge and help to create the modern-day computer. A pertinent example of such need is the history of the "analytical engine" conceived by Babbage about 150 years ago. The analytical engine was a perfectly viable design for a modern computer, albeit mechanical rather than electrical. Nevertheless, Charles Babbage was unable to implement his invention to any extent because of a poor technological base. In his day, precision machining was an extremely difficult task. A fully implemented analytical engine, would have required a roomful of clockwork, essentially impossible to construct.

Humanity had to wait until 1946 for the necessary coincidence of favorable factors among which two were decisive: (1) A sufficiently advanced electrical technology. This allowed first for relays, later for vacuum tubes, both of them vastly superior to mechanical devices. (2) A serious national-defense need for gun-control and code breaking. One might say that, in the 1940s, the power of whole countries backed the computer project. The combination of such favorable political and technical factors justified the major initial effort. It eventually culminated in the 30-ton ENIAC Using 200 KW of power to feed almost 20,000 vacuum tubes, the ENIAC, built at the University of Pennsylvania, was literally a monument of modern technology. It appeared only after the Second World War had ended, but the way for computer development had opened.[1]

The road from this point on was smoother, and the history of computers parallels the progress in electronics. Almost every year brought forth new and more sophisticated electronics at lower and lower prices. Thus the appearance of the transistor created a new generation of smaller and less expensive computers. Later when the transistors became integrated into ICs, the computers followed by going into their next generation of cheaper and yet more powerful constructions. Finally the large- and very large-scale integration of the past years has made it possible to produce full-featured, affordable units in the $1000 range. This can be regarded as the turning point where the computer finally became a consumer product.

The Microcomputer

In the early years the computer was invariably a large, powerful, and expensive **mainframe** shared between many users. The first time that a smaller alternative appeared, was in 1960, when Digital Equipment Corporation (DEC) produced the first of an illustrious series of **minicomputers**, the PDP−1. (Programmed Data Processor–1.) At its price of less than $100,000 it was something of a bargain, when compared with the alternatives. Being more affordable, the newly born minicomputers sold well, especially the subsequent PDP-8 which appeared five years later.

The PDP-8 represented an important development, because it was still lower in price, at less than half that of its predecessor. It permitted the economic implementation of a new type of use, **real time** operation. This mode implies dedicated use and instant availability for whatever task is requested. This is important if the computer must gather data, which might be available only at specific moments. More critical yet, real time is essential for control operations. For such uses the general purpose mainframe computer remains only a poor second best.

DEC and several other companies continued on the same path with considerable success. They created a long series of minicomput-

---

[1] This is not the first modern computer. It was predated by Konrad Zuse's machine in Germany in 1941, and the Colossus code-cracking machine in England in 1943, as well as others.

ers that became more and more powerful (and expensive). This resulted, however, in a gap at the low price end. For instance, in the early 1970s, the lowest priced PDP system (or an equivalent from other companies) might sell for $20,000, hardly an attractive price for a consumer item. In this vacuum the birth of small and cheap computers began to take place. In November of 1971, the three-year old company Intel, created the 4004, the first microprocessor. It consisted of a 2300-transistor integrated circuit that could be used as a general purpose four-bit processing unit. It was followed in rapid succession by more advanced variants, the 8008, 8080, and 8086/8088. Other microprocessors from various companies were soon available, among which we can mention the successful 6502 and the 6800 series. It must be realized, though, that these are micro*processors*, not micro*computers*, and they need considerable additional circuitry to become computers. The latter appeared soon enough, and various kits as well as ready-made microcomputers became available for the enthusiast. Three of them soon (1977) emerged in a satisfactory, ready-to-use form and at reasonable prices: Apple II, Commodore PET, and Radio Shack TRS-80. These models were immediately accepted by the public and were followed by a veritable explosion of computer offerings.

Many of the early names are long gone, for the industry has matured and consolidated. A major stabilizing factor was the involvement, in August of 1981, of IBM. A vast number of new companies soon followed the IBM lead; producing clones, compatible with the IBM command structure. IBM itself was unusually permissive in allowing such cloning. At the same time a few other standards, such as those of Apple Computer and Commodore, progressed as well.

Today microcomputers tend to be very powerful. In a few more years they will probably fulfill 90% of the scientist's requirements. Nevertheless, computing needs always seem to grow faster than hardware can keep up. The power of current personal computers today reaches in the millions of floating point operations per second, (**megaflops**).[2] Yet many applications are so demanding that they need supercomputers that can run gigaflops, a thousandfold superiority. Do

---

[2] Computer performance is often measured in Mflops. Among alternate criteria are the Mips and Vups. The Mips can be derived from the clock rate of the computer, say, 33 MHz, divided by the average number of clock cycles per instruction, say, 5.5. The Vups are multiples of the performance level of an older computer, VAX/780.

not be surprised to see, in a few years, *micro*computers boasting gigaflops as well, yet sold at reasonable prices. But expect, at the same time, to see supercomputers that can handle teraflops.

What Makes a Computer?

One might ask, at this point, a rather simple-minded question: What is really all that special about a computer? The characteristic that singles out the computer is not the fancy screen, nor the keyboard, nor the fancy price. In reality what makes this device unique is its philosophy, the manner in which information is handled. One can describe it by the following three properties:

1. *The information is stored in memory and can be replaced by new information when desired.* This is an important characteristic but not unique to the computer. Such devices as tape recorders or VCRs also store information that can be written over. The information stored by computers is generally described as data, even though it may consist of text rather than numbers.

2. *The data in the memory can be altered by the system itself.* In other words, operations can be performed on the data present in memory. The operations can be numerical, but text can also be handled, such as in copying and searching. Some of these features are encountered also in many hand calculators.

3. *The mode of operation itself can be stored in memory and altered depending on intermediate results.* This capability is unique to the computer, in fact it defines the computer as opposed to the calculator. Perhaps Babbage recognized it in a primitive form, but in modern times it was first described in 1936 by Alan Turing. He discussed the subject extensively in terms of a theoretical computer that could execute any mathematical or logical operation no matter how complicated. This theoretical computer had all the important characteristics of a modern real computer, except that it was relatively simplistic and its implementation would have been very slow. J. P. Eckert and J. W. Mauchly, the creators of the ENIAC, appear to have been the first to apply Turing's concepts to a practical general purpose computer.

In addition to the three elements discussed above, information must be entered and retrieved. Without such input/output channels

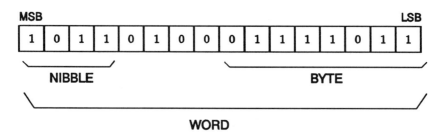

Figure 14-1 Relationship between nibble, byte, and word. MSB is the most significant bit (value 2^{16}), while LSB is the least significant bit (value 1).

(I/O), the operation would not have any human meaning. One can thus define the computer in terms of its ability to handle I/O, data, programs, and data processing

Computer Words

The information in the computer appears in form of collections of binary states, 0 and 1, either of which represents one bit. In most cases the information circulates in groups of bits treated as a unit. The smallest such unit is the **nibble**, a four-bit group. The old 4004 was a nibble processor, but today the concept is seldom used. Much more important is the **byte**, an 8-bit group.

A byte can take the values between zero = 0000 0000 and 255 = 1111 1111, for a total of 256 numbers, but this is not the only possible interpretation. For example, a byte can be made instead to represent the set of characters (letters, numbers, and such symbols as @ or #). The exact assignment of the 256 numbers to represent characters is a question of convention. The most common assignment, ASCII (pronounced Ask-ee), is shown in the Appendix. The assignments between 128 and 255 are subject to some variations between systems.

Contemporary computers operate with blocks of more than one byte at a time. Such a set of bytes is called a **word,** and we distinguish between 16-, 32-, and 64-bit machines in terms of the size of the words they use. Larger words are advantageous both because of the speedier moving of bits and because larger numbers can be defined by a single word. Figure 14-1 exemplifies the relationship between the various concepts discussed.

Consider a 16-bit word. In the computer, this collection of ones and zeroes may be assigned a variety of meanings such as:

1. The numbers between 0 and 65535, that is, $2^{16} - 1$.

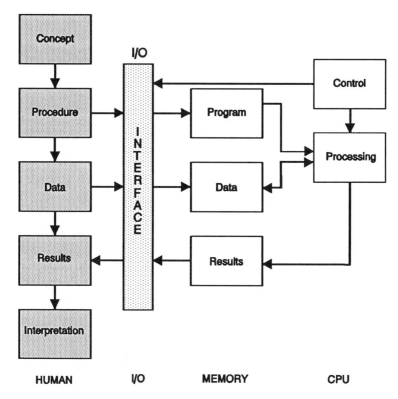

Figure 14-2 A general view of the computer as it interacts with humans.

2. Two ASCII coded characters of eight bits each.
3. A computer instruction.
4. An address in memory.
5. A dot (pixel) at a particular point on the screen.

In all cases they can be represented by the same set of high and low voltages. By just examining a byte, one cannot guess what it stands for. The meaning is a separate concept and must be specifically assigned to each word. This is done by the expedient of giving meanings to segments of computer memory. If a word appears in a segment assigned to contain programs, it will be interpreted as a command. If it is encountered in the data area, it will be interpreted as data. In the video area the word will help build an image. We can think of the computer as a gigantic collection of 0's and 1's that do not make any sense until the various memory areas are assigned specific meanings. Of particular importance is the program and the exact position (address) of the first instruction. Once this unique starting point is defined, everything falls in place. If we should use any other memory position as the first command, the meaning would become garbled; the computer would see only nonsense and would eventually crash.

Computer Architecture

The majority of human tools can be thought of as extensions of the body. Thus a camera can be considered an extension of the eye, while the bicycle augments the power of the legs. In this context, the computer is an extension of the *mind*, creating a division of labor between human and computer.

This thinking is illustrated in Figure 14-2. The section on the left represents the human part of the work. The boxes marked *concept* and *interpretation* represent assignments that will probably always remain with humans; they constitute the part that we like to call thinking. On the other hand, the procedure (programming) tends to be shifted more and more toward the machine. The tendency is to tell the computer what is to be done rather than to describe step by step how to do it. The computer is then allowed to choose the best way of implementing the particular task.

The boxes labeled *data* and *results* define the direct contact between human and machine. They represent an area of intense commercial effort. If a program fares well in this area, it is described as user-friendly, it reaches a larger proportion of users and it sells better.

Let us now refer to the computer itself. In addition to the I/O two parts can be identified: on one hand the control and processing unit whose collective name is the **Central Processing Unit,** CPU; and on the other hand the memory which contains the program, data, and results. The simplest hardware view of a computer is therefore in terms of I/O, CPU and memory. This is not including the human component at all. A more elaborate description is given in Figure 14-3, where the CPU itself is shown as consisting of various intercommunicating components. CPUs are extremely complex, with perhaps a million circuits, all in a single chip.

Computer Buses

The internal structure of computers involve several parallel communication paths, called **buses.** In Figure 14-3 the lines connecting various boxes represent such buses. At first sight, one might think that

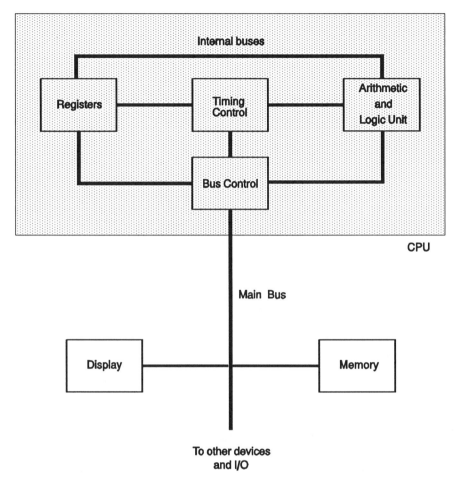

Figure 14-3 General block diagram of a computer.

a set of direct connections between each device and all the others would be superior. Consider 100 locations that need such intercommunications. If the connecting path is eight-lines wide, the total number of connections would be $8 \times 100 \times 99 = 79,200$ wires, certainly too cumbersome to be practical.

A typical bus system, shown in Figure 14-4, contains a large number of long parallel conductors in the following substructures: (1) A data bus carrying the sequence of 1's and 0's to be transmitted. At any given time, only one word can be accommodated. (2) An address bus containing a number representing the destination. It is assumed that to each possible destination has been assigned a unique

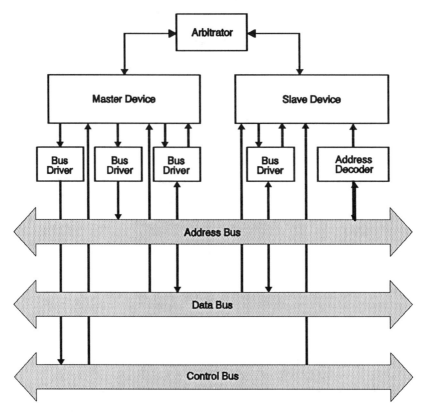

Figure 14-4 Bus control. A master and a slave device selected from a larger group are shown. Often there is only one master, the CPU, with several slaves.

number, an address. (3) A control bus, that serves as administrator of the process and also transmits instructions from a controller. Notice that the address of the originator is not present on the bus, only that of the receiver.

In the top part of the figure is shown the interplay between two devices, one that takes control (the **master**) and one that is controlled (the **slave**). The overall process normally consists of several phases:

1. A request from a would-be master is made, asking for control of the communication path (the bus). This is always the first step.

2. The request is received by an arbitrator, which examines the desired conditions and the presence of other requests, and then decides whether control is to be granted.

3. Upon receiving acceptance, the master takes control of the bus and loads on it the address of the slave, as well as specific instructions. Information is not yet sent.

4. The slave decodes the address on the bus and compares it with its assigned address. If they match, it responds by either receiving or sending a piece of information, depending on the instructions on the control bus. An acknowledging protocol ensures that the data are transmitted successfully.

5. The master remains in control and may repeat the operation, as needed.

6. Finally, the master informs the controller that it has completed its task. The bus then becomes free for other users.

In general provision is also made for **interrupts** that, as the name implies, interrupt an ongoing process to execute some urgent task.

A Simple Example of a Bus

We shall illustrate the electronic implementation of a bus, using a simplified scheme. It should be emphasized that actual bus constructions involve many thousands of circuits and are considerably more elaborate than our examples.

Consider a four-bit bus that interconnects eight devices. Let us address ourselves first to the problem of assigning the bus to one of the eight possible users as originator of the transmission (the **bus master**). An easy selection procedure would be simply to allot time to each device in sequence as determined by a clock. Figure 14-5 shows how such a clock line can be used to establish which is the bus master. A ring counter circulates a single logic 1 surrounded by 0's. The position of the lone 1 decides which device is given master status. The other devices are only allowed to respond to the master instructions and are all designated as **slaves**. It is assumed that all devices have received instructions as to where to send information and can recognize that their data is ready to be transmitted. They only have to await their respective turns.

The time-share procedure outlined above, is simple but impractical, because it is time-consuming. Even if only one device is active, it still must wait for seven idle periods before transmitting again. Also, a long message will be cut into many small pieces. In Figure 14-6 we give a portion of a more elaborate bus system that operates on the principle of **requests**. The box marked *device* generates the informa-

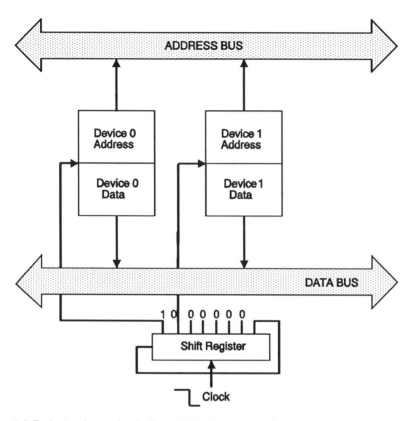

Figure 14-5 A simple method of assigning bus control.

tion, that emerges on data, address, and control lines. The device begins by requesting bus control with a HIGH on the *data ready* line. The bus traffic cop is the *bus busy* line. A HIGH on this line means that the bus is free. If the bus is indeed available, the AND gate permits the *data ready* signal to reach the flip-flop and set it. The output of the flip-flop is connected through an inverter to the *bus busy* line. As a result *bus busy* goes low, and inhibits any other device from setting its flip-flop. Only one flip-flop can be set at any given time.[3]

The output of the flip-flop has the dual function of maintaining the bus at zero and indicating that transmission is allowed. This latter

---

[3] This simple circuit is not foolproof; two bus requests at precisely the same instant could both be granted control. The problem would not appear when an arbitrator is present.

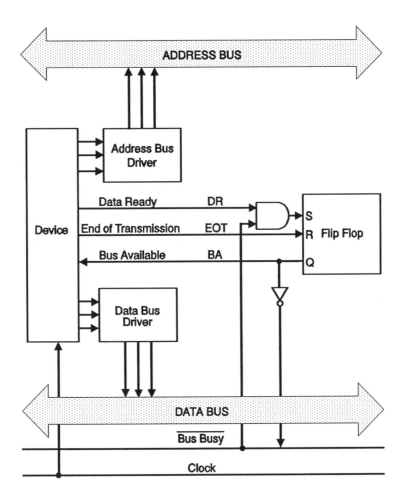

Figure 14-6 A request-based bus control system.

function is implemented by activating the *bus available* line, which enables the data and address drivers of the device, and permits data transmission to begin. *Bus available* acts as a feedback, to inform the master that the request was indeed granted; otherwise it might appear that a message had been sent, when in reality the bus was busy throughout. When the sender completes the message, the *end of transmission* line is used to reset the flip-flop, thus freeing the bus for other uses. The sequence of operation in our simplified system is thus: ask for control−receive control−send address−send data−release control. Evidently there must be a time delay between these phases.

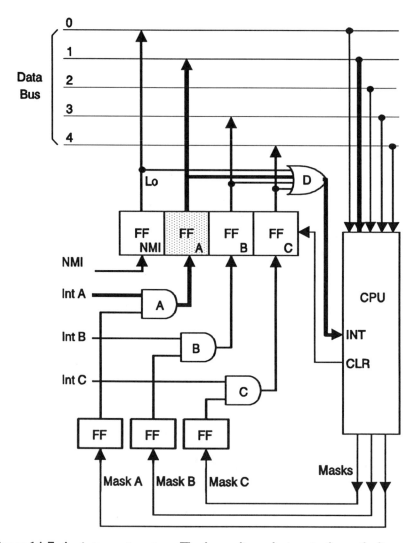

Figure 14-7 An interrupt system. The heavy lines designate the path discussed in the text.

This is taken care of by synchronizing the control request at state 0 of the clock. The data stream is then sent in synchronism with the HIGH state of the clock. The control request system in modern busses is very elaborate, with carefully selected levels of priority and facilities for interrupting an ongoing task.

The interrupt protocol is a provision for reassigning the bus control if a more urgent task appears. Some operations, such as incrementing the system clock, or various memory maintenance pro-

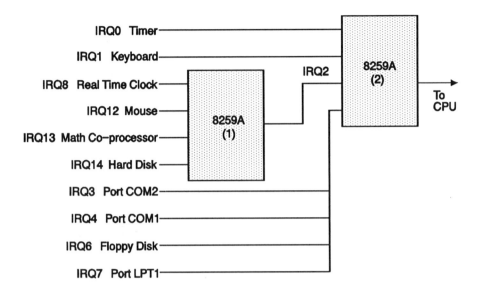

Figure 14-8 An interrupt system based on a cascade of two ICs, each capable of generating eight levels of priority. Some of the levels are not used in the IBM PC and are not shown.

cedures cannot be postponed without penalty. A good interrupt system takes into account the relative importance of various tasks, and assigns to each of the devices on the bus a **priority level** in the form of a number. A higher priority task can interrupt one at a lower level. In addition, some of the levels can be inhibited from interrupting, a procedure called **masking**. In contrast, some important requests such as RESET may be assigned non-maskable interrupt (NMI) status, which itself cannot be interrupted.

A simple implementation of such an interrupt system is shown in Figure 14-7. One of the levels (NMI in the figure) is nonmaskable. If only one interrupt request occurs, say, at Int A, it sets the corresponding flip-flop (FF-A). The HIGH output of FF-A passes through gate D and informs the CPU of the existence of an interrupt. The CPU then looks at the data bus which now has the line 1 HIGH, indicating that the interrupt was from line A. The ongoing process is halted and replaced by the interrupting process.

If several requests are received simultaneously, they will set several of the flip flops, but generate only one common INT signal from gate D. The CPU examines the values on the data bus and grants

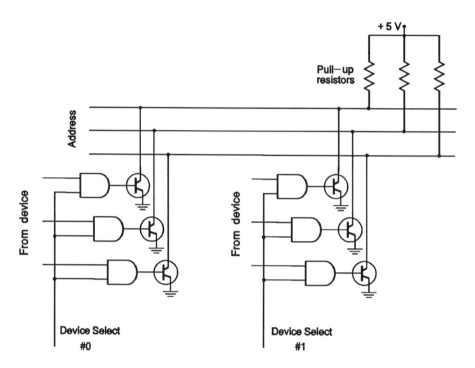

Figure 14-9 The electronics of bus drivers using pull-up resistors. We are showing lines from only 2 devices, with 3 address lines each, whereas in practice there
might be 10 devices and 20 address lines.

control to the highest priority demand. When the task is finished, the
bus is delivered to the next most important interrupt, and so on. The
four flip-flops A–D are cleared after the information is received by
the CPU. The system is then ready for another interrupt, even if the
previous ones were not yet serviced.

In practice a special IC such as the programmable interrupt
controller 8259A is used in place of the CPU to handle the interrupt
administration. Each device that might require bus control is provided
with a physical connection to it. A total of eight lines called interrupt
requests IRQ are provided. The lower numbered lines have priority.

It may be that there are more candidates for interrupts than the
eight available positions. The solution is to cascade two units as in the
example of Figure 14-8. IRQ2 is fanned-out into as many as eight
sublevels, by connecting a second 8259A chip; all of the new levels
have an intermediate priority between IRQ2 and IRQ3.

Various operations take place when the timer interrupt (IRQ0)
occurs. This repeats 18 times a second to advance the real time clock;

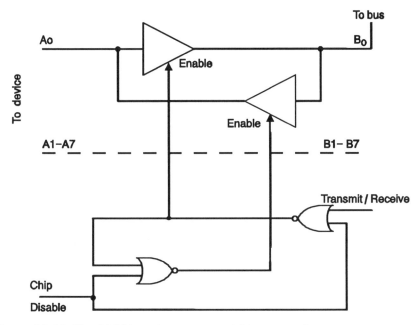

Figure 14-10 The 8304 bus driver, an octal three-state device.

since the main operation is interrupted anyway, it allows also various programs to service special needs. Notice the high priority given to the keyboard interrupts. Later in the book, we shall explain the meaning of various devices mentioned in the figure.

Bus Drivers

We shall now discuss the modes for loading information on the address bus. Figure 14-9 describes circuitry for specifying the desired destination (writing on the bus). It operates by forcing the corresponding bus lines to take logic 0 or 1. By themselves, the lines remain at logic 1, charged to +5 V by the pull-up resistors. The function of the transistors is to generate logic 0, where needed, by grounding the corresponding bus line; logic 1 is implemented by simply leaving the line as it is. Any of the groups of transistors can write on the bus, but only one device at a time is allowed to do so. The lucky master is selected by a HIGH on the corresponding *device select* line. All the other transistor inputs are blocked by a LOW at their AND gates.

A more widely used procedure employs three-state logic, discussed in Chapter 9. A major practical difference between the two methods is that the power consumption is much smaller for the three-state. In addition, in the absence of activation, the pull-up resistors automatically loads 1's into the line, while the three-state simply leaves the bus disconnected. Figure 14-10 shows an example of a three-state bus driver (model DP-8304 of National Semiconductor). This is actually a **transceiver**, as it can take care of both reading and writing on a set of eight lines. The *chip disable* line activates the whole device for either read or write. The *transmit/receive* line then decides if the transmission is to take place from left to right or in the opposite direction. The circuit formed by the two NOR gates ensures that only one direction is possible at any given time.

The Slave

At the receiving end of the bus is the slave, an example of which is shown in Figure 14-11. Two clock lines, ϕ_0 and ϕ_1 appear in the figure. The need for double timing arises from the fact that whenever a new address or data is sent, there is a short transient period when the information has not yet stabilized. To avoid data transfer during this unreliable period, the receiver can use a clock slightly delayed with respect to that used by the sending circuitry. The AND gate has now the function of inhibiting receipt of data until the bus has had time to stabilize. A signal is sent to the D flip-flops only if the correct address is present. In the example shown, the address is 5; the presence of the value 101 on the bus indicates that #5 is indeed the device spoken to. This selection function was described in an earlier chapter as *address decoding*.

To be useful, a bus must also include the facility for selecting between *read* and *write*. This implies two different modes of operation:

1. The master device selects the address and *sends* (writes) a message to the device with that particular address.

2. Alternatively, the originating device decides about an address, from which it will *receive* (read) data. A special R/W control wire that indicates *read* by a HIGH and *write* by a LOW can take care of this duality.

Real World Buses

There are many different buses in modern small computers. Among them we can mention:

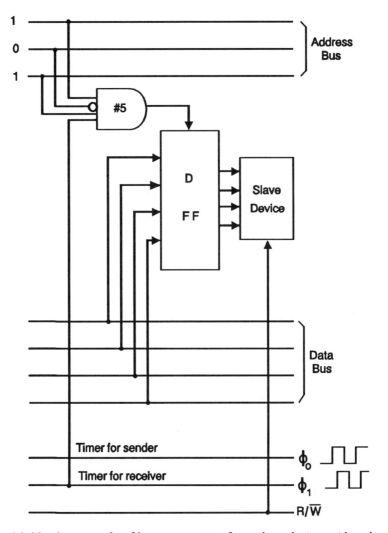

Figure 14-11 An example of bus connection for a slave device, with only a few bus lines shown.

- NuBus (also known as IEEE 1196) used by Macintosh and NeXt computers, among others.

- AT bus (also called ISA, for Industry Standard Architecture) used by a large numbers of PC compatibles. With only one master it is not very advanced, but is practical and widespread.

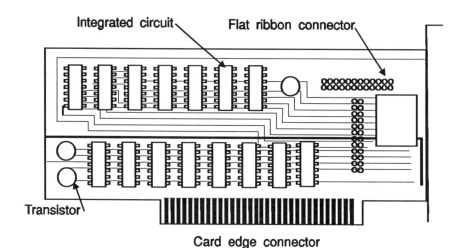

Figure 14-12 A typical plug-in card.

- EISA bus (Extended ISA), a 32-bit replacement for the aging AT. This multimaster bus will be described shortly.

- MCA (Micro Channel Architecture) a 16/32-bit bus used by IBM PS/2, IBM System/6000, and a few clones. It is also intended to replace the AT bus.

- S-100, an older bus (also called IEEE 696). Historically important, it had a good multimaster ability.

- VME bus (standard IEEE 996), a modern 32-bit bus used in certain super-microcomputers.

- Futurebus (IEEE 896), an advanced design not yet generally adopted.

- Apple II bus, a historically important design used in one of the first personal computers.

We shall describe the 32-bit EISA bus (ee-sa) in more detail. This is an extension of the widespread ISA bus, that has been adopted by a large number of clone makers.

As stated earlier, a bus consists physically of a large number of parallel connectors, terminated at the CPU on one end, and with a set of sockets at the other end. To the outside world, the bus is therefore observable as this set of sockets to accomodate various circuits built on cards (Figure 14-12) . The development of more efficient busses generally involves more elaborate sockets, and the EISA bus is no exception. The original PC cards required a 60-connector socket; the ISA bus added a 36-connector extension socket. Finally the EISA standard in order to allow compatibility includes both the PC and AT contacts. It uses a two-level socket; the lower part of the socket follows the old 60 + 36 = 96 contact AT bus setup, while the upper part connects with a third set of contacts.

The CPU

Among the devices that control the bus, the CPU or microprocessor is the master *par excellence*. It controls the operation of the whole computer, it processes data, and is the direct executor of the user's commands. The bus itself continues inside the CPU chip, but usually in a changed form. We can thus distinguish between the outside bus, sometimes called the **interface bus**, and the internal bus. This latter connects various components, among which are the **arithmetic and logic unit (ALU)** which has the function of data processing, and the **control unit (CU)**, in charge of the logistics. We shall concentrate more on the data processing functions of the ALU.

To discuss the operation of the CPU in more detail, it is essential to understand the basic principles of memory organization. The fundamental unit of storage is the word consisting of two or four bytes. Words are stored in a continuum of memory locations which are individually addressable. This is done by means of unique addresses, normally described by their hexadecimal values. Thus the memory space might extand from 00000 to FFFFF (i.e., one megabyte). Sections of this 1-Mbyte space are allocated to various needs, such as instructions and data.

In its simplest form, the operations of the CPU follow a scheme such as that in Figure 14-13. A register called the **instruction pointer** (IP) contains the address of the next instruction to be executed, 144FF in our example. The control unit uses the program counter address to fetch the instruction via the bus. The instruction thus retrieved is a collection of 1's and 0's and is sent to the **command decoder** (CD) for interpretation. The CD receives the instruction and transforms it

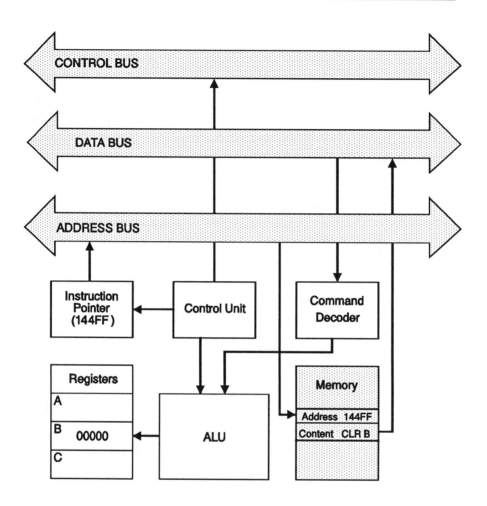

Figure 14-13 The general operation of the CPU.

into a series of individual electrical signals that will implement the required operation. The **arithmetic-logic unit** (ALU) executes this set of subcommands, as instructed by the decoder. A series of **registers** (very fast one-word memories) keeps track of partial results and operational conditions. In the figure the instruction "Clear Register B" or CLR B is fetched from the memory position 144FF, decoded in CD, and sent to the ALU. The ALU then clears the register (by loading into it a string of zeros).

After each instruction is executed, the IP provides the address of the next command. The cycle repeats and the operation continues until the end of the program. The ALU is capable of executing a variety

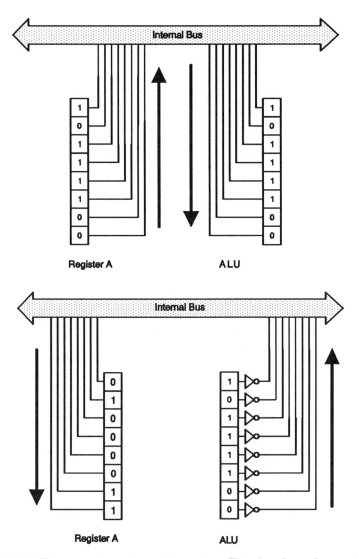

Figure 14-14 The operation of complementing. The data from the register A is transferred to the ALU, then transferred back in its complemented form. The vertical arrows indicate the direction of information flow, which is assumed to enter from the left and exit at the right. Only the active connections to the bus are shown in each case.

of perhaps several hundred commands. This is its most distinguishing feature. It receives data always at the same inputs and sends results always at the same output. There are no moving parts inside, yet it can process the data in hundreds of different ways. For example, if

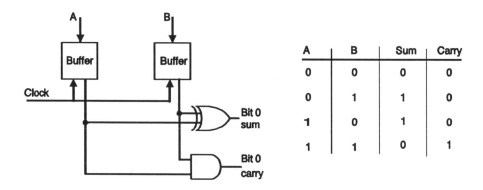

A	B	Sum	Carry
0	0	0	0
0	1	1	0
1	0	1	0
1	1	0	1

Figure 14-15 An example of half-adder.

the instruction is "ROR A—rotate to the right the contents of register A", the CPU behaves as if it were simply a ring counter. For all its 100,000 transistors, it operates just as if it were a 50-transistor circuit. As another example, the ALU might be asked to execute the command "Complement A, (NOT A)". In such a case all other functions are inhibited except for the action of complementing. The CPU becomes the modest circuit of Figure 14-14. Each command transforms the ALU into what amounts to a different object.

As another example, let us discuss the operation of the ALU when executing the ADD instruction. The two words to be added are brought in two ALU inputs, and addition is done bit-by-bit from right to left. If the carry were not needed, the operation would be simple, as it would involve just one XOR operation per bit. Remember that XOR gives 1 for the combinations 0, 1 and 1, 0 and zero for the combinations 0,0 and 1,1, exactly as needed by the addition.

The problem lies with the 1, 1 combination. A single XOR gives the correct result of 0, but without carrying into the next higher bit. Additional circuitry is needed to generate the **carry**, and a basic unit is shown in Figure 14-15. The two buffers retain the input until the process is completed. This circuit still cannot do the complete addition, since it can only provide carry-out. In a complete addition, all bits require both carry-in and carry-out, hence the need for a **full adder**. An example is given in Figure 14-16. Two stages are shown, for bit 0 (LSB) and bit 1. A total of 16 such stages are needed to add 16-bit numbers. The carry C_{aux} is provided to the first stage for implementing 32-bit addition, in which case a carry will be needed between the first and the second group of 16 bits.

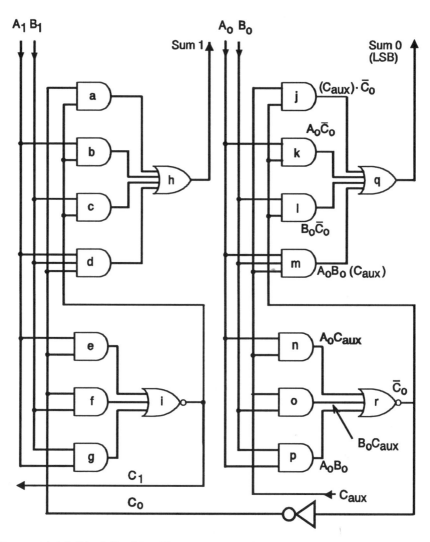

Figure 14-16 The full-adder. Two stages are shown, bit 0 on the right and bit 1 on the left. The symbols C_0 and C_1 refer to the carry from bit 0 and 1 respectively.

The signal C_0 is HIGH in the following situations: (1) the AND combination of A_0 and B_0 is 1; (2) the auxiliary carry, C_{aux}, is set while A_0 or B_0 or both, also have value of 1, taken care of by gates n and o. Gates j, k, l, and m generate the sum itself. The reader is urged to write the truth table for the Sum -0 and verify that an arithmetic sum is indeed generated. The set on the left takes care of the bit 1 and operates similarly, except that the carry is now internally connected. This situation remains the same all the way to bit 15, where the carry

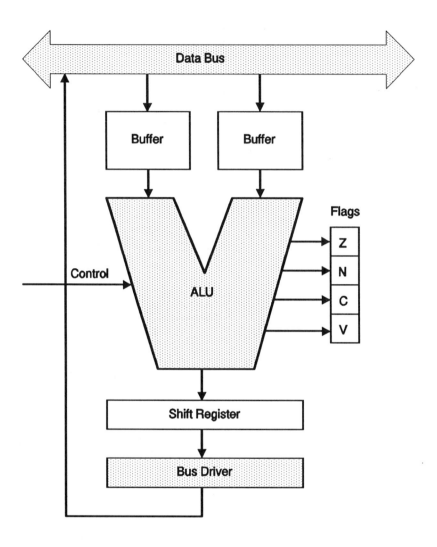

Figure 14-17 The ALU and its immediate accesories

becomes available as an output. Note that when signed additions are performed, bit 15 (MSB) needs to be treated differently, because it contains sign information and is not a digit.

In the previous examples, the reader was presented with a series of equivalent circuits for the CPU. Each instruction changes the equivalent circuit in a unique way. The situation, however, is a bit more

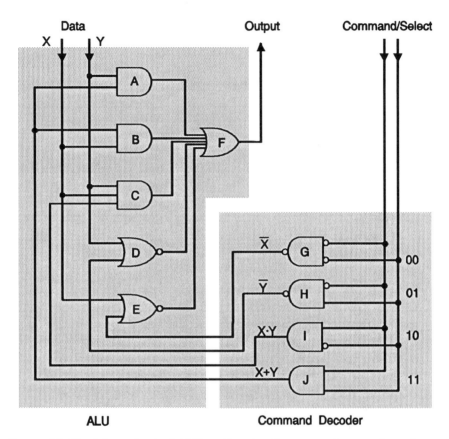

Figure 14-18   A hypothetical ALU performing four operations under software command.

complicated than is suggested by the examples. Even in the simplest CPU, several additional functions must be active at all times (Figure 14-17). The two buffers shown act as sample-and-hold (S/H) units needed because the result of each operation between two numbers is immediately written back on the bus, thus changing its contents. Additionally, the shift register is shown as an obligatory pass point. This part of the ALU is also always active, allowing multiplication and division by powers of two.

The flags are also continuously operational. They are driven by special circuits which signal with a high if the result of the previous operation was zero (Z), negative (N), or had a carry out (C). The overflow (V) refers to the following special condition. If both positive and negative (signed) numbers are represented, the first bit (MSB) is used for the sign. A carry into this bit would change the sign. Thus

Table 14-1

Typical ALU commands

Function	Mnemonic
Add	ADD
Add using carry	ADDC, ADDX, ADC
Subtract	SUB, SBB
Increment by one	INC
Decrease by one	DEC
Multiply, with sign	MUL, IMUL, MULS
Divide	DIV, IDIV, DIVS
Logical AND (bit by bit)	AND
Logical OR (bit by bit)	OR
Invert (logical complement)	NEG
Shift one bit to the right	SHR
Ring counter shift; rotate right	ROR
Rotate through carry left	RCL, ROLC
Transfer data; move	MOV
Clear (replace by zeros)	CLR
Jump (branch) if negative	JN, JNS
Jump if result was zero	JZ, BZ, BE, JE
No operation (waste one cycle)	NOP
Halt (operation stops)	HLT

two large positive numbers could give a negative sum, evidently an error. The V bit serves to identify this anomaly. Various actions are taken depending on the values of the flags. The number of flags in modern micros may be 16 to 24.

A listing of a small fraction of the typical commands that an ALU can perform are given in Table 14-1. The mnemonics vary with the make of the CPU, but several hundred are usually available.

It is somewhat difficult to conceive how a fixed circuit can take so many faces. The secret is in selectively activating specific parts by using **microinstructions**. We shall illustrate with a very simple implementation of an ALU that can operate, not in 100 ways, but only three: AND, OR, and INVERT. The basic circuit is shown in Figure 14-18. Only one bit is shown, eight such units are needed for each byte.

The decoder is constructed in such a way that only one of the *command select* lines can be activated at any moment. When *complement X* is decoded (from the command 00), gate G goes low. This, in turn, causes gate E to be activated, in the sense that its output becomes simply the complement of *X*. All the other gates, A to D, have 0 outputs. Consequently gate F will simply reproduce the output of gate E. The circuit as a whole ignores *Y* and gives as output the complement of *X*. Similarly when the instruction is 01, *Y* is complemented, by activating gate D and maintaining the other gates at 0 output. The output of the ALU as a whole will contain the complement of *Y*.

Gate C is activated by the instruction 10 at the input of gate I. It generates logical *XY*. Again gate F simply reproduces this value. Logical *X+Y* is implemented when the command is 11. Both gates A and B are activated; they reproduce the inputs *X* and *Y*, respectively, at their output. Gate F ORs the two values and generates *X+Y*.

This example shows that a single pair of inputs can generate a variety of outputs depending only on a "software" command. The quotes refer to the fact that, at this level, the software command is not a procedure but a set of actual voltages present in physical transistors.

A Modern Example of a CPU

The first integrated circuit CPUs were reasonably complex in design, with 2300 transistors in the pioneering 4004 of 1971. Today this number looks small indeed, with several companies producing million-transistor CPUs. The evolution has been dramatic. The graph of Figure 14-19 shows the progress of the "80" family of CPUs derived from the original 4004. The logarithm of the number of transistors comprising the chip is plotted against time. The straight line indicates a purely exponential growth: the number increases by an order of magnitude every six years or so. If the trend continues, by the year 2000 the CPUs will reach 100 million transistors. (Of course the curve has to taper off eventually, or else the CPUs will end up having more transistors than there are atoms in the universe.)

In this section we give a short description of one of the members of this family, the 275,000-transistor 80386. It is commonly referred to merely as the 386, and is built with CMOS technology. The general architecture is given in Figure 14-20.

A number of internal 32-bit buses are shown. This is typical of the fact that, as the complexity of the processors increases, the internal connections tend to become more significant than the transistors, both

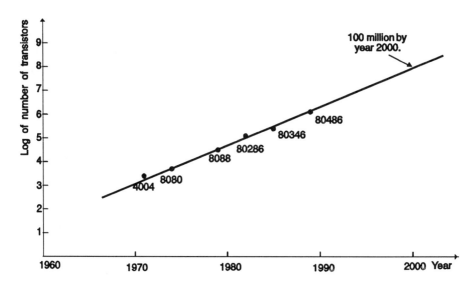

Figure 14-19 The growth in complexity experienced by the CPUs. Similar graphs could be made for memory capacity progress.

in terms of space and of cost. The *bus driver* and *control* section takes care of external communications. It also administers the bus control requests. It receives as input the physical address, data and control information.

The program code is brought in from the memory and processed simultaneously in several sections of a code stream. This type of parallel operation is called **pipelining**. The instructions are first loaded and kept in the 16-byte queue. This saves time, since idle moments can be used to load future instructions. From the queue, the code flows to the decoder. The decoder also proceeds in advance of the need and generates a second queue of up to three decoded commands. Next, the microcode section generates the set of commands needed to drive the ALU proper. The ALU executes the microinstructions and sends back information about the results in the form of one-bit flags, such as for parity, zero result, and overflow. The ALU has control of two dedicated buses, one that leads directly to the bus driver carrying the calculated data, and another one that communicates with the *segment unit*, in order to generate the actual, physical, memory address.

The memory addressing in the 386 is rather complex. The largest memory unit is the **task**. Each task uses an independent environment, with individual attributes attached to it. A task can contain up to 16,381

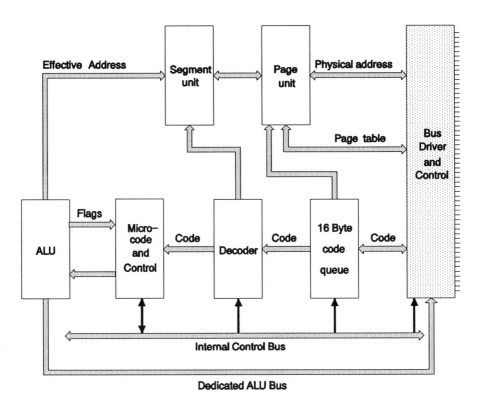

Figure 14-20 Simplified architecture of the INTEL 80386.

segments. The segments can be of variable size, but no larger that 4 Gbytes. In total there can be 16 k × 4 G = 32 Tbytes, that is 32 terabytes or trillions of bytes. The physical memory is invariably much smaller, perhaps 4 Mbytes, but it can be expanded by such external media as hard disks to create a **virtual memory** space. The virtual memory can be much larger than the physical computer memory, when an interchange of blocks of information with the disks is allowed. The CPU can address data as if it were in the memory proper, if the system can swap this information between the disk and unused blocks of internal memory.

The ALU generates a 32-bit address, usually called the **effective address**. The segmenting and paging circuits take care of fitting this together with the much larger virtual address, and both of them with the actual chips of memory present in the computer. In addition, the 386 is multitasking, so that several programs (tasks) can coexist. They are called up successively to use the same physical memory for different operations. Again the addressing must be adjusted before the memory can be reused for consecutive tasks. The paging unit takes care of another organization of the memory space, this time into a large set of interchangeable (swappable) 4 kbyte **pages**.

To give an idea of the intricacies of the 386, let us examine in more detail the register set, shown in Figure 14-21. Of the set of eight 32-bit general registers, the first four can be used separately as 16- and 8-bit registers if needed. This ensures compatibility with earlier versions in this series of CPUs, which used smaller-sized registers. The extended stack pointer (ESP) serves to define the stack, a portion of the memory that is organized in sequential manner. Any part of the memory space can be assigned to such sequential operation. The ESP register provides the current address within the stack. As this pointer moves, it allows access to other positions in the stack, in sequential rather than random order. The order in which the information is deposited (**pushed** onto the stack) is the reverse of the readout order (**popping** from the stack). A last-in-first-out rule prevails.

The stack is important for managing interrupts. The content of all the registers (environment) at the moment of the interrupt is preserved in the stack. When the interrupt process terminates, the CPU is returned to the exact previous condition by simply reading the status information in reverse order and restoring it. Several layers of interrupt can be accommodated, since one interrupt can, in turn, be interrupted by another. Do not forget that the stack does not represent a different type of hardware but only a manner of organizing the memory.

The **instruction pointer** (IP) contains the address of the next instruction. As each new command is fetched, the IP is incremented to contain the updated address of the next instruction. The flags register is quite similar to our example in Figure 14-17. The **segment registers** are elements that are involved in the physical address calculation. They each identify one of a large set of possible portions of the total memory space. Each segment contains a certain type of information. Separate registers are assigned to define the specific segment used for the commands (CS), and stack (SS), and also four different areas of data storage (DS−GS). The individual segments can be of various sizes. It is not sufficient to specify the location within the segment under consideration without providing information about the segment as a whole. This is taken care of by a set of large registers that are not accessible to the programmer. This lack of accessibility

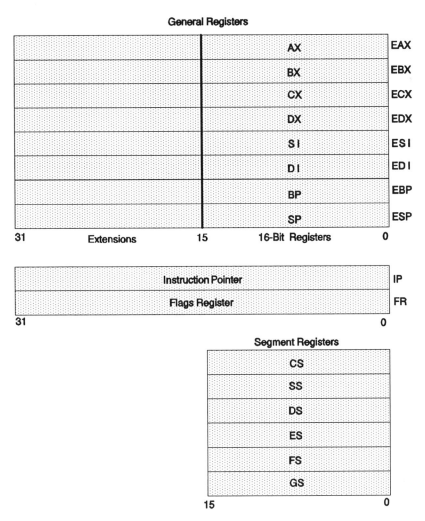

Figure 14-21 The register set for the 80386 CPU. The general registers can be used as 16-bit units (e.g. AX), or 32-bit units (EAX).

might seem to be a great disadvantage, but in reality the majority of CPU internal operations are invisible to the user, just as the majority of inner electronics cannot be reached. This class of activity is called **transparent**. In modern designs the computer operation is becoming increasingly invisible. The transparent registers mentioned above contain the beginning and end addresses assigned to a given segment; this is decided by the computer without any programmer intervention.

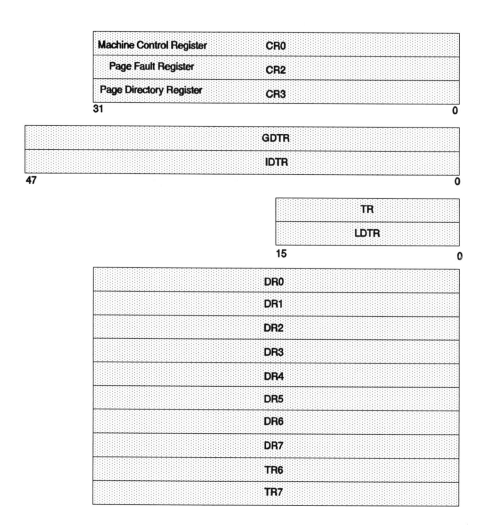

Figure 14-22 Internally operated registers in the 80386.

More administrative information is contained in the control registers CR0–CR3 (Figure 14-22). Note that CR1 is missing. It is common practice to reserve unused portions of the CPU for future development.

The other registers store the address where descriptive information about the status of the task at a particular time is to be found (GDTR, IDTR, TR, LDTR). Finally, a whole set of debugging and

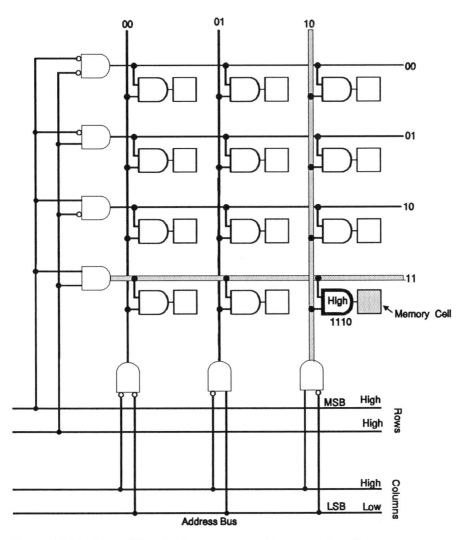

Figure 14-23 A small bank of memory positions used to illustrate memory address. The bus carries the value 11 for the rows and 10 for the columns.

testing registers is thoughtfully provided. in DR0−DR7, TR6, and TR7. These registers offer the programmer a degree of control and insight into the deeper layers of internal operation.

Memory

The previous discussion did not clarify one point: How does the memory know that it has been addressed? How does one implement,

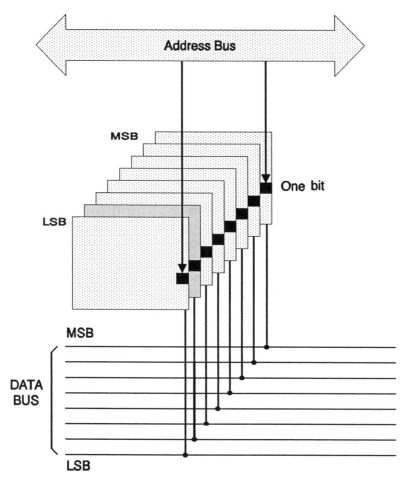

Figure 14-24 The storage of a single word, with each bit in a different bank. Each bank consists of a matrix like that of Figure 14-23. All banks are activated by the same address.

for instance, the query: "Memory 365,655, what's the value that you are storing?" If the computer has 16 million memory positions (16 Mbytes) it is not easy to interrogate every single position and expect an answer only from number 365,655. For this reason, in practice, the address space is divided into a number of sections and only one at a time is active. Within a section all memory locations are addressed at the same time, but only one answers the call.

We shall illustrate by an example the manner in which this could be done. Figure 14-23 shows part of a small bank of memory cells

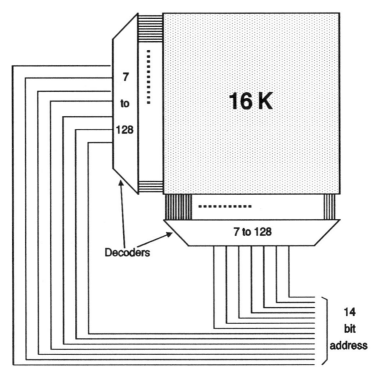

Figure 14-25   A 16 kbit memory chip and its addressing, using a 14-bit address. Most chips are larger (1 or 4 Mbits).

containing gates and flip-flops, organized in a 4 × 4 matrix for a total of 16 cells of which only 12 are shown. The four rows are coded 00 to 11, the same is true of the columns. If the row code is combined with the column code, a series of four-bit numbers result 0000, 0001, 0010, etc., where the first two digits in each group refer to the column and the last two to the row. Inspection of the figure shows that every value of the four-digit number corresponds to one and only one cell. We are thus able to refer to each cell by simply feeding to the address bus the correct combination of row and column selectors. A set of only four wires permit us to address 16 components. This is an example of **address decoding**. Address decoding becomes especially economical for larger matrices. One million cells can be addressed by only 10 rows and 10 columns. (Note that a total of 20 bits are used, giving 2^{20} distinct addresses.)

A given cell knows that it is addressed by the presence of a HIGH at the output of its associated AND gate. In the example of the figure, only row 11 receives a HIGH. Similarly among columns, only 10 receives a HIGH. The only gate that will receive HIGHs in both inputs is therefore gate 1110. This decoded signal activates the 1110 flip flop, which then reads from or writes to the data bus. Of course the scheme discussed takes care of only one bit. A set of eight such matrices of memory cells is needed to operate on a byte. This is illustrated in Figure 14-24 for the case where a number of banks or **pages** of memory are activated simultaneously. Once an address is given, it is received by all the pages at the same time. In each page, all memory cells receive the command, but only one memory cell is activated. The set of eight enabled cells, one from each page, combine to form a byte.

In a more realistic example there might be two sets of seven control lines decoded into 128 rows and 128 columns, as seen in Figure 14-25. Only 14 wires are sufficient to take care of $128 \times 128 = 16,384$ memory positions, so they can be individually addressed. Eight such units are required to form a complete byte.

Also important is the ability to select one or another of the memory banks (**bank select**). The capacity of a single chip is generally smaller than the total memory needed. Hence many chips must be used to complete the memory. The memory banks must be provided with additional circuitry to allow a selection between the read and write states, as well as synchronization with the clock .

The read/write selection is needed in all cases, except for **read only memories,** (ROM), that are loaded only once for permanent storage. The ROM acronym is used in contradistinction to **random access memory** (RAM). This latter term, meaning the conventional read/write type, is a misnomer; both ROM and RAM use random access. It is best to ignore the meaning of the initials and think of RAM just as conventional read/write memory.

Types of Memory

For many years, memory was implemented by **cores**, which are hardly ever used nowadays. Introduced in the early 1950s, core memory consisted of small magnetic rings, that could receive two states of magnetization. A collection of wires threaded through the rings served for both reading and writing. The core had the disadvantage of requiring expensive mechanical assembly, and it has been abandoned in favor of semiconductor memories. Unlike cores, modern RAM memories are volatile. This means that once the power is turned off, the information is lost. There are many varieties of RAM, two major classes being the **dynamic RAM** (DRAM) and the **static RAM** (SRAM).

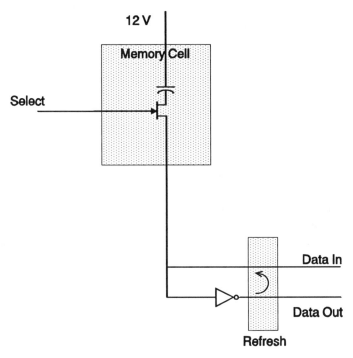

Figure 14-26   A simplified DRAM element.

Each cell of a typical DRAM has as active elements one transistor and one capacitor in an arrangement like that of Figure 14-26. The capacitor is physically very small, and its capacitance is minimal, perhaps 0.05 pF, so that it cannot maintain its charge more than a few milliseconds. This situation might look hopeless, but actually a millisecond is a long time by computer standards. To compensate for this discharge, the DRAM memory is provided with a **refresh** circuit that goes through a recharge cycle for all the cells that store a 1. The refresh circuit makes use of the buffer shown in Figure 14-26 to sense the residual charge and send the output back through *data in* to recharge the capacitor. The process is extremely reliabil, so that even after billions of refresh cycles errors are rare. The buffer is also needed to amplify the energy in the capacitor; the charge is so tiny that without buffering it cannot propagate the signal into any significant length of conductor. SRAMs, contain sets of flip-flops, one for every bit. They are faster, more expensive, and occupy more physical space than DRAMs. They are used only in key locations where their speed is important. They do not require refreshing.

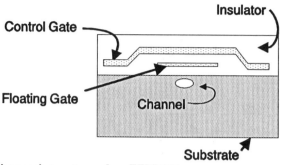

Figure 14-27 Internal structure of an EEPROM.

In addition to the two types mentioned, there are other types akin to shift registers. They include **charge coupled devices** (CCD), and **bubble memories**. These are serial devices, and consequently are slower, but compensate for this disatvantage by their a more compact size and fewer control requirements.

PROMs

Numerous types of **programmable read-only memory** exist. The original type, the PROM is programmed by fusing tiny connectors inside the chip. More recent types (late 1970s) allow bulk erasing with ultraviolet light, they are called erasable PROM (EPROM). Electrically erasable PROMs are also available (EEPROM). One such bit is shown in Figure 14-27. It consists of a FET whose gate is completely isolated. A momentary high voltage (30 V) on the control gate can load the voltage of the floating gate permanently (for years). This establishes a 1 in that position The floating gate can later be altered by a second high voltage signal applied to the control gate. Many other varieties of memory are continually being developed. Other forms of memory such as magnetic disks and floppies will be discussed in a later chapter.

Error Detection

The number of processes occurring in a computer in each second is staggering, yet by its very nature the computer demands rigorously error-free operation. For this reason a great deal of effort has been spent to improve the reliability of computer operation with a variety of error detection and error correction procedures. Note that all errors in binary systems reduce to 0 and 1 interchanges.

A widespread method of error detection is the **parity check**. This makes use of a ninth bit added to each byte. This bit is programmed

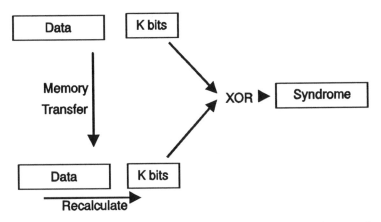

Figure 14-28 Use of the Hamming code to effect an error correction scheme.

with special circuitry to ensure that the total number of 1's (the **parity**) in the nine-bit group, is odd. If the byte itself had an odd number of 1's (an odd **parity**), the ninth bit becomes 0. If the byte has an even number of 1, then the ninth bit is set to 1, thus causing the total number to be odd. Similar procedures can maintain, alternatively, an invariable **even parity**. In either case the parity is checked whenever a memory operation is performed. If the parity is found to be changed, corrective measures are taken. If a satisfactory correction is not feasible, you receive a dreaded *parity check error* message, and the system halts. Most computer designers elect a complete stop when this occurs, as opposed to allowing the operation to proceed with an error present. Whether this is right or wrong at the user level is debatable, but it usually means that the computer must be restarted (rebooted).

The parity check is a good detection method when there is only one error present, which is normally the case. The error is detected but not automatically corrected. For example, if the byte sent is 1111 0101, and the parity convention is odd, then (since the number of 1's is even) the parity bit will be 1 ensuring an overall odd parity. If now a simultaneous error of two bits occurs, for instance generating the erroneous 0011 0101, the number of bits remains even and consequently the required parity bit remains 1. The error is therefore not recognized, because there is no discrepancy between the parity bit and the parity of the byte. More elaborate methods are needed to take care of multiple-bit errors. Let us concentrate first on single-bit errors.

The number of error events is conveniently expressed by the **device failure rate** (DFR). The term failure refers to both **hard** errors, where the damage is permanent, and **soft** errors that are transient, nonrepetitive. Both cases involve failure to behave as expected. For modern equipment DFR runs at the level of 0.02% in an interval of 1000 hours for a single device, an impressively small number. From it we can calculate the mean time between failures

MBTF = 100 × 1000/DFR hours

This corresponds in our example to 5 million hours, or close to 600 years. Unfortunately, this optimistic value refers only to single devices. If, for example, there are 1000 devices, the MTBF is lowered in direct proportion, to about one error in 5000 hours. This larger frequency could mean that error detection is not sufficient, a situation requiring the use of error correction procedures.

A simple example of an error correction scheme applicable to data transfer is illustrated in Figure 14-28. Each eight-bit byte is extended with four extra **redundant** bits (also called K bits) that are generated by a special circuit following a procedure known as the Hamming code. Immediately following the transfer the byte is submitted to a second code generation. The two sets of four bits, one generated at the source, the other at the destination, are xored to give what is called the syndrome. If 0 is obtained, there was no single-bit error. Any other value for the syndrome indicates an error. The error might have been in the byte itself, or in the redundant K bits. A total of twelve possibilities exist. The syndrome can take any values between 0000 and 1111. It indicates which bit was in error; in our example the value of 0011 indicates that the third bit is in error, and a correction is automatically made. (Recall that a bit can only be 1 or 0, so correction means simply complementing the bit.)

Chapter 15

Computer Peripherals

In this chapter we discuss, largely from the hardware point of view, the electronic aspects of such peripheral equipment as keyboards and screens. As in preceding chapters, most of the examples will be taken from the PC compatible world.

The Keyboard

The most immediate contact between user and computer is through the keyboard. What really happens when we press a key say, a Q, on the keyboard? It turns out that the actual process is quite complicated and involves many individual steps. In a simplified view, the action is to connect two conductors in some unique way and send the generated code to the CPU.

Consider the original IBM PC keyboard, which has 83 positions. As seen in Figure 15-1, pressing a key causes a specific connection of one row wire with one column wire. As seen from the figure, a total of $4 \times 23 = 92$ distinct possibilities exist (out of which only 83 are used).

The identification of a contact between two wires is made by a circuit that monitors continuously all the 83 possible switches, to see if any contact has occurred since the last check. This is illustrated in Figure 15-2. The pressing, for example, of the Q key causes lines 21 and B to become connected. The controller identifies this event by the expedient of grounding lines 1 to 23 one by one in a fast repetitive

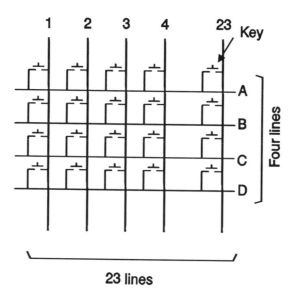

Figure 15-1 Simplified keyboard. The pressing of a key results in a contact between one horizontal and one vertical line.

sequence, while monitoring lines A–D. When the Q key is pressed, there is no immediate effect until the sequence of grounding reaches line 21. At this point, the grounding propagates into line B and is immediately identified as a closure at position 21–B (the position associated with Q). The controller translates this information into a special **scan code** which is unique for each key. The scan code is then sent serially to the CPU.

When the key is released, a second code is sent to indicate end-of-actuation (it consists numerically of just the first code plus 128). The reason that two codes must be used is that the system must know the *duration* of each keypress to be able to identify multiple key actuations. For example, pressing simultaneously the *Shift* and the Q keys transmits the sequence:

{press *Shift*} {press Q} {release Q} {release *Shift*}

which translates as capital Q, as opposed to the sequence:

{press Q} {release Q}

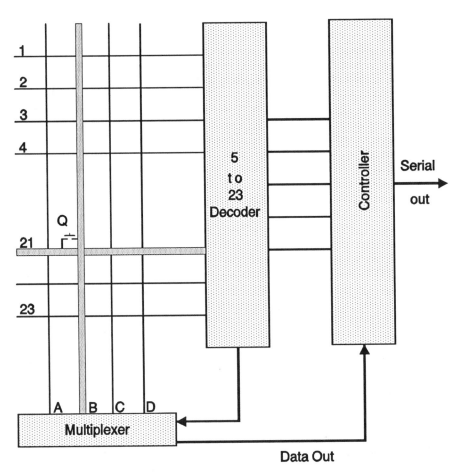

Figure 15-2 The result of pressing letter Q.

that is interpreted as lower case q. In many such keyboards, diodes are wired in series with each switch, to prevent the propagation of the closure signal if several key are pressed simultaneously.

When the scan code is received, a special type of routine, called **interrupt** (INT)[1] takes place. It receives the transmission and decides about the action to be taken. Thus a reset command receives immediate attention, but other keypresses receive a lower priority or are simply be ignored.

---

[1] Do not confuse these interrupt routines, which are *programs*, with the hardware interrupts of Chapter 14.

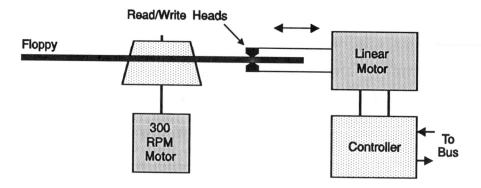

Figure 15-3 Schematic representation of a floppy disk drive.

In general it is better to have a single controller (e.g. Intel 8279) for both the keyboard and the monitor, since most of the time the information entered on the keyboard is reproduced on the display. Modern keyboards often have 100 keys or more.

Diskettes

As a rule, a large portion of the microcomputer memory is **volatile,** meaning that the information disappears when the power is turned off. This can be a blessing if there is trouble with a jammed computer: all one needs to do is to restart the computer (reboot); there will be no reminiscence of the past trouble. On the negative side, volatile memories require the operating system and programs to be reloaded each time the computer is powered up. (Memory located in ROMs is not affected by loss of power.)

It is essential to have some form of separate, permanent storage device in which programs and data can be safely kept. Nonvolatile storage devices mostly employ magnetic and optical media. Magnetic storage can make use of tape, but the more common storage devices are removable disks, called **diskettes** or **floppies,** and **hard disks.**

Floppies were originally introduced in the early 1970s with a diameter of 8 in., Today 8 in. floppies are still occasionally used, but have mostly been replaced by the 5.25- and 3.5-in. sizes. Both types consist of a plastic disk coated with a very smooth magnetic layer, and sandwiched between two cleaning tissues. The ensemble is enclosed in a protective jacket. The 3.5-in. variety has a more rigid and better sealed jacket. It seems destined to replace the larger type.

Table 15-1

Representative floppy disks

Type	Tracks/side	Sectors/track	Capacity
PC 5.25	40	9	360 K (DD)
PC 5.25	80	15	1.2 M (HD)
PC 3.5	80	9	720 K (DD)
PC 3.5	80	18	1.44 M (HD)
Mac 3.5	80	8–12	800 K (DD)

Floppies employ a drive that rotates the magnetic disk at about 5 rotations per second (300 rpm) while lightly pinched between two read/write heads (Figure 15-3). The heads operate individually and record a series of magnetic dots, on concentric annular paths called **tracks** or **cylinders**. The heads are able to read the dots back, meaning that they can reconstitute the original train of pulses and thus the sequence of 1 and 0 that produced the dots in the first place. The system uses a **controller** to locate the head over the correct track and read or write magnetic information.

The basic unit of information storage is the sector. A fixed number of sectors, consisting usually of 512 bytes each are assigned to each track. The organization of the more common types of floppies is given in Table 15-1. Not shown are the less common high capacity diskettes such as the 2.8-Mbyte type. The total capacity is the product of the number of tracks on the two sides, multiplied by the number of sectors in each track and by the number of bytes per sector. In the case of Macintosh computers, the tracks have a variable number of sectors: the first 16 tracks are the longest and are divided into 12 sectors; the next 16 contain 11 sectors, and so on. The innermost tracks, being shortest, are assigned only 8 sectors each. This somewhat complicates the bookkeeping but allows a higher capacity (800 K as opposed to 720 K for PC diskettes), without compromising data integrity.

In the table the DD (double density) and HD (high density, also called quad density) differ not only in the sector organization but also in the width and magnetization of magnetic dots produced on the track. Thus the PC 360-K disks (DD type) have a track width of 0.33

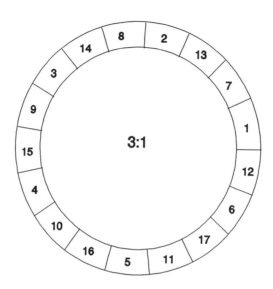

Figure 15-4 Example of 3:1 interleaving. One track is shown, indicating the sequence in which the sectors are read. It takes three complete turns to read the whole track.

mm, while the HD 1.2-M floppies, only use 0.16 mm. The disks are physically interchangeable, yet many HD heads on PC computers cannot erase properly the entire 0.33 mm path on the DD disks. Although the HD drive may initially read the lesser density disks perfectly well, the accumulation of new information, together with the half-erased tracks, could end up generating errors. Newer 1.2-M drives seem to have the problem solved, but the caveat remains never to use a 360-K disk at the 1.2-M density. This is also true of the DD and HD 3.5-in. disks. They have the same track width but different magnetic coating formulation. Both types can be used safely in the same drive, but neither should be written in the wrong density, even if the floppy itself tolerates it. Higher density diskettes, of up to about 40 Mbytes are also available, but they have not become too popular as yet.

The sectors on a track are not written on consecutively, in order to allow some housekeeping time between sectors. This is called **sector interleaving**, and it is totally invisible to the user (in computer parlance it is **user-transparent**). Interleaving can be 1:1, meaning no interleaving at all, or 2:1, 3:1, etc., meaning that every other or every third sector is read in physical sequence. An example of 3:1 interleaved track is shown in Figure 15-4. The disk controller ensures that the data are gathered from the correct sectors into a smooth flow of information.

A new disk cannot be used directly as fabricated; it must be organized in accordance to a proper **format**. The format is largely a problem of software, and a given drive and controller may accommodate a variety of formats. The process of formatting consists of making the disk follow a given writing scheme by establishing the tracks and sectors. Each sector receives at its very beginning an internal code (seven-byte number) followed by a delimiter (gap) consisting of the number in hex FF00, needed for synchronization purposes. Data (up to 512 bytes) can be written from this point on. The end of the sector is marked again by another gap to separate it from the following sector.

Additional housekeeping information is written on the outermost (first) track. At the start is the **boot**[2] **record**, which contains general information about the operating system used and the type of formatting. Next is the **file allocation table** (FAT), which contains the position of data sectors for each file. When the disk is new, the files are simply stored in sequence. However, after erasing and rewriting data a few times, any new file will have to make use of whatever sectors are available and will not be able to ensure contiguous positions any more. This is where the FAT helps. The files are written in pairs of consecutive sectors called **clusters**. Each cluster is numbered, and the FAT simply contains the successive addresses of clusters on which the file continues (not necessarily in numeric sequence). When the file is completed, an **end-of-file** (EOF) is flagged. Figure 15-5 gives an example. The file directory contains the address of the first cluster, 12, while the FAT indicates the continuing sequence of clusters 13 − 14 − 15 − 27 − 28.

A third formatting structure is the **directory**, which is a listing of the file names, together with some pertinent information, such as date,

---

[2] The term "boot" originates with a German story about the legendary Baron Münchausen, who was so strong that he could lift himself by his own bootstraps, hence the original computerese **bootstrapping** for computer self-start. The word was later shortened to **boot**.

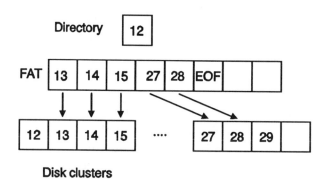

Disk clusters

Figure 15-5 The function of the FAT. The first cluster of the file is indicated by the directory, following this the FAT takes over and guides the head through the correct sequence in which the data are to be read. Often the files are fragmented over many areas of the disk with no ill effects, except for a slower response.

and size. This information is displayed when a directory listing is called.

As already mentioned, the FAT, directory, and boot normally occupy the outer positions on the disk. In some Apple disks the directory is located at an intermediate position, on track 17, so it can be accessed more speedily from various points on the floppy disk.

The floppy drive itself is a relatively simple device. The heads receive and generate information in the form of a series of pulses, that are transmitted by means of two wires. There are, in addition, a line that gives the location of the read/write heads, and a few control lines.

Single-density disks are recorded by the older frequency modulation (FM) method, shown in Figure 15-6. A four magnetic-element cell is assigned to each mathematical bit. If we call the two states of magnetization **dot** and **space** logical zero is described by the sequence dot−space−space−space, while logical one is represented as dot−space−dot−space. In both cases, the area marked C carries the information, the rest of the occupied area is simply used as a buffer zone. In Figure 15-6c the sequence 1100 is shown represented by magnetic elements. From the figure, it is evident that the number of dots per unit time is higher in the region of 1 than in the region of 0, justifying the name of frequency modulation.

A total of 16 magnetic elements are needed to describe only four bits. This is a waste of disk space, but is mandated by the need to have all magnetic dots surrounded by spaces in order to prevent interaction between neighbors. There is another alternative, however: a proce-

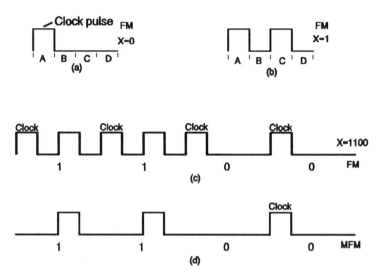

Figure 15-6 Magnetic recording on the diskettes. Each pulse is actually a magnetic field inversion. (a) The representation of a logical 0 and 1 in the FM method. (b) The sequence 1100 in FM. (c) The same sequence in MFM. The clock pulses are eliminated in all cases except between the two zeros, where the clock pulse is needed.

dure that decreases the density of dots, called **modified frequency modulation** (MFM) method. This procedure removes all dots that are not absolutely necessary for intelligibility. Thus the first magnetic dot, marked A in Figure 15-6, is removed from all 1s, since the pulse at position C can take care of the information content. This is also true of single 0 flanked on both sides by 1. The clock pulse is allowed to remain only when there are two or more consecutive zeroes. As a result, the density of magnetic dots reduces to about half. This permits writing twice as many bits for a given track length, hence the name of double density given to this technique.

The information written on the floppy is a combination of data and timing information. The two are separated by the controller, and the data are sent byte by byte through the bus directly to memory. The controller also records on the track a 16-bit word that is calculated from the data by an error detection technique known as **cyclic redundancy check** (CRC). When the data are read again the CRC is recomputed and compared with the original. The controller orders a reread in case of discrepancy. After several unsuccessful tries, the feared "unrecoverable error" message is eventually displayed.

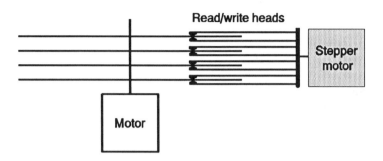

Figure 15-7 Schematic of a hard disk drive. The arm holding the heads is usually not straight, as shown for simplicity, but in the form of an elbow.

Hard Disks

The hard disk is so named in contrast to the flexible floppy. It is principally characterized by the fact that it is enclosed and usually not removable. Being hermetically sealed, it can be made to operate under much more efficient conditions. Its rotation speed is on the order or 3600 RPM (as opposed to 300−360 RPM for the floppy), and the data density is considerably larger. Hard disks are sometimes referred to as **Winchesters**.[3]

Disk storage has been in use for a long time. Thus the successful RAMAC 350 hard disk of 1953 contained packs of 50 immense disks (22-in. diameter), for a stored capacity of only 5 Mbytes. One can appreciate the progress in data storage by considering that today the smallest hard disk drive stores 20 Mbytes, on only one 3.5-in. disk. The improvement in data density is perhaps 10,000 to one.

Microcomputer hard disks come mostly in 3.5- and 5.25-in. diameters, and have several disks called platters, on the same axis. (Figure 15-7). Each disk has a read/write head on each side. All the heads move together, activated by a single motor. The head motor and

---

[3] The name comes from the first small (30 Mbytes) hard disk introduced by IBM under the designation 3030. The moniker is reminiscent of the famous Winchester 30-30 rifle.

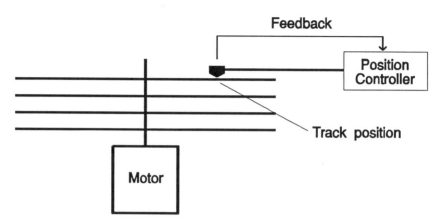

Figure 15-8 Servo-positioning for the read/write heads. One disk side is sacrificed for the purpose, leaving an odd number of surfaces for data.

the rotation motor are usually located outside the main casing of the disk pack.

A very common method of head location uses a stepper motor, with magnetically defined angular rest positions. Every time that it receives a pulse, it advances to the next position by exactly the same angle. The controller knows which is the present track, and sends the correct number of pulses, to move the heads exactly to the next needed track. This procedure looks elegant but is not perfect. There might be 1000 tracks on each disk side, at very small separations from each other. Even such a small effect as thermal expansion changes the track position sufficiently to generate errors and requires special thermal compensation devices.

An alternative procedure is to locate the heads using position feedback. The motion of the heads is now continuous rather than stepwise, and one of the platters must contain special tracks which generate a signal that measures alignment with the track (Figure 15-8). This signal generates an error command and actuates the motor to readjust the head position as needed. The motor must be of very smooth motion and of low inertia. A linear motor is similar in principle to the voice coil of a speaker but physically longer. The motor remains stationary in the absence of current and moves in either direction when a current of proper sense passes through its coil.

The read/write heads touch the platter surface when at rest, but the air carried by the rotation of the disk lifts them off by a few micrometers. Since the magnetic surface is extremely smooth, this small distance is sufficient to guarantee lack of *mechanical* contact, while permitting essentially perfect magnetic contact between the head and the magnetic film on the platter surface.

The disks are enclosed in a housing that is assembled in a clean room to ensure an inner atmosphere totally free of dust. An internal filter maintains this purity throughout the life of the device. The case has another ultrafilter connected to the outside atmosphere to allow for changes in the atmospheric pressure. This is needed not so much for avoiding stresses in the structure as for minimizing the effect of small openings such as pinholes in the casing, through which the atmospheric pressure variations would drive impure air into the enclosure.

Formatting a hard disk drive involves several steps. The first step, usually done before delivery, is **low-level formatting**. This marks the tracks and sectors (typically 17 sectors per track and 1024 tracks) on each platter. The identifying information ensures that the correct sector is being read by the controller. The next step is **partition**. In this process, a disk drive (physical drive) can be divided into several **logical drives**. The CPU regards each partition as if it were a different disk drive, with its own directory structure and even different operating systems. The next step is **formatting**. It generates a boot record with identification information, one or two copies of the FAT and the main (root) directory for each partition. The FAT contains the successive addresses of the clusters where the files are written. The clusters could be larger than for floppies, perhaps of four sectors of 2048 bytes each.

Hard disks usually contain some unavoidable surface defects which are identified in manufacture. The whole sector containing them is marked as bad and sequestered to avoid errors. As an additional measure to prevent corruption of information, the data are recorded together with an error correction code, which allows the recovery of most reading errors even if several bits are flawed.

Recording Modes

As is the case with floppies, hard disks need a controller to administer the data stream. Hard disks exhibit a much higher storage capacity and also a faster data flow. The direct management by the CPU would therefore involve too much overhead. Newer designs tend to relieve the CPU of most of the task, by implementation of so-called **intelligent drive interfaces**. Nevertheless, unsophisticated controllers are still in considerable use. The most common type of standard for older controllers is the ST-506 (usually coupled with a newer standard and described as ST-506/412).

The ST-506 is used in one of two different modes: modified frequency modulation (MFM) and run length limited (RLL). The MFM was discussed in relation with double density floppies. Hard disks typically use 17 sectors per track of 512 bytes each. At a rotational speed of 3600 RPM or 60 rotations per second, the data become available to the head at a rate given by $60 \times 17 \times 512 \times 8 =$ approx. 4 Mbit/s.If interleaving is used, the rate must be divided by the interleaving factor. If, for example, every fourth sector is read, the data transfer rate cannot be larger than 1 Mbit/s. This rate might look large, but if frequent disk reads are done, such as in database searches, the computer's overall speed could be slowed down to unacceptable levels. The computer is said to be I/O bound.

The route to faster data transfer for a given rotation speed lies in higher recorded densities. The **run length limited** (RLL) technique, described below, allows an increase in density of about 50% over MFM, without requiring any physical changes to the magnetic surface. Nevertheless, it places more stringent requirements on the device as a whole, by using 26 rather than 17 sectors per track. Both MFM and RLL use the presence of a magnetic flux transition to identify a 1 and its absence to identify a 0.

MFM procedures code individual bits into magnetic transitions, whereas RLL codes the data flow one byte at a time. Each byte is coded into a 16-element magnetic sequence. Since there is much more information in the 16-element than required by the 8-bit byte (256 times more), it is possible to avoid the specific combinations that are difficult to record magnetically. Only sequences with magnetic transitions separated by at least two blank spaces are allowed. In the most common scheme, the number of consecutive spaces ranges from two to seven, hence the complete name of the method 2,7 RLL.[4]

The fact that the pulses are more widely separated in RLL than in MFM permits more dense data recording, without bringing the magnetic dots so close together as to cause interference. Even though in this coding scheme the number of magnetic positions is actually increased by extra spaces, the data can be squeezed together tighter, resulting in an overall gain of about 50% in both data storage and

---

[4] Other choices such as 3,9 RLL are also used.

transfer. The maximum theoretical rate of data transfer for RLL is increased from 5 to 7.5 million bits per second (Mbit/s).

ESDI, SCSI and IDE Disk Controllers

In the ST-506 standard mentioned above, the disk drive does not perform any data processing (it is "dumb"); it only interchanges raw data with a controller, where a data separator extracts the information from the magnetic sequence of transitions, and sends it to the bus.

In contrast to the old ST-506 standard, two more recent types of controllers are considerably more "intelligent": they are known as ESDI and SCSI.

The **enhanced small device interface**, ESDI (pronounce Ess-dee) was designed by a large group of manufacturers seeking a higher efficiency standard. The most important characteristic of ESDI drives is that the MFM pulses are processed on the drive itself, cleaned of timing and formatting information, and sent to the controller. ESDI drives also offer efficient administration of bad-sector rejection. The drives operating under this standard are capable of writing 34 sectors per track and permit a data transfer of 10 to 20 Mbit/s.

Also in widespread use is the small **computer system interface** SCSI (pronounced Skuh-zee, not Sexy). It extends to other peripheral devices in addition to disk drives. In its more recent form elaborated by the ANSI standards organization, it can use up to 32-wire parallel communication path. The parallel data transmission of SCSI is much more efficient than the serial path of the other standards including ESDI. It permits fast data transfers up to 100 Mbit/s. The controller is also smarter, assuming such tasks as participation in the arbitration of bus requests and direct transmission of data from device to device without CPU intervention. In the SCSI protocol, each device is assigned an identification number (ID). Each device can be assigned the function of either initiator (controller) or target. The higher the ID, the higher will be the priority. The hard disk, has a low ID and is consistently cast as a target. If two devices request control simultaneously, an arbitration procedure assigns the initiator-target status, based on the ID. Overall the interconnection is more or less the same as the classical bus discussed above, except that there are no interrupts. Often the SCSI standard extends to external ports by means of a connector (see Table 15-2). Only the even-numbered pins are listed; all the odd pins serve as grounds. All logic is inverted (TRUE = ground),

Table 15-2

ANSI standard for SCSI connections

Pin number	Name	Description
2−18	DB0−DB7,DBP	Data, odd parity
20−24, 28, 30, 34	GND	Grounds
26	TER, PWR	+5 V supply
32	ATN	Attention
36	BSY	Busy
38	ACK	Acknowledge
40	RST	Reset
42	MSG	Message
44	SEL	Select
46	C/D	Command/Data
48	REQ	Request
50	I/O	In/Out

Note. Macintosh computers use the same protocol, with a 25-pin socket.

since open-collector WIRED-OR configuration is used throughout, with 330-ohms termination resistors.

The process begins with an initiator pulling to zero (**asserting**) line BSY and simultaneously identifying itself by asserting the data line with the same number as its ID. After about 2 μs the initiator looks on the data lines to see if there is no other request from a higher priority device; in the affirmative case it retires its request and waits for the next opportunity. This scheme makes every device into an arbitrator. If the initiator is successful, it asserts line SEL (by pulling it to zero volts) and signals on the data bus with which device it desires communication. After some additional handshaking, the transmission can begin. Data are sent on eight lines and an odd parity line, and the transmission is controlled by the initiator by means of the control lines listed in the table. At the end of the transmission, line BSY is de-asserted (meaning that it is left disconnected and allowed to go by itself to +5 V). This signals that the bus is free for the next operation. A disadvantage of SCSI is that there are several variants of it such as SCSI I, SCSI II, Fast SCSI, Wide SCSI, etc. that are not totally compatible.

Of widspread use is also the **integrated drive electronics** (IDE) type of controller. It has the advantage of being less expensive than the other types, while maintaining a reasoanbly high level of performance.

The speed of accessing a given track is important in data retrieval. For microcomputers it is in the tens of milliseconds. If the program calls for frequent disk searches of small blocks of data, track access might be considerably more time-consuming than the actual data transmission. (Keep in mind that in a period equal to a typical 15 ms access time, full-speed operation could transmit several megabytes of information.) In such a case a memory buffer, called a cache, will considerably enhance the overall operation. Depending on its organization, the cache (typically 64 to 512 Kbytes), may retain the data previously accessed, with the idea that it might be requested again. Alternatively, the cache could store the rest of the current disk track, to make available the data immediately following. It is not uncommon for a cache to contain the material needed in 80% of the requested disk reads (80% hit rate).

Typically, hard disk drive systems consist of the drive itself, cabling and a controller card that is plugged into the main computer bus. Figure 15-9 shows the various arrangements of the more popular systems.

Other Forms of Mass Storage

The development of mass-storage technology for microcomputers follows two main directions.

1. High-capacity removable storage offers good security and unlimited total capacity. Removable floppies can be brought to high capacity by simply employing the high precision techniques for head location already in use for hard disks. The slow rotation speeds of floppies is difficult to increase, especially since the head touches the disk. Even if the head is made to fly close to the disk, surface wear remains a problem. Nevertheless capacities of 20 to 80 Mbytes are

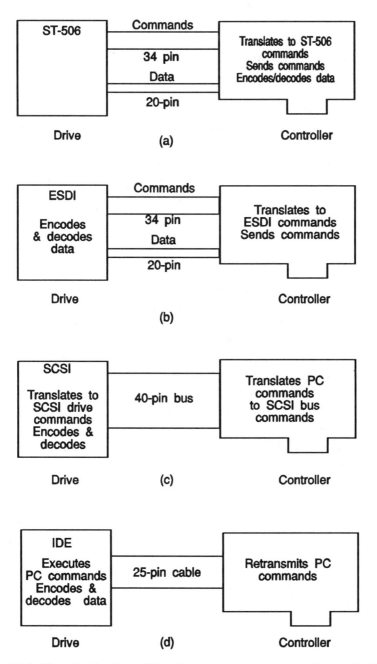

Figure 15-9 The distribution of functions among the controller and drive for ST-506, ESDI, SCSI, and IDE standards.

easily obtainable. Also promising is the removable Winchester, that performs admirably in terms of data transfer and wear, but still requires careful handling. A very efficient technique for removable media employs the Bernoulli effect which raises a rotating disk by aerodynamic force against a head. Commercial units store 90 Mbytes per disk. Another elegant procedure employs optical tracks to guide the floppy head with great precision. They are called **floptical** disks and have typically about 20 Mbytes capacity.

2. Development of very high capacity storage. If large enough, it can obviate the necessity for rewriting on the media, just as to-day very few people erase and reuse sheets of paper. This would eliminate a major weakness in the computer data, the lack of a reliable audit trail.

The several types of optical disks have a great future. The read-only disk (CD-ROM) is derived from the compact audio disk. It has a capacity on the order of gigabytes and can be filled with the content of hundreds of books. Whole encyclopedias can be accomodated in a CD-ROM. Its three-mile long track can store a billion bytes, yet any byte is accessible within a second or two. The data transfer is reasonably rapid at about 0.1 Mbyte/s. These optical disks can be inexpensively duplicated in mass production which is a great advantage if the distribution is widespread. At this stage however, the CD operates only the *read* mode.

In contrast, there are various forms of **read/write** optical media which allow writing by the user. The write−once−read−many−times (WORM) drives appear to have a bright future. It is basically a disk of recording material, usually a tellurium alloy, covered on both faces with plastic films. The record consists of microscopic dots burned on the disk with a laser. In this way an optically distinguishable spot is generated for every bit of information.

One erasable optical medium, more exactly an opto-magnetic medium, is described in Figure 15-10. The disk is coated with a thin (10 nm) layer of a magnetized material that can lose magnetization locally when heated above its Curie temperature.[5] A sharply focused laser beam heats a spot momentarily while a weak magnetic field is

---

[5] The Curie point is the temperature at which a material loses its coercive magnetic field and can no longer retain magnetization. It is about 200 °C.

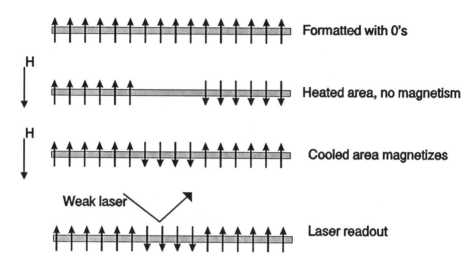

Figure 15-10   Erasable optomagnetic medium. A very small portion of a track is shown.

applied. The field is not strong enough to change the magnetization of the disk. The heated spot, however, becomes first nonmagnetic and, upon cooling, assumes the magnetic direction of the applied field. If, for example, 0's represent upward magnetization, the disk can written with 0's using an upward field, and with 1's by using a downward field. The readout depends on the interaction between a given spot and a second polarized laser beam which is made to reflect from that particular spot. The so-called Kerr effect causes a small change in polarization of the light, dependent upon the sense of magnetization. This change in polarization can be identified by a special optical detector. The materials used to coat the disk surface, comprising gadolinium or terbium alloyed with cobalt, exhibit a strong Kerr effect. They also show high magnetic coercitivities which are higher than those of magnetic tapes. This suggests a very long data life for this medium. The very high capacities of magneto-optical drives makes them important in archiving. Calculations show that the price per million words is about four times smaller than for a paper archive and about one tenth of the cost of recording on high-capacity hard disks. The space occupied is about 150 times smaller than that of a paper archive. Tapes, to be discussed in the next section, have about the same advantage in cost as optoelectrical media, but the latter are expected to becomes still less expensive as the technology matures.

Magnetic Tapes

Historically tape has been strongly connected with computer progress. A variety of tapes are available. The half-inch (nine-track) tape is the most common, and it is often used as a data-exchange medium between computers. It is not used much by micros, where one is more likely to encounter 0.25 in., 8 mm, or 4 mm tapes. In all cases the access is sequential because of the tape's geometry. For this reason tapes are more useful for backup or distribution of blocks of data, than for normal random-accesss use. The capacity leaves nothing to be desired, with 40−80 Mbytes for the simpler quarter-inch tape and over a gigabyte for 4- and 8-mm tapes using helical recording (similar to the tape handling technique used in VCRs).

Displays

Not too long ago, small computers used **teletypes** (TTY) to enter and display information. Those were glorified typewriters with communication facilities and punched paper tape memory. This slow and noisy method has been universally replaced by screen displays or **monitors**.

The cathode-ray display operates on the same principle as the television screen, but it has a higher resolution. It uses a focused electron beam that is guided to strike a phosphor-coated screen. The point of impact generates a luminous dot, that has some degree permanence, decaying exponentially. The half lives for the phosphorescence vary between 1 ms for very fast special purpose displays and 100 ms for slow displays which generally are more pleasant to look at. A disadvantage is that bright moving points leave tailing traces behind them.

The color of the light emitted depends on the chemistry of the phosphor. It is often yellow-green or orange for monochrome (one color) monitors. Alternatively, paper-white displays that generate black characters on a white background are also available. Color displays combine three colors: blue, green, and red, in tight clusters to synthesize points of all the visible colors. For instance, the sensation of yellow is produced when the green and red are mixed in equal amounts, and the blue is turned off. In this case, the eye perceives yellow even though there is no yellow actually present as such.

In order to produce the image, the various points on the screen are struck by the electron beam in an organized sequence, such as that

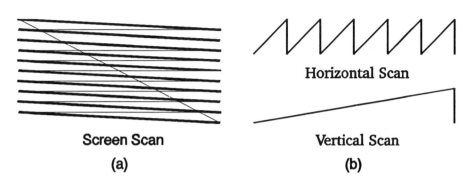

Horizontal Scan

Screen Scan   **Vertical Scan**

(a)   **(b)**

Figure 15-11 The screen scan (raster). The lines in (a) fuse in vision to form a continuous image. (b) The electrical control signals. The frequencies are perhaps 35 kHz for the horizontal scan and 60 Hz for the vertical scan. The return traces, shown as fine lines, are normally visible.

shown in Figure 15-11. The electron beam is deflected by electric or magnetic fields in two orthogonal directions. A slow vertical, downward deflection is combined with a fast repetitive horizontal sweep. This results in a series of parallel horizontal lines slightly sloping because of the vertical sweep. At the end of each line, a synchronizing signal blanks out the beam to allow it to return to the beginning of the next line without tracing a diagonal on the screen.

After all lines have been traced, both scans reset, while the beam is blanked for the longer time necessary to return all the way to the origin. The frequency of the horizontal scan is between 15 and 45 kHz. Note that during each sweep, information about the status of all individual points has to be transmitted. This information requires a bandwidth in the megahertz range. The vertical sweep takes usually 1/60 s, so that the vertical frequency is 60 Hz. At this frequency, the effects of image persistence of vision and of the phosphor combine to produce a complete image rather than a moving dot. Note that the requirements for computer monitors are more stringent than for television, where the screen is normally viewed at a much larger distance.

A design trick is sometimes used to improve the image stability by alternating between even- and odd-numbered lines in consecutive scans. Thus a single scan covers only half of the lines, say, lines 1, 3, 5,... The next scan takes care of the lines that have been left out, numbers 2, 4, 6,... This is called **interlacing**; with this procedure even a slow 40 Hz scan is perceived as steady by most people.

In a typical display there may be 400 horizontal lines, each containing 720 horizontal points, called picture elements or **pixels**.[6] There would be, in this case, a total of 720 × 400 = 288,000 pixels. This is a staggering amount of information, to be transmitted 60 times a second, and it explains why the progress in displays has lagged behind for a long time.

To generate text, the screen is divided into rows of cells, or **matrices**. The matrix might be, for example, composed of 9 × 16 pixels. Within the matrix, bright and dark points generate a semblance of the character as seen in Figure 15-12. The characters are kept in memory by numbers, according to some convention, such as the ASCII code of the Appendix. A special decoder translates this code into the pixel map shown in the figure.

A row of 80 characters requires 80 matrices or 80 × 9 = 720 pixels in a line. Vertically there are 25 lines of text in a standard full PC screen, requiring 80 × 16 = 400 pixels, producing a 720 × 400 screen structure. The characters are synthesized line by line, illuminating the proper pixels along one row, then those for the second row, and so on. The situation is complicated by the fact that a given horizontal scan cuts through the whole row of 80 characters at the same time, and the electronics must keep track, of which pixels are lit and which ones are dark in each cell. This is illustrated in Figure 15-13, which reproduces a small portion of text. It is hard to imagine the amount of activity behind a screen that looks to you so placid and motionless. In addition to the display of text, most monitors are able to represent graphics. In the **graphic mode** each pixel is addressed individually, without making use of matrices. Returning to our example, 288,000 pixels need to be specifically controlled. This requires a large block of memory, in order to form a **bit map**, containing the ON or OFF assignment for each bit. This need is fulfilled with additional video memory, specifically assigned to the screen graphics, rather than with normal computing memory. Typically there might be 256 Kbytes of memory supported by a separate controller such as the 82706, which can be programmed to operate in many different modes, text and graphics of various resolutions.

---

[6] IBM calls them "pels".

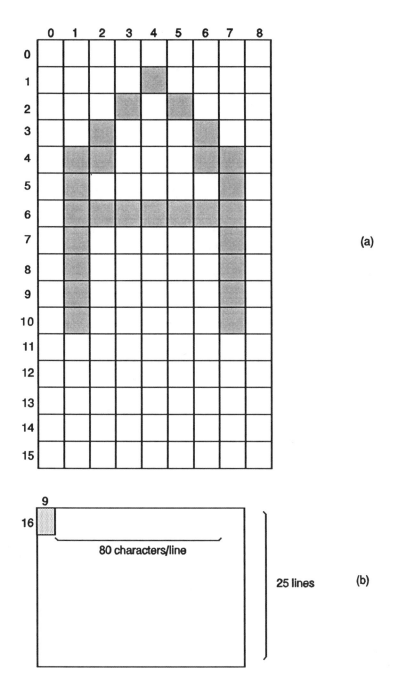

Figure 15-12 (a) An example of a 9 by 16 matrix for the representation of characters, (b) the relation of the matrix to the whole screen.

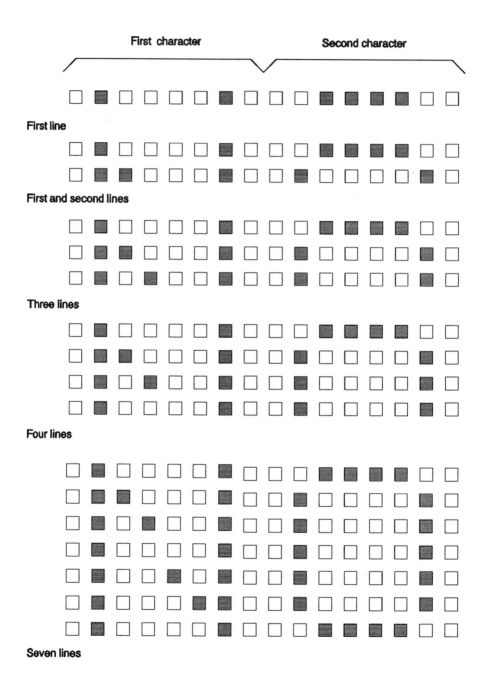

Figure 15-13 The process of writing a line in a character-only display. The letters
NO are shown out of an entire line of characters using an 8 × 7 matrix.

Table 15-3

Examples of display modes

Type	Number of pixels (typical)
MDA	720 × 350 monochrome
CGA	320 × 200, 4 colors
EGA	640 × 350, 16 colors
VGA	640 × 480, 16 colors
Super VGA	800 × 600, 16 colors
8514/A	1024 × 768, 256 colors
XGA	1024 × 768, 256 colors

Note: The meaning of the initials is as follows: MDA—Monochrome Display Adapter, CGA—Color Graphics Adapter, EGA—Enhanced Graphics Adapter, VGA—Video Graphics Array, XGA—Extended Graphics Array. 8514/A is also called Super VGA.

A great variety of display modes are employed by microcomputers. Table 15-3 includes a few from the large selection available. In each type there are variations.

Other Forms of Displays

Portable and other low-power applications often use liquid crystal displays which need negligible power, but must be illuminated from outside. The liquid crystal compound fills the space between a transparent polarizing plate and a mirror. The light penetrates the plate and is reflected by the mirror, rendering a characteristic homogeneous silvery appearence. A network of conductors on the plate, applies an electric field to specific locations, causing the liquid crystals to align in the electric field at those specific spots. The aligned liquid crystal interacts with the polarized light and changes the plane of polarization by the correct amount to generate crossed polarization when the light re-emerges through the polarizing plate. This in turn causes the light to be absorbed. Graphics and text are formed by the combination of such light and dark pixels. The energy needed to align the liquid crystals is very small, hence the low-power operation. Liquid crystal displays tend to be darkish in ambient light. Some designs use backlighting to improve legibility. Color liquid crystal displays can also be made.

Another type of display, the **gas plasma,** contains a target composed of a number of minuscule neon tubes, one per pixel. It uses more energy than the liquid crystal type, but is self-illuminating. Also generating its own light is the **electroluminescent** panel that replaces the neon tubes by dots of electrically excited solids which emit a pleasant greenish light.

Scanners

Images from outside sources can be converted into bit maps by means of optical scanners. The image from a sheet of paper — text, drawing or photograph — is illuminated by a thin horizontal LED light source. A charge coupled device (CCD) array receives the reflected light. More light is reflected by the blank parts than by the text or lines, resulting in different readings from the elements of the array. The line is digitized into 200 to 300 dots per inch. Each dot is interpreted as either dark or light and this information is stored in one bit. The resulting bit maps are useful for line drawings and for text. Photographs and other continuous tone materials cannot use this procedure but can be reproduced by special methods such as **dithering**. This uses a matrix of dark and bright dots to generate the sensation of grey. Typically 16 shades of grey are mimicked, by using a matrix of 16 dots, varying from 16 white dots, to one black and fifteen white dots and so on. The technique is useful with laser printers, that can only generate black and white. Note that resolution is considerably diminished since a cluster of points is used where only one was used before.

For scanning text, special **optical character recognition** (OCR) software is available. It is used to identify each letter and convert it to ASCII, but with loss of information about size and shape. Pattern recognition programs are often fooled by unusual letter types.

Mechanical Printers

There are two types of mechanical printers: preformed character and dot matrix printers:

1. **Preformed characters**, invented by Johann Gutenberg in the 15th century, are still in use to-day. The principle consists of constructing each character to form a movable letter-printing unit, and then combining such units together to print the whole text. One of the modern implementations, the daisywheel printer, has the outline of each symbol (letters, numbers etc.) placed in relief at the end of a flexible finger. The fingers are struck by a hammer which causes them to press a ribbon onto the paper and create a typed character. With a good ribbon, the daisywheel generates printing at a resolution equivalent to 500 dots per inch (dpi).

It is in the nature of such printers, to assign to all letters the same space, as a result an I has too much space left around it, while an M looks crowded. This uniform spacing gives the text a "typewriter look". Most books and magazines use **proportional spacing**, where the letters are assigned spaces proportional to their width, from the narrowest for I the widest for M.

2. **Dot-synthesized characters** represent a newer technique in computer printers. It works on a principle similar that of screen displays. The characters are described by a collection of 1's and 0's in a matrix, for example, of 12×18 dots. The bitmap for each character can be stored in the printer's meory or in the computer. It may even be generated as needed. Graphics can also be printed by activating the proper collection of dots.

The resolution of a printing system is defined as the number of distinguishable dots per inch (dpi). Books and magazines are printed at 1000 to 2000 dots per inch. This is approximately the smallest detail that can be resolved by the human eye upon close inspection.[7] The minimum dot resolution needed for enjoyable prolonged reading is about 300 dpi, but text still looks acceptable at 125 dpi. Slightly rough paper, as normally used in copiers, cannot do justice to printing at better than about 500 dpi. This is why art magazines use special, very smooth paper.

Implementation of the character bit maps to generate text can be done with a set of short rigid wires, that are driven by solenoids and

---

[7] The resolution of the human eye is normally defined in terms of line pairs per milimeter, rather than dots and spaces. At reading distance, the eye acuity is on the order of 125 lines to the inch, correponding to about 250 dots per inch; at closer distances this may increase to perhaps 1000 dpi.

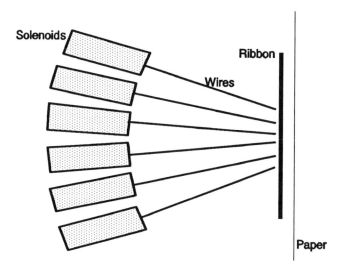

Figure 15-14 The principle of dot-matrix printing. The solenoids are grouped in a "head." Measures must be taken to dissipate the considerable heat produced. Not to scale; the wires actually extend no more than about 1 mm from each tiny swolenoid.

strike a ribbon facing the paper (Figure 15-14). The set of wires, the **head** moves across the paper and generates a collection of dots in accordance to the map stored in memory. Each wire generates a dot. If the printer has six wires vertically spaced at 1/144 in., it will print a track of 6/144 in[8]. Consequently it needs several passes to complete a row of letters. This multipass printing is a common procedure, because it allows a variety of letter sizes to be printed on the same page by simply shifting among a collection of maps, and implementing them by the necessary number of passes. A additional, overprinting set of passes, done after offseting the head by a half-dot, considerably improves the quality.

Laser Printers

The principle of the laser printer is the same as that used in Xerox copiers, which were first introduced over 30 years ago. In its modern form, **xerography** uses a photosensitive drum to register any high

---

[8] In typographic terms, 1/72 in. is a **point** In normal text, the heights of capital letters are 10 to 12 points, requiring three to four passes of the printing head.

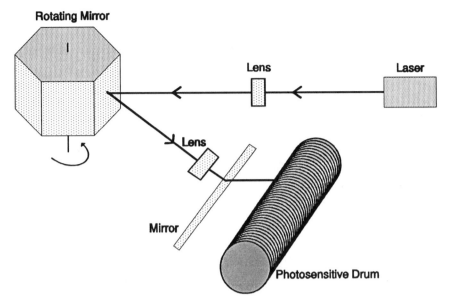

Figure 15-15 Simplified diagram of the optics in a laser printer. Each face of the rotating mirror generates one horizontal scan. The rotation of the drum takes care of the vertical scan.

contrast image in terms of areas that are electrostatically charged and areas that are not. The drum is then treated with a pigment, called **toner,** that adheres to the electrostatically charged areas and not to the neutral areas. The toner is subsequently transferred to a sheet of paper by applying an even stronger electric field, with a "transfer corona wire." The electrostatically held image is made permanent by contact with nonstick heated rollers, which melt the toner into the paper.

In conventional copiers the image is generated by illuminating the document to be copied, and projecting it onto the drum. In contrast, laserwriters do not use an original; the image has to be generated and written on the drum directly. The most common procedure is shown in Figure 15-15. The small dot written by the laser beam must be positioned on the drum with a precision at least equal to the desired resolution of the printer, usually 1/300 in. This requires a sophisticated optical system. The bit maps are similar to those used in dot-matrix printers, but with the difference that more dots are needed to generate each character. Laser printers can accept bit maps for a variety of fonts, which are **downloaded,** meaning that they are transferred from the main computer to the printer memory. In some printers, notably those following the **PostScript** protocol, the description of the letter contours is sent to the printer which generates the actual character.

In addition to the laser system described above, some devices write on the drum by means of a row of many LEDs (perhaps 2400 of them) in line or by a strong light source illuminating a long row of 2400 liquid crystal microshutters, that are electronically activated as needed. The three methods seem to produce equivalent results. In each case, the drum is written on, dot-by-dot, in a repetitive, painstaking manner, until the whole active area is covered with dots or blanks. There is no significant difference between writing graphics and text except that a full page of graphics needs an tremendous amount of memory. If the image is 8×10 in., it needs $8 \times 300 \times 10 \times 300 = 7.2$ Mbits, or about a megabyte. Gray is more difficult to represent than black and white. It can be synthesized, as discussed previously, by forming the image out of a collection of matrices of, say, 4×4 dots. This causes the resolution to drop by a factor of four, to 75 dpi.

Inkjet Printers

A technique that will probably be very succesfull in the future is the ink-jet printing. It allows for inexpensive color as well as black and white printing. It consists of a set of nozzles built into a moving head, similat to that of dot-matrix printers. In the color printer, each nozzle ejects drops of a specific color to combine into any desired hue. The resolution is comparable to that of a laser printer, except that the ink may spread slightly onto the paper fibers.

Chapter 16

Data Communications

All microcomputers need to communicate and interact with printers and other devices. Sophisticated computers can extract immense benefit from sharing data with other computers. In the future we will, no doubt, see much stronger interaction between individual micros in data exchange, personal communications, marketing, and computations. The role of the mainframe as the hub of all serious computing appears to be shifting away and is being replaced by networks of small computers. While it is not really conceivable that big computers will disappear, a well designed network of small computers can cope with most of the jobs encountered in practice. The next chapter will go into more detail about such networks; this chapter describes the fundamentals of communication circuitry.

The computer may communicate in a variety of ways; some of them are illustrated in Figure 16-1. Most communications consist of streams of HIGH and LOW bits, transmitted through wires, radio waves, or optical fibers.

The digital transmission of information can be serial, one bit at a time as in cases *a* and *c*, or it can be parallel, in units of one byte at the same time sent on eight wires as seen in *b*. Case *d* describes a variety of interaction types such as by digital and analog voltages or closures of contacts.

For a communication to be intelligible, both sender and receiver must use compatible hardware and adhere to the same set of conven-

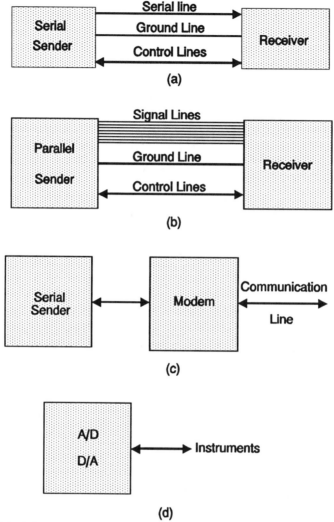

Figure 16-1 A few modalities of computer communications.

tions, commonly referred to as the **communication protocol**. A few
standards involving such protocols are listed in Table 16-1.

Line Drivers and Receivers

As a rule, transmission originates and terminates in specially designed
integrated circuits that are optimized for good noise immunity. A
simple example of such a driver/receiver pair, the ICs DS1488/1489 is
shown in Figure 16-2. This pair is quite satisfactory for short distances,

Table 16-1

Examples of communication standards

Type	Comments
RS-232	Serial, Short distances (50 ft)
RS-423	Serial, faster
RS-422	Serial, differential signal
RS-485	Serial, as 422 but multiuser
IEEE-488	Used in instrumentation
CAMAC	Parallel or serial
RS-357	For facsimile transmission
RS-366	Automatic calling
IEEE 802.3	Network communications

for instance across a room, for moderate transmission speeds. However, the reliability deteriorates quite rapidly when operating at longer distances resulting from of a variety of causes:

1. The signal attenuates with the distance and loses its high-frequency components. This creates a distortion that manifests itself as a decrease in both rise time and amplitude.

2. If the impedance of the transmission line is not matched by a suitable resistive termination at the receiving end, reflections may occur, which are troublesome for longer lines. On the other hand, if impedance matching is done using an appropriate low-resistance termination (perhaps 100 ohms), high currents are generated that in turn, may require a counterproductive lowering of the signal amplitude to keep the power dissipation within limits.

3. Electromagnetic interference (EMI) increases with the length of the wires. Often, though, the major interference sources are not far away on the line. The biggest contributors are switching devices such as relays.

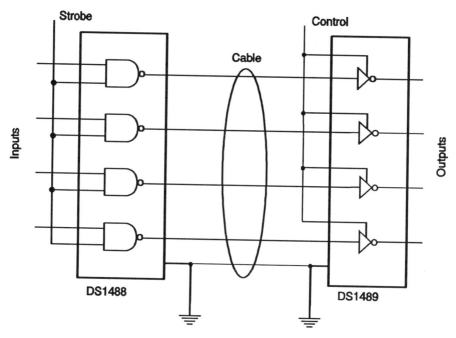

Figure 16-2 Simple line driver/line receiver pair. The two grounds should be interconnected to ensure the same reference level.

4. The transmission lines themselves may also emit radiation, that can be picked up by other transmission wires within the same cable (**crosstalk**).

5. Finally, the ground connections at the two ends might be at different voltages, producing ground-loop currents in the wiring.

The circuit mentioned above (Figure 16-2) tries to fight interference by using a rather large voltage swing between the 0 and 1 states (perhaps -7 to $+7$ V, compared with 0 to $+5$ V for TTL). This increased difference ensures that the signal is larger than the noise by a sufficient margin. On the receiving end, noise rejection is implemented by hysteresis at the input. Typically once a transition, say, $0-1$, has occurred, the opposite transition is inhibited by the hysteresis until the signal has dropped by a considerable amount (perhaps 1 V).

The simple system of Figure 16-2 above is satisfactory under favorable conditions, but for more demanding applications (higher frequency and longer distance), a differential mode of data transmission gives better results; the schematic of it appears in Figure 16-3.

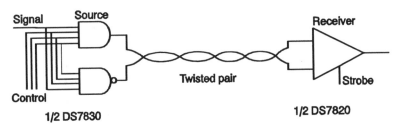

Figure 16-3 A longer distance communication using twisted pair wire.

The source has the additional provision of several AND-ed input lines and contains both AND and NAND gates on the same chip. The signal is sent simultaneously, both unchanged and inverted, on the two wires. In this system, logic signals are carried not by conventional voltages but as inversions of the relative polarity of the two lines. To each of the two possible relative polarities is assigned a logic value of 1 or 0. The receiver is capable of interpreting reliably a signal swing of as little as 0.2 V in the presence of strong noise (10 V), but only if the noise is equal in both wires (common mode). Along their path, the lines interchange positions as they twist around and tend to suffer an identical effect from electromagnetic interferences. The resulting current is then canceled out by the differential receiver. The twisted arrangement also produces a very low level of radiation and tends to interfere little with other devices.

Coaxial Cables

Coaxial cable can be used for medium distance communication, and is generally superior to the twisted pair. It comes in two different types, the **baseband**, and the **broadband** cable. The baseband type is used for direct transmission of digital data at rates of up to 10 Mbits/s. It consists of a central wire, surrounded by a high-quality insulator and a wire mesh outer conductor. The ensemble is isolated by a protective jacket. In contrast, the broadband type uses a solid aluminum tube for the outside connector. Broadband cables are used with modulated signals; they can transmit at about ten times the rate of the baseband cables (about 100 Mbits/s).

An example of baseband cable transmission, the so-called **Cheapernet**, is shown in Figure 16-4. The transceiver unit, 82502, requires additional circuitry before connection to the computer. The cable used is the same one often encountered in laboratories under

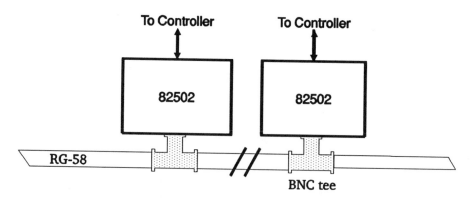

Figure 16-4 Example of communication using coaxial wire (coax).

the designation RG-58. Conventional BNC connectors are employed. An open-ended cable is shown; in fact these cables are limited to about 600 ft, and must be terminated by 78-ohm resistors. Many transceivers can be connected on the same line. Each of them must be relatively powerful; at a voltage level of 11 V the current required can be as large as 0.5 A (several watts of power).

Fiber Optics Technology

Fiber optics permits the transmission of information by means of light pulses conducted through transparent filaments. It is challenging copper wiring in essentially all long distance communications. Seldom has a new technology shown so many dominant advantages. Essentially the only drawback of this type of transmission is the difficulty of connecting and splicing, and even in this area the progress is rapid. On the other hand, the list of advantages is long indeed:

1. Lower attenuation with distance (by an order of magnitude).
2. Higher bandwidth (100 MHz to 1 GHz)
3. Essentially total immunity from electromagnetic interference, ground loops, voltage transients, even lightning strikes (unless they melt the cable).
4. Lower weight than coaxial wiring (10 lb/1000 ft versus 50 lb/1000 ft)
5. Much lower possibility of damage to instruments or humans if accidently severed.
6. Much lower power dissipation (milliwatts versus watts)

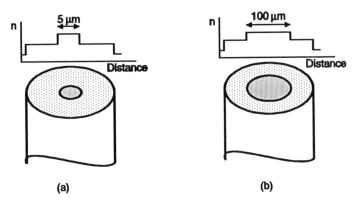

Figure 16-5 Two types of fiber optics conduits. (a) unimode (b) multimode.

Two typical fiber optic structures are shown in Figure 16-5. In both the light beam is contained in a core of high refractive index material surrounded by a **cladding** of lower index. The type in Figure 16-5*b* is easier to couple optically because of its larger core diameter, but it allows a multitude of light paths (**multimode**) because the light reflects internally from the cladding. A logical transition, say from light OFF to ON, takes a certain time to complete. At the receiving end, the signal begins on the moment that light arrives following the shortest of all the reflective paths. The light intensity continues to grow until that following the longest path arrives. After this moment the light intensity remains at a constant level for the duration of the light pulse. The whole process is analogous to the rounding off of square waves by RC time constants. This optical effect is called **modal dispersion**. The delay occurs both at the beginning and at the end of a pulse of light, and increases with the distance travelled. A typical delay might be 20 ns/km.

The single-mode fiber, shown in Figure 16-5*a*, has a more homogeneous light path and it suffers a dispersion of only about 2ns/km. Another popular type of fiber uses a graded refractive index structure, rather than two discrete indices; it can also provide low modal dispersion. In all cases the glass used has an almost incredible transparency. If the ocean were as clear, you could see its bottom.

As illustrated in Figure 16-6, the light can be provided by LEDs that are directly driven by TTL levels or by the faster (2ns) laser diode. The receiving end is serviced by a PIN photodiode, a phototransistor or a photodarlington, that receives the optical signals and regenerates the electrical logic levels. The optoelectronic circuit is relatively

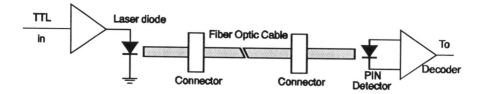

Figure 16-6 A fiber optic data link.

simple, but there are difficulties in splicing into such a fiber. Optical "circuits" are mostly arranged in series and very seldom in parallel.

Modems

Direct transmission of digital data over telephone lines is not desirable. Yet telephone lines can transmit sinusoidal tones just as easily as they can accommodate voice. In addition until recently no direct electrical connection into the telephone system was permitted. As a result digital communication by telephone required a **modem (mod**ulator-**demod**ulator). This is a device that transforms digital data into tones to be transmitted by telephone or other analog media. Another modem is required to perform the reverse conversion at the other end. The frequency band that the telephone systems makes available is 300 to 3000 Hz.[1]

A basic example of modem is the classic Bell 103 standard. It uses two carrier frequencies, one for each direction of communication, in a system called **duplex**. The 1's and 0's are represented by two pairs of frequencies (**frequency-shift keying**, FSK):

1070 Hz = forward 0   1270 Hz = forward 1
2025 Hz = return 0   2225 Hz = return 1

Bandwidth considerations limit the transmission in this system to about 300 transitions per second, or 300 baud.[2]

---

[1] The higher frequencies are limited by the transmission bandwith; the lower frequencies are eliminated because they require a disproportionate amount of energy, especially for male voices, without gain in intelligibility.

[2] Baud is derived from the name of the communication pioneer Emile Baudot.

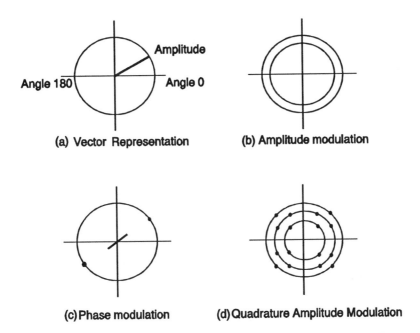

Figure 16-7 Vectorial representation of phase and frequency modulation. (a) Vectorial representation of a steady state, **single** frequency sine wave. (Multiple ferquency schemes such as FSK cannot be represented in this manner). (b) Amplitude modulation. (c) Digital phase modulation. (d) Amplitude quadrature modulation.

Considerable amount of creativity has been spent to develop more efficient data transmission, even though it would appear impossible to transcend the bandwidth limitation. A variety of solutions have been found using better **encoding schemes**. They are summarized in Figure 16-7. In this representation, a pure sine wave is described by a rotating vector of length equal to its amplitude. This is essentially the same as the trigonometric representation of the sine by a moving point on a circle. Part *a* of the figure shows such a vector with a phase angle of about 40° referred to the origin. (Note that frequency-shift keying cannot be represented in this way because it involves more than one frequency.) Since the radius of the circle represents the amplitude, changing the size of the circle represents **amplitude modulation**. If the modulation is analog, such as in AM signals, the

It is the number of transitions per second, usually close to the number of bits per second, **bps**. Baud is hardware-oriented; bps is a software concept.

circle swells and shrinks in a continuous manner. On the other hand, if the modulation is digital, the amplitude jumps between two values and the corresponding vector jumps between two circles, as is the case in Figure 16-7*b*. Another procedure, **phase modulation** leaves the amplitude constant (same circle) but causes sudden changes in the phase; it is described in Figure 16-7*c* as a vector jumping to a point diametrically opposite on the circle (for a 180° modulation). This is the principle used in **phase-shift keying**, PSK; the frequency is unique for each of the two channels (forward and return). A HIGH bit is identified not by a change in frequency but by a 180° phase jump. The absence of such a jump within the predefined time period is interpreted as a 0, and the presence by a 1. In this procedure the flow of data is considerably compressed, since the pair of information: *frequency + phase* is now replaced by *phase* alone. Twice as much speed is possible (600 baud). The circuits required are complex and the probability of error is greater hence better error correction is required in this scheme.

Finally, **quadrature amplitude modulation** (QAM) shown in *d* of the figure. It uses a combination of phase and amplitude jumps. The standard allows a total of 12 permissible phases and three permissible amplitudes. Out of the 36 possible combinations, only a **constellation** of 16 points is actually used. A complex encoding scheme transforms the digital data into simultaneous jumps of phase and amplitude, as the point jumps between the 16 permissible positions. This technique allows transmission at 2400 baud. Careful error correction schemes are needed.

Other protocols allow even higher communications speeds. Thus the CCITT V.29 standard provides for 9600 baud. This high speed is obtained by the artifice of **inverse multiplexing**. Four channels of 2400 baud share the transmission of a single channel between them, resulting in an effective rate of 9600 baud. By the same token, two 9600 baud can themselves be combined to give 19,200 baud.

Parallel Transmission

Parallel communication is based on the use of a set of eight (or more) wires conducting separately modulated signals. Evidently, a method must be provided to let the sender know that the message has been received. This could be resolved by the conventional human expedient of asking the listener to repeat the message. In this case two-directional transmission and an additional read/write control line would be

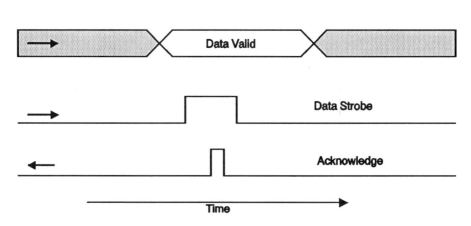

Figure 16-8 A simple example of handshaking procedure. The upper part of the figure follows a widespread convention for representing generalized digital signals that could be either 1 or 0's. The X-shaped zones describe data changes. The sender uses the *strobe* to indicate that valid data are present on the bus, while the receiver indicates with the *acknowledge* that the data were received.

needed (for example using three-state devices). However, this type of answer-back procedure proves to be inefficient since it cannot be determined if the forward or the answer-back transmission was the cause of an error.

In practice more efficient error detection schemes such as parity check, are used together with some handshaking procedures. Figure 16-8 shows such a case where a **strobe** signal on one of the control lines indicates valid data sent and an **acknowledge** (ACK) pulse sent back by the receiver indicates that the data have been accepted. The handshake is defined as just such a pairing of control messages. A special parity bit takes care of error detection, as discussed in a previous chapter. If a parity error is recognized, the acknowledge pulse will simply not be sent. After a short pause, the sender identifies this lack of response as an erroneous reception and sends the same byte again, hopefully without error. The procedure is reasonably safe, except that if the acknowledge pulse sent by the receiver is lost somehow on the way, a repeated byte might appear in the message.

The sender will not proceed until it receives the acknowledge pulse, at which time the strobe signal returns to zero. It follows that the trailing edge of the data strobe does not carry information, except to generate a secondary handshake, which indicates that the acknowledge pulse has been received. This means that the important strobe information is concentrated at the ZERO-TO-ONE initial transition.

Table 16-2

Assignments of the Centronics pins

28-pin	36-pin	Signal
From computer:		
1	1	$\overline{\text{Strobe}}$
2−9	2−9	Data (bits 0−7)
16	31	$\overline{\text{Initialize}}$)
14	14	Logical ground
16,19−25	16,19−30,33	Grounds
18	18	5 V
To computer:		
10	10	$\overline{\text{Acknowledge}}$)
11	11	Busy
12	12	Out of paper
13	13	Select
15	32	$\overline{\text{Error}}$

This is taken advantage of in the scheme of Figure 16-9, where *transitions* rather than *states* are used to carry the handshake. This has the advantage that both upward and downward transitions carry messages. In the figure the transition used for strobing the first byte is ZERO-to-ONE. The next byte is strobed with a ONE-to-ZERO transition, and so on. The process is called **non-return to zero** (NRZ) and is largely a DC procedure, since either HIGH or LOW might remain on for long periods of time.

The Centronics Parallel Interface

Centronics is a printer manufacturer that has developed a protocol followed by a number of other companies (often in simplified form) for connecting printers to computers. This protocol (Table 16-2) uses eight data lines, and a few control lines, of which the most used (Data

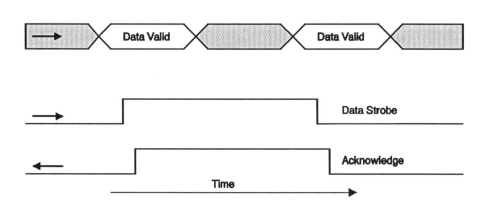

Figure 16-9   The nonreturn to zero (NRZ) handshaking protocol.

Strobe, Acknowledge, and Busy) are shown in Figure 16-10. In action the data are first sent on the eight lines, and allowed to stabilize; then the strobe goes low for a few microseconds, indicating the availability of reliable data. The response could be a busy signal or an acknowledge. If the printer is not busy executing some other operation, the acknowledge will come in about 10 μs, and the next cycle of transmission can follow. If the printer in unable to accept data, it sends a busy signal, freezing the transmission until it is ready; then the acknowledge signal is sent, and the cycle is completed. Note that the acknowledge signal is somewhat redundant. Some manufacturers do not

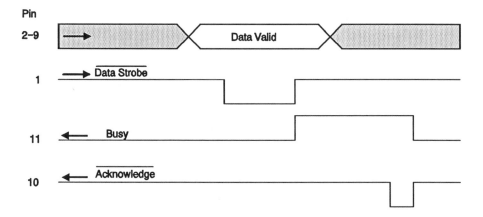

Figure 16-10 The Centronics handshaking procedure.

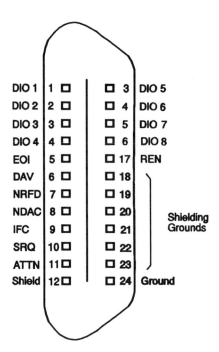

Figure 16-11 The socket used in the IEEE-488 (GPIB) devices. Both male and female connectors are present. The "shielding grounds" are ground wires that are twisted together with the control lines on the opposite side of the socket, to improve noise rejection and diminish interference.

employ this signal at all, limiting themselves to the eight data lines, the ground and the busy.

The maximum cable length is about 10 ft, limited by crosstalk, while the maximum transmission rate is of about 1000 characters per second.

The IEEE-488 Standard

A well-documented and complete protocol for the interconnection of various instruments with computers has been introduced by Hewlett-Packard. It was originally called HPIB, later known as GPIB, and eventually adopted by the Institute of Electrical and Electronic Engineers in 1975, as the IEEE-488 standard. This standard permits the direct connection of any number of properly equipped devices to form systems of great complexity (up to 15 devices: computers, printers, plotters, voltmeters, voltage sources, counters, and many other instru-

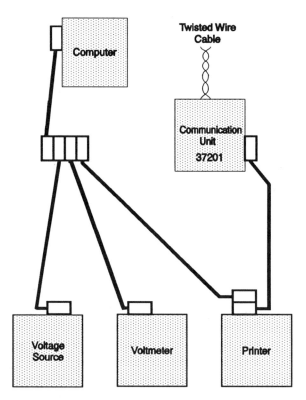

Figure 16-12 An IEEE-488 setup. The cables can be connected in any order, because they always form a unique bus.

ments). In Table 16-3 we give a few important characteristics of this standard.

The basic connector is shown in Figure 16-11. The major control lines are twisted together with ground lines. A short description of their functions is as follows:

ATN(attention); the data lines contain a command.
SRQ(service request); calls the controlling device.
IFC(interface clear); a reset of controlling functions.
NDAC(data not accepted); not yet done with reading data.
NRFD(not ready for data).
DAV(data valid); a byte has been successfully received.
EOI(end or identify); last byte of a data block.
REN(remote enable); activates a device for command.
DIO(data lines).

Table 16-3

Characteristic parameters of IEEE-488

Data transmission	Flow of parallel words
Word width	8-bit
Speed of transmission	Typically 0.2 Mbyte/s
Separation between devices	2 m maximum
Max. total cable length	20 m
Maximum no. of devices	15 (typically 8)
Connectors	Both genders at each end

Figure 16-12 shows a typical setup. The connectors can be stacked, since they are provided with double male/female termination at each end. A star configuration, as well as series arrangements are possible, but in all cases the devices are electrically connected in parallel to a bus. Functionally there is no appreciable difference between various manners of connecting, as long as the 20-meter length limitation is not exceeded. The communication unit shown in the figure allows longer transmission spans. For example the Hewlett-Packard 37201 extender, can communicate at great distances, with a twisted pair line. At the other end another 37201 extender connects to another IEEE-488 device.

Although all devices are connected to the bus in the same manner, their communication functions can be different. The IEEE-488 standard accepts several type of functions, the most important being control, sending data (talker), and receiving data (listener) in various combinations. Figure 16-13 shows an example of a typical configuration, where the controller/talker/listener can be a computer. Other devices, such as a voltmeter, might only receive orders and send data (talker/listener). Still another type of device might be only a listener (for example, a programmable power supply) or only a talker.

The controller is in charge of the operation via the **management bus**, as well as of the actual transfer of data, via the **transfer control bus.** Before the actual transfer, the controller begins with a call to **attention** (ATT) and stops all data movement. It then designates the

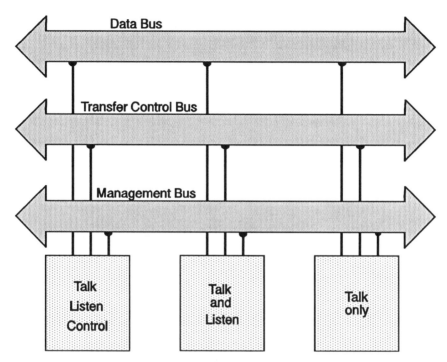

Figure 16-13 Configuration of an IEEE-488 system connected to the I/O bus.

active talker and the listeners. When line ATT returns to normal, the controller is said to deassert attention, and the data transfer can start.

An example of data transfer is given in Figure 16-14. Three control lines are involved: **data valid,** DAV, **not ready for data,** NRFD, and **data not accepted,** NDAC. First the listeners must be ready, and they indicate this condition via the NRFD line. This line is set up in a WIRED-OR configuration, so that as long as one or more devices are *not* ready, the line will be pulled down to LOW. When all devices are ready, the pull-up resistor brings the line HIGH, and the talker writes data to the bus. It then signals that this is the situation using the data available line DAV. The talker then waits for an answer from the listeners. When the slowest listener finally gets the message, the NDAC, also operating in WIRED-OR configuration, gives a pulse, and new cycle can start. In addition to this simplified data transmission, there are a great variety of sophisticated IEEE-488 protocols available, for instance, procedures using more than one controller.

In practice the use of IEEE-488 devices involves a board to be plugged into the microcomputer. Cabling is then used to form a

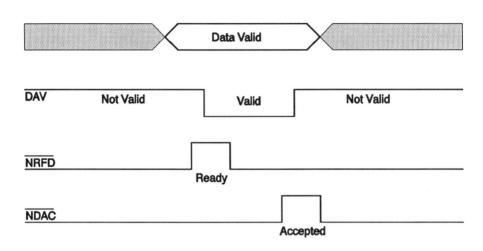

Figure 16-14 Basic signal exchange in a IEEE-488 communication.

configuration with instruments such as those in our previous example. The devices must be specifically provided with the IEEE-488 capability. The software must also be specifically designed, and is often quite complex, but it is possible to operate it at a simpler level, such as in BASIC.

Serial Communications

In serial transmission the characters are normally sent in an asynchronous manner, meaning that there is no special rhythm to the transmission. The characters are sent in packets of bits called **frames**. Such a frame is shown in Figure 16-15. It is seen to consist of a series of transitions between a positive voltage (usually +12 V) called **space**, corresponding to logic 0, and a negative voltage (−12 V) called **mark**, corresponding to logic 1. The nomenclature originates with telegraphy.

Since there is no special time when a frame can be sent, a method must exist to signal the beginning of a word. This is ensured by keeping the line at a negative voltage whenever no transmission takes place (this corresponds to a mark or logic 1). This situation lasts until a frame begins. The first action within the new frame is to signal with a change to 0 that the transmission of the word has begun. This is the **start bit**. Even though the transmission is asynchronous in the large picture, it is carefully timed within each frame. The duration of each bit, as

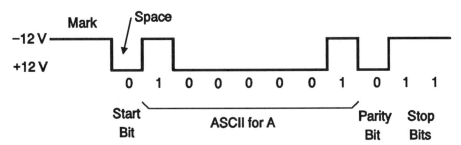

Figure 16-15 An example of serial transmission. The −12 V is referred to as **mark**, and corresponds to logic 1, while the +12 V is a **space** and is assigned the value of 0.

defined by the Baud rate, is kept rigorously equal. The start bit, and all the following ones must have the correct duration—for instance, at 300 baud they are 1/300 s or 3.33 ms long.

Next the data bits are transmitted; in the figure this is the letter A or 1000001 followed by the parity bit. In this example we have assumed that the protocol asks for even parity, so that the parity bit is 0. Finally, one or two marks (logic 1) are sent as **stop bits** in order to completely separate this frame from the next. The baud rate, the number of bits (7 or 8), the parity (odd, even or none), and the number of stop bits (1 or 2) must be established identically at both ends of the transmission line. Sometimes the parity bit is sent always as a 0 or always as a 1, even though it is not needed, for purposes of compatibility. The most common selection is 7 data bits, even parity and 1 stop bit. In some cases one encounters 8 data bits, no parity, and one stop bit. When no frame is being sent, the line is kept at mark condition (logic 1). If a long sequence of spaces (logic 0) is transmitted (more than about five frames), the series is interpreted as a break command, and this has the effect of a reset or stop in the operation.

The RS-232 Standard

In 1969 the Electronic Industries Associations (EIA), in collaboration with Bell Laboratories and others, devised a standard for serial transmission known as RS-232. It has several improved versions, notably RS-232C and the more recent RS-232D. Present use is somewhat different from the original intention. The title of the standard reads: "Interface between data terminal equipment and data communication equipment employing serial binary data interchange." The standard establishes two types of devices: the **data terminal equipment** (DTE), which is the source, and the **data communication equipment**

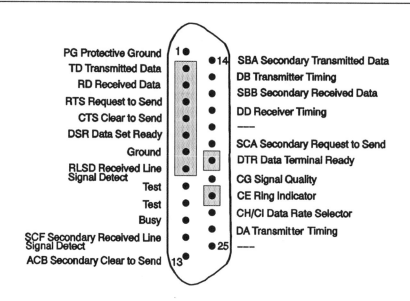

Figure 16-16 Socket connection for RS-232; a DB-25 connector is used. The shaded part represents the portion that is commonly used.

(DCE), which is the receiver. The standard is used for communications between terminals and modems, but it is not restricted to such applications. In the numerous non-modem applications, many pin connections are ignored and may even be reassigned, sometimes in a rather random manner. As a result one must be cautious to ensure compatibility of connection. The standard defines a specific connector,[3] the DB-25, shown in Figure 16-16. Not all the pins are needed in normal communications, at the most nine lines are used: 2–8, 20, and 22. A nine-pin connector is sufficient to accommodate all the needed lines and is occasionally used, in spite of the standard.

Wiring Diagrams for RS-232

Computers and printers do not readily fall into either of the two categories, DTE or DCE. Yet it is important to know the exact assignment of each, since the pin connections are different. For the moment let us assume that we have a well defined pair of DTE/DCE devices,

---

[3] In principle DTEs should be provided with male connectors, while DCEs should use female connectors. The rule is not universally respected.

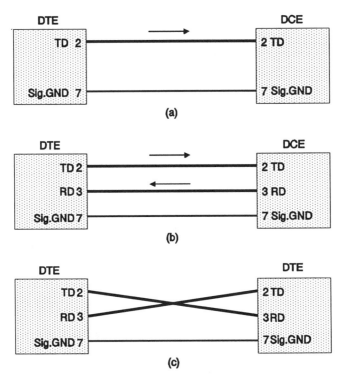

Figure 16-17 Examples of basic RS-232 connections.

and let us proceed first with simple examples, then with progressively more complex ones.

The minimum amount of wiring consists of just two lines, numbers 2 and 7, as seen in Figure 16-17a. The data flows in one direction only, toward the DCE, using line 2 (labelled *Transmit Data*, TD or TXD). The common line (number 7) is essential, to close the signal circuit, not only as a ground. We can assume that the data are formatted according to a protocol, such as that considered above, and sent in frames of 9 to 11 bits.

If bidirectional transmission is needed, a second line can be activated, namely line 3 (*Receive Data*, RD or RXD), it transmits in the opposite direction, as shown in Figure 16-17b. Note that the DCE and DTE devices have different pin assignments:

DCE receives on pin 2 and transmits on pin 3.
DTE receives on pin 3 and transmits on pin 2.

The procedure breaks down if two devices of the same type need to communicate. Unlike the IEEE-488 communication where a bidirectional bus is involved, the RS-232 is designed to be unidirectional. The solution is to make the connection artificially symmetric by crossing lines 2 and 3. Now the signal is sent from either end on TD and received on RD, as in Figure 16-17c. Pin 7, the reference ground, must always be connected to pin 7 at both ends. The arrangement is called a null-modem connection; it allows two DTE devices, such as two computers, to communicate with each other. In general one can assume that printers are DTE, and that modems are DCE. Computers are mostly DTE.

Software Handshaking

The circuits in Figure 16-17 are not provided with means of controlling the stream of data. However, even with this simple system, there are means of controlling the transmission by software handshaking. One device sends the information and the other sends acknowledgment signals.

An example of this type of procedure is called the **XON/XOFF protocol**, in which the receiver sends ASCII 17 (so-called XON) whenever it is ready to receive data in its buffer. As the transmission continues, usually at high speed, the buffer gradually fills. When the buffer is almost full, the receiver sends an ASCII 19, (XOFF) requesting a stop in the transmission until the next XON. Thus the transmission takes place in bursts. There are other such handshaking protocols, among them the ETX/ACK, which is very similar in nature but uses the ASCII 3/ASCII 6 pair.

Hardware Handshaking

Considerably better control is obtained if handshaking is implemented in hardware. The only disadvantage is that it requires dedicated wires, For this reason hardware handshaking cannot be used with modem-to-modem communications when only one pair of wires is available.

A complete set of handshaking lines is illustrated in Figure 16-18α. The connection on pin 20, **data terminal teady**, DTR, indicates, just as the name suggests, that the DTE is ready to send. Its counterpart is the **data set ready** (DSR), sent by the DCE to indicate that it is ready to receive. There is really no obligatory sequence as to which of the two precedes the other, since no communication can take place unless both are asserted. Normally throughout the transmission both lines

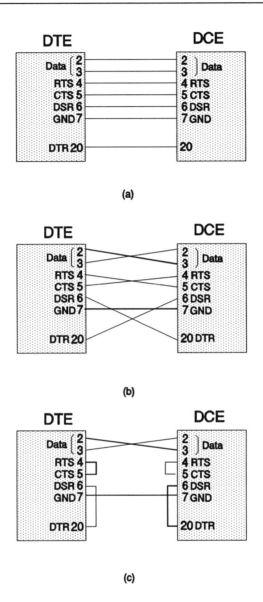

Figure 16-18 Commonly used RS-232 connections: (a) straight connection; (b) connection between partners of the same type; (c) simplified connection between similar partners, without handshaking. The simpler connections shown in Figure 16-17 can also be used.

remain active (logical 0, the positive voltage). The actual handshaking is implemented on another pair of conductors, **request to send**, RTS, and **clear to send** CTS, indicating that the two participants are both ready for transmission. If either of the two lines is inactive (logic 1), the corresponding communication does not take place. Not shown in the figure are two more lines used mostly by modems: the **carrier detect** CD, which is active LOW if the modem communication is stabilized, and the **ring indicator** RI, which is used to indicate that "the phone is ringing."

The ring indicator serves to tell the computer that a communication is in the process of being established, but no action is taken until a carrier is detected; at that point a CD signal is sent. Then follows the activation of the two lines RTS and CTS, and the communication can begin.

This arrangement can also be adapted for symmetrical operation, as indicated in Figure 16-18b. All lines are crossed except for line 7, the signal ground. The cables used may look externally identical to the previous case, yet they have quite different connections. In case of doubt, it is a good idea to check such cables with an ohmmeter to avoid surprizes.

Still another cable connection is shown in Figure 16-18c. It is used when two computers are interconnected, and in many other circumstances where the conventional straight cable does not work. It fools the computer into believing that some of the needed handshakes are present.

In addition to RS-232, there are several other similar standards which allow for differential operation, permitting longer transmission distances.

The UART

The RS-232 signals can be generated by a variety of devices, one of which is the **universal asynchronous receiver/transmitter,** UART (pronounced *you art*). It has enjoyed a long and successful life. We shall use the model 8250, shown in Figure 16-19 as an example. The address lines $A_0 - A_2$ are not used by the decoder. The unused address lines serve to transmit operation-mode instructions. Internally the UART is quite sophisticated. It contains ten registers and can effect both serial-to-parallel and parallel-to-serial conversion. It can also generate the various RS-232 signals mentioned earlier. It can control start, stop, and parity bits in a variety of configurations, and operates at up to 56 000 baud.

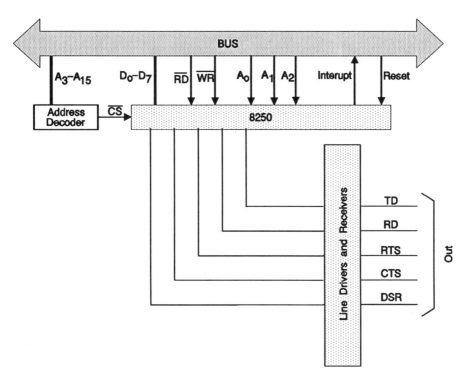

Figure 16-19 Simplified example of connections for a UART. Only the principal output connections are shown. The interrupt INTR serves to request service from the CPU. The master reset line, RST, is used to clear the internal registers. Line drivers and line receivers are needed since the UART can only generate TTL voltages, which are too small for RS-232 use.

Laboratory Interfacing

As a rule, when used for laboratory measurements, the computer requires the following elements:

1. A means to apply an excitation signal to the system under study and to measure the resulting effect. For example the application of a magnetic field, and the measurement of a resulting electrical voltage.

2. Local analog signal processing. For example preamplification and filtering of the small electric signal.

3. A means of entering the information into the computer.

4. Data processing by software and the output of data in a desirable format.

5. Software for treatment of blocks of data, such as in spreadsheets.

A great number of measurement projects can be carried out with GPIB instruments. The IEEE-488 standard is implemented by numerous instrument manufacturers as options to their products. The required software is also available. Other instruments have RS-232 connections, and are rather straightforward. Another very flexible option is to interface the computer directly with the analog signals, available at the recorder output of many instruments. Interface cards are available for various computer buses providing a variety of digital and analog inputs and outputs. Among them are the following:

1. Several channels of A/D conversion.

Thanks to modern sampling and multiplexing techniques, the interface cards usually have an ample number of analog input channels. More important than the number of channels is the speed in terms of samples per second; often a large number of points stem from an event that has short duration. Another figure of merit is the resolution, measured in terms of the number of bits in the digital word. An eight-bit word is just about the minimum, since it provides only 256 different values. For a signal of about half-scale, the resolution is close to 1%. In general, the larger the number of bits, the slower the conversion.

2. Several channels of D/A conversion.

Voltage outputs are less commonly needed, and two 10-bit channels will usually suffice. They can be used to generate excitation signals and also to drive recorders, or to control instruments parameters.

3. Digital input/output.

If events must be monitored by the computer, the experimental set-up must contain channels that can communicate with the computer by means of logic transitions. The computer, in turn, must process such information in **real time**, meaning that it must accept the information at the moment it arrives, even if it must interrupt an ongoing task.

On the other hand, digital *outputs* can be used to control ON–OFF events such as relay closures or position actuators. It is highly desirable that the computer digital grounds be independent of the outside ground. This can be ensured by using opto-isolators, discussed earlier, one in each of the digital lines.

Chapter 17

Computer Networks

Mainframes and Micros

When mainframes and micros are compared, a major factor is the difference in the overall power. We can think of power as an arbitrary combination of speed and memory size. The mainframe computing power is perhaps 50 times larger than that of a PC, about 100 Mips and 128 Mbyte of main memory as opposed to about 2 Mips and 2 Mbytes. This seems to put mainframes and PCs in entirely different classes; yet we must consider that the former may service simultaneously a large number of terminals. If we divide the mainframe power by the number of active terminals, it turns out that each of the terminal might claim perhaps 1 Mips and 1 Mbyte. In other words the power per terminal is brought in the range of or even below that of the PC. It might appear therefore that there is no special advantage in using a mainframe.

In reality this comparison is valid only for small programs or applications. The equation changes when really demanding uses are contemplated. Now, the mainframe-terminal combination comes into its own, on account of the following advantages:

- The mainframes have much larger total power and can handle large programs that may be unmanageable for a micro.

353

- The mainframes allow extensive sharing of information; for example, a database can be accessed simultaneously by many terminals.

- The mainframes have powerful collections of peripherals and high storage capacity (hundreds of gigabytes).

On the other hand, microcomputers benefit from:

- Much lower price, even when calculated in dollars/Mips.

- A vast collection of low-cost application programs.

- A user-friendly interface, and good graphics.

- The feeling that the user is in control of the computer.

Hence it is advantageous to combine the strengths of each of the two types and to connect microcomputers as terminals to the mainframe. In such implementations the micros should retain the ability of reverting to local operation whenever the application does not need the muscle of the mainframe.

At the present time we are witnessing a new generation of mainframes which can be considered to have originated in 1964 with the IBM 360 series[1], that offered the novel element of a common architecture and carefully controlled software compatibility between various models across the entire line of computers. This compatibility is probably the main reason for the enormous success of the 360 series In most cases, compatibility with microcomputers was not included, as the operating systems of microcomputers were too restrictive, and not normally compatible with mainframes.

The IBM mainframes are typically surrounded by a collection of type-3270 terminals all administered by a **cluster controller**, as seen in Figure 17-1. Some terminals itself are totally "dumb," acting as a keyboard-screen pair, while the cluster controller takes care of all the

---

[1] It is interesting to note that a typical 360 mainframe computer of the 1960s is comparable in many respect to a micro of today: 16 registers, 16 Mbytes of memory, about 1 Mips. The old mainframe had much more elaborate I/O facilities.

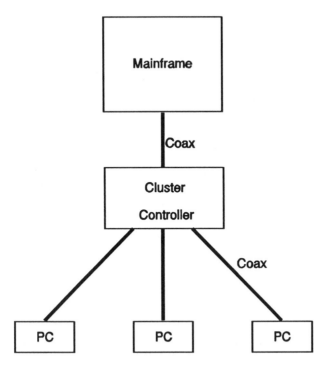

Figure 17-1 A mainframe-to-terminal connection. Either coaxial cable or twisted pair wires could be used.

communications. In contrast, the distributed function terminal (DFT) is "intelligent" which relieves the controller of most communication details. It has advanced features such as divided screens and multiprogramming but still falls short of a good PC, especially in terms of graphics.

In order to use a micro in place of a 3270 series terminal of either type, it must appear to the mainframe as identical to such a terminal. This requires special software and a plug-in board such as the Digital Communication Associates IRMA. These boards have all the necessary electronics and are provided with connections for a coaxial cable for data transmission. Several thousand feet of cable can safely be used. The cluster controller itself is generally situated close to the mainframe.

Instead of coaxial cables, inexpensive twisted-wire pairs are often used for shorter distances. They must be provided with a small transformer at each end (a **balun**) to simulate the impedance of a coaxial wire. One advantage of twisted-pair wires is that they are often already present in a building, along the telephone-line paths.

Networks

With the increased power of computers came the need to be able to interchange data across large distances. The first major development in this area was the Defense Department's **ARPANET**, in 1969. It connected large mainframes by means of smaller **interface message processors** (IMP), and used the existing telecommunication facilities. The major innovation of ARPANET was transmission by the segmentation of the message into **packets**. This method is fundamentally different in philosophy from conventional telephone operation. Normally a telephone connection, once established, remains the caller's "property" for the duration of the call. Unfortunately, this approach is not economical with computers whch transmit in high-speed bursts interspersed with long periods of silence.

The packet system breaks the message into many segments to which pertinent information is added to allow the packets to travel independently, much the same as parcel-post packages. At the destination the several segments of the message are reconnected, and the complete message regenerated. In this approach, fewer telephone lines are needed, since they are now shared by many packets of different origins. The result is a more manageable transmission cost. A side effect is that the messages are reasonably secure from unauthorized interception, being divided out on various paths and sent at various times.

The ARPANET was soon expanded to involve several hundred computers coast-to-coast. The procedure was eventually standardized as CCITT X.25, and was commercially adopted by a number of networks. At present there are hundreds of large networks world wide. In contrast to such large-scale **wide area networks** (WAN), smaller networks can also be set up within a reduced geographical area, such as a building or a group of buildings. They are known as **local area networks** (LAN).

The first efficient LAN, known as **Ethernet,** was pioneered by Xerox Corporation in 1973. The mission of LANs from the beginning was to provide communication between computers and thus allow sharing of expensive equipment as well as of files. For example, a LAN might connect various authors and allow them to collaborate in publishing a magazine. The complete text is stored in a central hard disk. At each station the user can take care of all word processing and spelling corrections. When satisfied with the text, the user can send

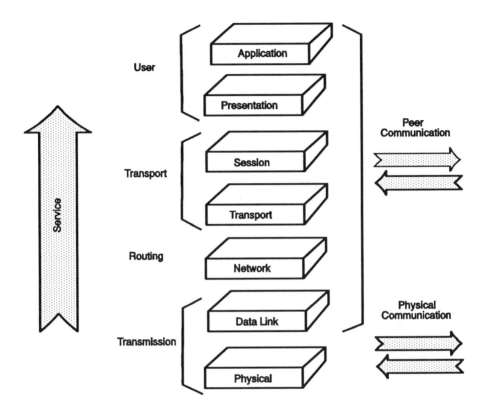

Figure 17-2 A general picture of the ISO model. The layers are conceptual elements, some represent hardware and some software.

it to central storage, to be combined with the efforts of other collaborators, processed further, printed, and so forth.

For many such uses the PC network can efficiently replace the mainframe, which would have been an overkill in the first place. On the other hand, for major applications the mainframe is always needed. In principle, the set of PCs in a LAN could be combined to form a powerful multiprocessor problem-solving unit, a sort of synthetic mainframe. When satisfactory multiprocessing software becomes available, it is conceivable that such sets of PCs might replace the mainframe altogether, but for the time being this is not the case.

Network Structures

Networks contain several components, both software and hardware. A model that organizes and classifies these elements was presented in 1977 by the International Standards Organization (ISO) and is known as the **ISO Model**.[2] This master plan divides the communication process into seven layers and establishes standards for both hardware and software. Layers of the same order communicate with each other and establish local protocols for the data transmission process, which can then proceed unhindered by the differences between the two participants. Figure 17-2 shows an example of the ISO conceptual representation of a computer with its communication gear and software. The communication is sent down through the tower of layers, to the **physical layer** where it is transmitted to the other computers in the network. Two stages are involved:

1. First is the peer-to peer communication. Each layer communicates with the corresponding one on the other computer. (The communication is not direct since the physical layer is always used as a medium.) This stage serves to generate agreement about the coding and encrypting (**presentation**) and about the exact strategy of transport. The **session** layer takes care of the mode of communication, and the **transport** layer takes care of the handshaking protocol.

2. Once the first stage finished, the actual message transmission can take place. Only the lower three layers are involved. They take care of the routing (**network layer**) and physical transmission (**data link** and **physical layers**). In LANs all nodes can communicate directly with each other, and the network layer has little work to do. In contrast, in WANs the routing is very important, and special nodes consisting of only the three lower layers are used to direct traffic.

Transmission always contains both the actual message (generated in the **application layer**), and additional "headers" from each layer, identifying the destination, type of coding, and other pertinent information. At the receiving end, the transmitted sequences are stripped of the headers to reconstitute the complete message.

Partial implementation of the ISO model can be encountered in a variety of systems, among which are the IBM **system network architecture** (SNA) and the DEC **digital network architecture** (DNA).

[2] This model is sometimes referred to as **open system interonnect** (OSI)

In general the tendency is toward gradual increase in compliance with the OSI model, except perhaps for the specifications of the network layer, which have received considerable criticism. We shall concentrate mostly on the two bottom layers of the ISO model, beginning with the major element of the physical layer, the actual cabling.

Interconnection Media

As introduced in the previous chapter, the principal transmission media for computer data are coaxial cable, twisted-wire pairs and optical fibers. A moderate level of standardization exists. Thus IBM offers several types of standard cabling, among which we mention:

1. *Type 1 Cable*. A double pair of twisted wires made of solid #22 wire twisted six times per inch and protected by a braided shield. It has a nominal impedance of 150 ohms.[3]

2. *Type 5 Cable*. Two optical fibers within the same sheath.

On the other hand, DEC offers a different set of cabling favoring coaxial types:

3. *Standard Ethernet cable*. A relatively rigid 0.5-in. coaxial cable. (50 ohms).

4. *Thin wire Ethernet cable*. A 0.25-in. thick coaxial wire similar to that used for cable television, but with BNC connectors.

5. *Twisted-wire pair*. Four twisted pairs, unshielded (about 100 ohms impedance).

Twisted pair cables tend to displace the other types for LAN use, because they are much less bulky and less expensive. Communications can be run at as much as 10 Mbit/s on such wiring. Eventually, fiber optics will probably displace all the other types, but at this point the technology is still not completely developed.

---

[3] This particular impedance has a specific meaning. It refers to the impedance of an infinite cable, measured at the operating frequency. A resistor of this value, which is connected at the end of a shorter line, makes it behave as if it were infinite and this is usually a desirable feature.

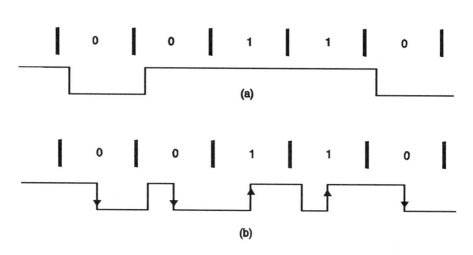

Figure 17-3 Examples of encoding methods: (a) SLDC; (b) one of the several varieties of Manchester encoding. The arrows indicate the bit-carrying transitions.

Data Links

The data are transmitted through the cables in accordance with some agreed upon protocol. For small distances, one can use RS-232C techniques. In general, LANs use more complex schemes, for example the IBM's **synchronous data link control** (SLDC). In this coding method, the presence of a transition either upward or downward represents a 0, and the *absence* of any transition within the allocated time is interpreted as a 1. This is exemplified in Figure 17-3a.

In the SLDC scheme when long sets of 1's are transmitted, the voltage level remains constant for long periods of time (the equivalent of a low frequency). Some transmission procedures have difficultly in accommodating such slow changes. In such cases one can use the Manchester encoding, shown in Figure 17-3b, which exhibits much more regular transitions and is often preferred. In this encoding scheme, upward transitions represent 1's, and downward transitions 0's. Consecutive 1's pose no problem, since the voltage must drop first before rising again to signal the 1's, as can be seen in the figure. In each period there must be either one or two transitions present. This guarantees a narrow bandwidth. Special periods with no transitions at all can be used as delimiters, before and after a frame of transmission.

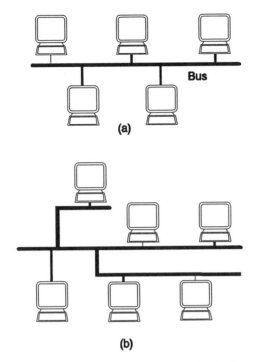

Figure 17-4 The bus topology: (a) simple; (b) branched, also called **tree topology.**

Network Topologies

The term "topology" is used to describe the type of connectivity encountered in a network. The most important network topologies are the **bus**, the **star** and the **ring**.

Bus Topology

This case refers to a single uninterrupted connection, such as the straight line of Figure 17-4a. Also under the same topology falls any branched structure as long as it is of one piece. An example is the **tree** of Figure 17-4b. In bus topologies, all stations have equal access to the communication channel at all times.

Special care must be taken to avoid simultaneous use by two stations (referred to as a **collision**). The most common collision avoidance method is the **carrier sense multiple access with collision detection** commonly known as CSMA/CD. In this procedure, each station listens to the bus to see if anybody is transmitting a message at that

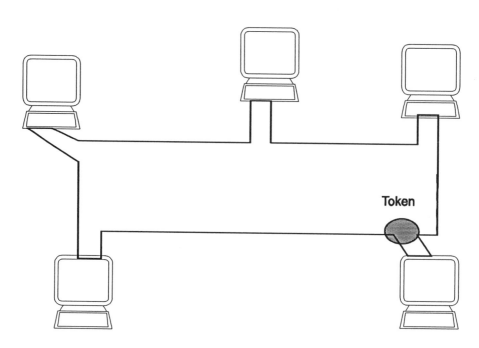

Figure 17-5   Ring-topology network operating as token ring. The token circulates around the ring in a fixed direction.

particular moment. If all is silent, transmision start. All stations receive the message, and begin by examining the header, where the destination is stated. Only the intended receiver accepts the message, the others ignore it. If two station transmit simultaneously, the transitions superpose and create nonsense. This is identified by both stations as a collision, and both transmissions cease. The situation is analogous to the case of two polite persons starting to speak simultaneously. They both stop as soon as "collision" is apparent. Sometimes they will start to speak again at the same time. The members of a LAN family take care of this eventuality by waiting for a *random* time before attempting retransmission. A second collision is therefore rare. Another method for collision avoidance, used occasionally, is the rather inefficient procedure of simply assigning consecutive time slots to all stations on the bus, whether they need it or not.

Ring Topology

The ring topology is illustrated in Figure 17-5. Each station receives the transmission via one port and sends it again through another port.

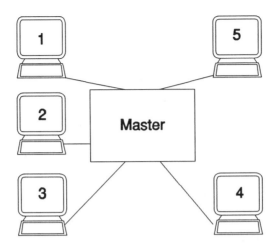

Figure 17-6 Star topology.

Each station reads the header as the message passes through. Messages addressed to the station itself are accepted, the others are retransmitted unchanged around the ring. Collision avoidance is implemented either by the **slotted ring** or the **token ring** technique. The slotted ring uses a series of time transmission periods, called **slots**, which circulate around the ring. Any station that wants to transmit uses a slot to send the message. If the message is larger than a slot, it is divided between consecutive available slots and reconstituted at the receiving station. The receiving station has the responsability of emtpying the slot.

The token ring is similar in design. A small message called a **token** indicates the FREE or BUSY condition. When a FREE token passes through a station, a message can be attached to it, while its condition is changed to BUSY. The pair circulates around the ring until it arrives at the destination. At this point, the message is accepted. However, the token+message is not stopped but sent ahead on the ring. In this way all messages must make the complete circle to the original sending station, which returns the token to FREE status, and sends it ahead alone. The return message can be compared with the original for error detection.

Tokens can also be used with bus topologies, in which case it is said that the *logical* topology is that of a of ring while the *physical* topology is that of a bus.

Star Topology

The star topology, illustrated in Figure 17-6, is not unlike a mainframe surrounded by terminals. In this system it is the responsibility of the central master to respond to requests from the stations and make the connections between them. The connection is done by a direct electrical path, in a way similar to an old-time telephone operator plugging in a patchcord to make a connection. More often, the master arbitrates between communication demands from various stations, and services them one at a time, according to some pre-established scheme.

Network Implementations

There are a great variety of LAN implementations, of which we will describe a few typical examples.

1. **Ethernet,** the pioneering LAN, was developed at the Xerox Palo Alto Research Center (PARC). It uses coaxial cables in bus configurations, operates with the Manchester encoding scheme, and uses CSMA/CD for collision avoidance. Ethernet transmits the messages encapsulated into frames that contain a rather large overhead of operating instructions (26 bytes). Typically a frame begins with a preamble of a large number of alternating 1's and 0's which serve for synchronization, followed by the destination and source addresses. This is all the information needed by the lower layers of the ISO model to effect the transmission. There follows a type field of two bytes to be used by the higher ISO layers, and finally the message itself, which could be up to 1500 bytes. The frame closes with an error-checking sequence of four bytes.

At each station an implementation consists of a card plugged into the computer bus and provided with a cable connection. A short cable connects it with a **medium attachment unit**, MAU. This small electronic circuit couples the signal into the Ethernet cable. The physical coupling is made by a special tap that perforates the thick cable with a minimum of disturbance. Other Ethernet implementations use thin coaxial cable and twisted-wire pairs.

2. IBM offers several procedures and thorough implementation in both hardware and software. They support the token ring architecture which was originally developed at IBM as well as the PC

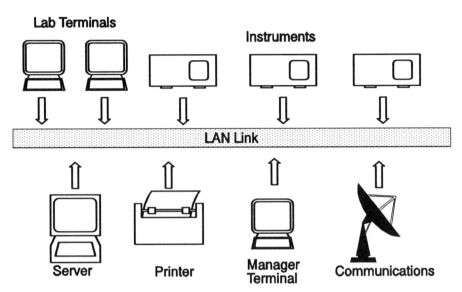

Figure 17-7   An example of LIMS.

Network that uses CSMA/CD architecture. At the higher ISO levels it also provides a variety of options for each layer.

3. Integrated Services Digital Network (ISDN) offers a standardized solution to long-distance networking. It is a comprehensive standard developed by CCITT under the aegis of the United Nations. The salient feature of this system is the presence of dedicated digital lines, as contrasted with conventional telephone lines that are optimized for analog transmission. It is expected that eventually nearly all telephone services will be converted to ISDN, world wide.

LIMS

Under the name of **laboratory information management systems**, LIMS, are found a variety of hardware-software combinations, serving the needs of laboratory measurements. The major components of such systems are:

1. A communication network, usually a small LAN.
2. A main computing and storage center, called a **server.**
3. Instruments directly connected to the network so that measurements are automatically entered in the system.
4. Terminals for manual entry of data and commands as well as for receiving calculation results, work assignments and so on.
5. A manager's terminal.

6. Software whose important functions are to identify the measurements needed, to calculate the results, to compare with requirements and standards, to make the report, and to archive the data.

Figure 17-7 shows an example of LIMS applied to quality control in industry. In general LIMS can take care of most operations that previously needed paper. Following is a possible sequence of operations in a LIMS in an industrial laboratory concerned with spectrophotometric determinations:

1. Sign on, call the proper program and give analyst's ID
2. Give sample ID, or read a bar code
3. Enter analysis type
4. Enter other data (sample weight, tray position, etc.)
5. Enter ID of calculation type and form to be filled
6. Instrument starts and sets up automatically
7. Correct number of replicate measurements are made
8. Data are stored in a database
9. Report is printed and archived
10. Manager may ask for specific data and statistical calculations
11. Manager may ask for an evaluation of analysis costs
12. Database can be used for establishing trends and other quality control estimations
13. Data can be communicated to headquarters

Bar Codes

The description of individual items in a large inventory can benefit considerably from automatic item identification by bar codes affixed to the object. Bar codes are representations of text or numerical data by means of an array of thin and thick bars. It is the ratio of bar thickness and not the absolute value that distinguishes between a 0 and a 1. This feature is important because it allows reading at various angles, resulting in variations in absolute but not relative thickness. Bar codes are widely used in supermarkets to convey item identification and pricing information. In LIMS their use is growing rapidly.

Several encoding methods are used, among them the **interleaved 2 of 5** and the **interleaved 3 of 9** named for the number of thin and thick elements used. The fixed number of elements can help in error detection. For example, reading four narrow elements when the code allows only three immediately indicates an error. Other error-detection methods are normally added, such as the **checksum** where the sum of all the members in a set is used to check the accuracy of the reading. It is claimed that bar code entry of data is about 20 times less error prone than manual entry.

Chapter 18

Construction and Troubleshooting

A well-designed electronic instrument should last for many years without any need for servicing. Often an instrument is discarded in favor of a more modern design with added features, even though it is still working perfectly. In the days of vacuum-tube electronics the situation was different. Tubes would deteriorate with use and eventually burn out, but with present-day solid-state devices, failure of active components is unusual. This explains why it is economical to solder integrated circuits directly onto the circuit board rather than to use sockets.

Nevertheless, instruments *do* fail occasionally. The substantial improvement in reliability of transistors over vacuum tubes has been lessened by a violent increase in instrument complexity. Counting all the transistors in the ICs, we find hundreds of thousands or perhaps millions of component devices in a single instrument. Even though the failure rate *per device* may be vanishingly small, a complex of a million of them will fail eventually. Consequently, it is well to be able to carry out at least the first stages of troubleshooting and repair.

Furthermore, if construction of a circuit is attempted, one must be able to distinguish between correct functioning and malfuctioning. In this chapter we shall give an introduction to the techniques for analysis and diagnosis, as well as some information about the methods of constructing small circuits. Diagnostic techniques are also needed for home-made devices that might not be operational as constructed. One often needs to "repair a new instrument into existence."

The Scientist's Job

One might ask: What is the portion of electronics work that is the responsibility of the scientist? The answer depends considerably on whether a qualified engineering staff is available, but one can assume that a knowledgeable scientist will occasionally attempt both construction and troubleshooting. Frequently the need is not so much to repair a device by oneself, as to be able to describe intelligently the nature of the problem. This chapter should help in both situation.

Tools and Facilities

If electronic work is to be attempted in the laboratory, it is desirable that space be specifically assigned for the purpose, with space for tools, test instruments and a small store of materials and components. The tools needed include the normal small mechanical work tools such as screw drivers, hex wrenches, and nut drivers. One of the tool kits available from electronic mail-order houses would suffice for other needs. A few special tools and instruments are needed in addition. The most important are:

1. A high-powered work light, with attached magnifier.

2. A temperature-controlled soldering station (about 25 W). Rosin-cored solder should be used exclusively; other types may be corrosive or have the wrong melting point. The solder should ideally be used within a narrow temperature range to ensure that the copper is wetted by the solder without damage to the components.

3. A temperature-controlled desoldering station, with grounded tip. A major cause of problems in printed circuit boards is the use of excessively hot irons for soldering and especially for desoldering.

4. Antistatic wrist strap and foot mat. Essential in dry weather, it avoids the endless frustration of beginning with a slightly defective circuit and ending up with a thoroughly inoperative one. CMOS and MOSFET circuits are particularly prone to electrostatic damage.

5. Isolation transformer. Some equipment, such as monitors and oscilloscopes, usually have the chassis connected to the neutral wire of the line. If the hot and neutral power lines are accidentally interchanged, the chassis may be charged to the line voltage, with danger

of shock and short circuit. This is alleviated by the use of an isolation transformer.

6. Digital multimeter.

7. Oscilloscope, preferably dual-trace, 10 MHz maximum frequency, 2 mV sensitivity.

8. Frequency counter, 6 or 7 digits (optional).

9. Audio oscillator, to 100 kHz, sine- and square-wave.

10. DC power supply with multiple outputs, one of them adjustable to 50 V.

11. Variable power transformer, to 150 V, 5A.

12. Logic pulser, an inexpensive tool to inject signals into circuits.

13. Logic probe, an inexpensive tool to identify the presence of pulses in a logic circuit.

14. Connecting wires, general-purpose flexible wires with insulated alligator clips at both ends. Check them frequently; after much use clip leads often break *inside* the insulation.

15. Instrument oil and degreasing spray.

16. Electric hand drill with a set of bits.

17. IC removers.

Component Set

It is very difficult to suggest a list of components appropriate for the great variety of scientific uses that may be encountered. The following list may serve as a rough guide.

1. Solderless bread boards (10). These are blocks of connectors for wires and components. They accept wire of 20 to 29 gauge (AWG). The 600-hole unit is about 2×6 inches. Two bread boards should have built-in power supplies. Not suitable for megaherz region.

2.. Wire jumpers for the solderless boards (at least 3 sets).

3. Resistors, 1/2-W. Buy a ready-made set.

4. Resistors, 2-W. Buy a ready-made set.

5. Precision resistors, 1/2 W, 1% or better tolerance, metal film types; these usually come in even values rather than standard resistor values.

6. Potentiometers, 22-turn. Useful values are: 100, 500, 1000, 5000 and 10,000 ohms.

7. Potentiometers, precision, 10-turn, with precision dials, 2 each of 100, 1000, and 10,000 ohms.

8. Capacitors, film type, 1 set.

9. Aluminum electrolytic capacitors, 1 set.

10. Modular power supplies to be included in instruments (2 of each kind): (a) 5 V, 500 mA, and (b) dual, positive and negative, 15 V.

11. Perforated board for wire-wrap pins (20).

12. Wire-wrap pins (1000). To be used with the perf board above and a wire-wrap tool. Do not use round pins for wire wrap.

13. Ribbon cable, multicolored, with about 16 to 24 wires, 50 ft. Alternatively 16-wire cables with socket plug headers on both ends (6).

14. Aluminum chassis and panels, as needed.

15. Wire-wrap sockets, 10 each of 8, 14, 16 and 20 pins. Also 10 each with solder pins.

16. Assorted TTL integrated circuits, preferably of the 74LS series.

17. Assorted 4000-series CMOS digital ICs.

18. Transistors, 25 each of 2N2222A, 2N2905, TIP29 and TIP30 for a basic supply; others as needed.

19. Diodes, 1N4004, 1N4448 or 1N914, and a few 1/2 W Zener diodes (4.7, 10, and 15 V).

20. Voltage regulators for +5 and ± 15 V, such as 7805, 7815, and 7915.

21. Operational amplifiers, including types 741 and LF354 or equivalents (25 each) and OP-07 (10).

22. Miscellaneous hardware, such as machine screws and nuts, especially 3/8-in. and 1/2-in. #6. Heat-shrink tubing, hook-up wire various colors (solid), wire-wrap wire, electronic grade solder, fuses, fuse holders, etc., to be acquired as needed.

Construction Methods

There are many ways in which a circuit can be implemented. In all cases the order of development should approximate the following:

1. Determine the need, the type of use, and the exact specifications of the device to be built.

2. Design the circuit and build a prototype on a solderless board.

3. Check the prototype against the specifications, and if feasible, check it also under the actual conditions of use. Make any needed corrections.

4. Build the final permanent circuit. Suggestions will be given below.

5. Test the final circuit and modify if necessary.

Circuit Boards

The best circuit boards are the so-called printed-circuit (PC) boards made by chemically etching copper-clad insulating boards, removing all copper except that comprising the desired circuit. These boards are useful for mass production but seldom practical for a single project. If it is desired to make the PC board, we suggest one of the kits that are commercially available.

For hand-made boards it is advisable to use sockets for all ICs. The standard board has a matrix of holes spaced at 0.1-in. intervals in both directions. The IC sockets can be lined up along one side of the board and the added discrete components mounted on the rest of the board. It is recommended that the output of each op amp or other IC be provided with a special pin to act as a convenient tie point for clip leads used in circuit analysis and testing.

The interconnections between sockets and discrete components can be made by direct point-to-point soldered wiring, which is satisfactory for small boards with only four or five sockets. For larger circuits, this method becomes very cumbersome, and wire-wrap or printed circuit techniques should be used.

The ground (or "common") from the power supply and all the other grounds must be arranged in a star connection, as illustrated in Figure 18-1.

Soldering

To solder wires to socket pins and other devices, small diameter insulated solid (not stranded) wire is the best to use. The wire should first be cleaned of insulation for about 1/4 in., then wrapped once tightly around the pin. Touch the joint with a moderately hot soldering iron and, at the same time, with the end of a solder wire. Hold it there until the joint is hot enough to melt the solder. Apply enough solder to wet thoroughly both parts of the joint. The solder will solidify quickly when the heat is removed. The finished junction should not show a big blob of solder, but rather just enough to coat the wires. Great care must be taken in soldering to two adjacent pins of a socket, as they are only a tenth of an inch apart. Try to avoid soldering two wires to the same pin; this can be done, but it is tricky.

Soldering heavier wire, such as those from the power line, is better done with a soldering gun rather than a light-duty iron, but the principle is the same. Don't *ever* use acid-core solder or soldering

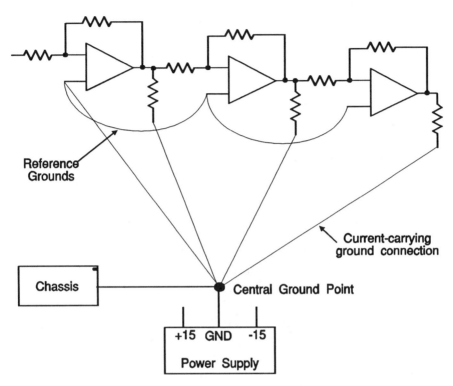

Figure 18-1 Strategy for ground connections. Ideally *all* ground connections should meet at a central common point. However, the lines that carry negligible current (such as the positive inputs of op-amps) may be connected in series as shown, with very little loss of signal quality. In noncritical applications, it is also admissible to daisy-chain the power-carrying grounds but a second, separate path must be used.

paste. These are fine for plumbing purposes but not for delicate electronic soldering.

Wire-Wrap Techniques

The wire-wrap method requires the use of special sockets provided with square pins about 5/8 in. long which extend through the holes in the perforated board. Special pins are available for discrete components. Also required is a special tool, either hand-operated or electrically driven, that contains a small reel of wire and means for twisting it around a pin. The wire used should be that sold explicitly for this purpose (usually AWG #30). The tool twists the wire tightly around the square pin so that the angles on the pin cuts through the insulation to make a permanent low-resistance junction. Hence it is not necessary to strip the insulation from the wire.

All the wire-wrap pins extend below the board so that the inter-connecting wiring can be out of sight in the completed instrument. A connection is started by twisting a wire around one pin, then feeding out enough wire to reach the second pin, twisting on that, then continuing to a third pin if desired, and so on.

Troubleshooting a Faulty Instrument

The first principle is always to check the simplest things first! If the instrument shows no response whatever to the controls, check first that the power cord is not faulty, that it is plugged in to a live outlet, and that the fuse is not blown. In a battery-operated instrument, check that the batteries are actually present and are not dead.

The next step should be to measure the power supply voltages, using a high-impedance voltmeter. (Do not attempt this for the high-voltage systems found in televisions and computer monitors; leave these strictly for professionals.) Remember that line voltage is dangerous. In fact any voltage greater than about 50 V should be treated with caution. If possible, measure the power-supply voltage without load; then momentarily connect the load and observe the effect. If the voltage drops drastically when the load is connected, then suspect a short to ground or failure of some component, such as an electrolytic capacitor.

If no voltage is seen, even without the load, look for an internal fuse. Sometimes commercial instruments will have a fuse inside, so the cabinet has to be opened. A blown fuse should not be replaced repeatedly until the reason for its demise has been established and corrected.

At this point the entire circuit should be inspected visually, in order to locate components such as resistors or transformers that appear to have been "fried." Many components, such as transistors and ICs, seldom show any external sign of internal suffering. Also look for any bits of wire, solder, or other foreign material that may have fallen on the board causing a short circuit.

If the circuit is operational but gives incorrect response, it is time to make detailed measurements. The strategy varies from case to case. If a schematic with marked voltages or waveforms is available, these should be checked with the high-impedance voltmeter or with an oscilloscope. If no such test-point data are available, the following sequence should provide sufficient information to localize the trouble.

First unplug the instrument (or remove the batteries), and measure the resistance between the power-supply common and all circuit points that are supposed to be at ground potential. No such values should be greater than a few ohms. Repeat for each of the power-supply outputs.

Next reconnect the power source, and check the voltages at all points that are supposed to be at power-supply voltages (particularly +5 V, +15 V, and −15 V, as appropriate). A difference between the expected and measured voltages of more than a few tens of millivolts should be considered suspicious. If there is an AC or logic signal normally present in the circuit, it should be checked at various strategic points in the circuit, using an oscilloscope.

Two possible circumstances that can cause confusion to the trouble-shooter are (1) multiple faults and (2) intermittent faults. It may seem unlikely that two or more faults would appear at the same time, but this happens more often than statistics would predict. It may be due to the "domino effect," where one component failing puts undue stress on its neighbors, producing a chain of destruction, or it might be the result of a sudden voltage surge destroying several components simultaneously. The operator, having located and corrected one fault, is apt to assume that all is now well and reassemble the instrument, only to find it still inoperative.

Intermittent faults are difficult to locate because they may not be present during the actual testing. This may be because the fault is the result of a warming-up process that may take considerable time to become evident. The temperature should be checked near various components to identify the ones that are overheating. The temperature can be measured easily with a simple thermistor circuit as described in an earlier chapter. A hot-air gun or a blast of cold air can sometimes be useful to make the fault appear or disappear.

A very useful procedure is to measure voltages and waveforms at key points when the instrument is operating normally, to be used as reference when a fault appears.

Testing Specific Components

Some components can be tested while still connected in the circuit, but the effect of other portions of the circuit may alter the readings. For this kind of test it is advisable to remove all components mounted in sockets. Before proceeding, check your test equipment, especially the leads, which should give zero reading on the ohmmeter. Wiggle the leads to discover intermittant contacts.

Resistors

It is easy to clip an ohmmeter across a suspected resistor. If the reading is significantly *higher* than the marked value, something is wrong, perhaps a poorly soldered terminal. On the other hand, if the reading is too *low*, it may represent the parallel resistance of two or more current paths. If one of the possible paths includes a diode or other semiconductor, reversing the leads to the ohmmeter may give differing readings that will help in diagnosis. If the reading continues to be unacceptable, it will be necessary to remove (unsolder) one end of the resistor. If unsoldering is impractical, the only alternative may be to clip out the offending resistor entirely and replace it with a new one. Unless the diagram specifies otherwise, select the nearest standard value resistor with 5% tolerance.

Capacitors

Nonelectrolytic capacitors seldom fail unless subjected to much higher voltages than rated. Electrolytic capacitors can fail either by opening up or by shorting the circuit. In either event, the only cure is replacement. Measuring a capacitance in a circuit is not likely to yield enough useful information to justify the work. Good capacitors (in operating circuits) often have a DC voltage across them, while defective ones don't. A medium value capacitor can be tested out-of-circuit with an ohmmeter. The reading should be low initially, and increase progressively to a very large value

Discrete Bipolar Transistors

If the transistor to be tested is socketed, remove it and test with a commercial transistor checker, following the manufacturer's directions. An volt-ohmmeter can be also used, testing the following:

1. Collector-to-emitter resistance should be very high for both polarities. (Test by inverting the ohmmeter leads.)
2. Base-to-collector or base-to-emitter resistances should be high for one polarity and low for the other. In operating circuits V_{BE} should be small (under one volt) while V_{BC} and V_{CE} should be large (tens of volts). Perhaps the best way to observe the output of soldered-in

transistors is on an oscilloscope while injecting an AC signal at the input. With a JFET, make certain that forward bias cannot be applied to the gate.

Operational Amplifiers

If the output of an op amp is neither zero nor at saturation (close to the power-supply voltage), it is probably good. Another way to test an op amp (aside from commercial op amp testers) is to set up a basic circuit on the solderless prototype board, and then vary the input while monitoring the output. Some experimenters keep such a circuit permanently, to facilitate routine tests on op amps. In most cases, if an op amp works at all it works well.

A very useful series of twelve articles entitled "Troubleshooting Analog Circuits," by Robert A. Pease, was published in EDN magazine during 1989, and republished as a combined volume. This forms a highly readable and reliable guide to its subject.

Testing Logic Circuits

This is where the logic pulser and logic probe become important. These are both small, fountain-pen-like units, each terminating in a small retractable clip that can be hooked onto component wires or tie points as needed. The operating power is sometimes taken from the circuit under test, sometimes from self-contained batteries. The pulser is connected to the input of the section is to be tested, and the probe to its output. Pressing a button on the pulser gives a momentary pulse of voltage, the result of which can be seen with the probe. For TTL and CMOS logic the pulse is 5 V. The probe usually has two LED indicators, one showing a logic HIGH, the other a LOW. In many cases the pulse can be seen propagating through the logic system, as the probe is moved from place to place. Alternatively, comparable tests can be made with a 5-V square-wave generator and an oscilloscope.

Testing Computers

Because of their complexity, computers require more specialized test methods, often provided in software. In many computers test routines are built in. For the IBM PCs these are called **power-on self-test** (POST) facilities. Each performs the indicated test on command, with a series of beeps and screen messages as output. A few typical signals are given in Table 18-1.

Table 18-1

Typical POST signals

Beep signals	Error messages
Short	Okay signal
(None)	Power absent
Many short	Power absent
Continuous	Power absent
Short/long	Display
Short/short	(See Table 18-2)
Long/short	System board
Long/short/short	Display
Long/short/short/short	Display

The short/short beep signals indicates that a coded message will be displayed. A few example of such displayed codes are shown in Table 18-2.

There is so much variation from one computer to another that little can be said about the details of trouble-shooting and repair. Whatever information can be gleaned from diagnostic software is all to the good and can serve as a guide for replacing boards or components, but the actual repair will normally be entrusted to a professional. For those with a Macintosh computer, the book *Macintosh Repair & Upgrade Secrets*, while somewhat outdated, is highly recommended. The author describes actual repair techniques, starting with how to open the case (nontrivial with the Mac) through such easy items as how to clean the mouse, to the more difficult areas of replacing components of the master computer board. One thing to remember with the Mac, and other manufacturers, is that merely opening the case voids the warranty.

Upgrading the Computer

There are two ways in which the performance of some models of microcomputers can be improved short of buying a new one. The

Table 18-2

Examples of PC error codes

First two digits	Location of fault
01	System needs "setup"
02	Power, memory
03	Keyboard
04	Monitor (monochrome)
05	Monitor (color)
06	Diskette
07	Math coprocessor
09, 10	Printer
11, 12	Asynchronous adapter
13	Game port
17	Hard disk
18	Expansion unit
24, 25	EGA
28	3270 adapter
29	Color printer
30, 31	LAN adapter
36	IEEE-488
38	Data acquisition adapter
48, 49	Modem
73	3.5 in. Diskette

The important digits are the first two. Note that 00 in the last two digits means "no error."

speed of numerical calculations can be increased significantly by adding a second processor that is designed specifically as a **mathematical coprocessor.** This unit can handle double-precision and floating-point calculations while the main processor is handling other operations.

The other improvement lies in adding additional RAM memory. In some computers this merely involves the insertion of additional memory chips into sockets provided for expansion. Nine such chips must be added at the same time. They can be had in strips called SIMMs (**single-in-line memory modules**) that fit into corresponding sockets. In other computers it is necessary to purchase an additional memory board that fits over or plugs into the preexisting mother board.

Viruses

The computer virus is an interesting development, in that it brings the computer close to the hazards of being alive. A virus is a piece of code, usually but not necessarily harmful, that can be inserted by stealth, into a computer's programs. Once in the computer, it reproduces itself and attaches onto various files. It is at first invisible but can be triggered at some future time and continues to replicate itself (like any live virus) and have a devastating effect on the host computer. The trigger may be the computer's real-time clock so that, for instance, the activity might take place the next time the day of the month is 18. The viral action might be merely to increase the burden on the computer, choking off its normal routines, or the virus may be programmed to erase data files.

Viruses are most commonly acquired through a contaminated floppy disk, or by importing files via a modem. There are several good virus-fighting software programs. Inherently such programs are tested on old viruses, and a new virus that cannot be detected may appear at any time. Many viruses attach themselves to the COMMAND.COM file in the operating system. One way of detecting them is to note the size of that file. The correct sizes for various versions of this file are (in bytes): DOS 1.0: 3231; DOS 2.0: 17664; DOS 3.0: 22040; DOS 3.1: 23210; DOS 3.2: 23791; DOS 3.3: 25307; DOS 4.01: 37557; DR DOS 5.0: 37557; DR DOS 6.0: 50456. If an increase in size is observed, one can suspect a virus.

A simple and partly successfull antivirus strategy is to set the read-only tab for all floppies not needing to be written on (e.g. the distribution disks of application software) and also to assign all the EXE and COM files on all disks as *read-only*.

Several programs can be obtained that will detect and help to kill any known viruses. It is well to have such a program permanently installed in any computer. Many monitor the size of all executable files and alert if any has changed. This is effective against the present viruses, and also conceivably useful against future ones.

Chapter 19

A Glimpse of the Future

It is always risky to speculate about the future of technology, but we shall offer some opinions about the developments to be expected in the next ten to fifteen years. We apologize in advance for all wrong predictions.

Electronic Components

While surprises always occur, we do not see the likelihood of any fundamentally different new devices. More probably, the existing ones will develop further, and more will be available as integrated circuits. Battery operation, for example in portable computers, is expected to improve when large capacity lithium cells are developed. The continuous replacement of analog by digital circuits should stabilize at about the present level.

Integrated Circuits

Integrated circuits over the years have shown a logarithmic growth curve, as shown in Figure 14-19. From this curve alone, we can expect hundreds of millions of transistors in a single IC. It is conceivable that the power of today's mainframe (hundreds of MIPS) will very soon be available for personal computers. By the same token, essentially all major equipment will contain a reduced set of only three ICs: a processor, a memory, and a power stage. The specifics of each application will be programmed in ROM. This should be true of Hi-Fi equipment, radios, televisions, etc., and whatever other instruments will be invented in the future. The implications point to devices with vast numbers of features and low prices.

The question is how to produce the powerful ICs neeeded. The most significant problems are the electrical interactions between internal elements and the difficulty of heat dissipation. Both increase rapidly with the number of active devices in an IC.

How far can we proceed in the direction of miniaturization and higher frequency? There appears to be a natural limit. When internal dimensions shrink to the same order of magnitude as molecular spacing (below 0.1 nm), the quantum-mechanical properties of electrons become important, and interfere with the operation. We still have a long way to go until we reach this natural limit By the year 2000 the lithographic methods, using far-UV or X-rays could achieve feature detail as small as 0.12 μm, allowing the production of gigabit ICs. (Such small details require elimination of contaminating particles as small as 0.03 μm.)

Another limitation that seems to be rather unforgiving is the amount of heat produced in the IC. This depends on the number of devices switching at any given moment. Even the present million-transistor ICs run quite hot and require careful engineering to avoid self-destruction. The heat puts a limit on both the circuit complexity and on its maximum operating frequency.

If superconducting devices are successfully developed that will operate at room temperature and resistant to magnetic fields, a complete reconsideration of electronics will be in order. Superconducting paths in integrated circuits will allow implementation of ICs containing billions of transistors and operating at gigahertz frequencies. Note, though, that heat effects will not be completely avoided, since superconductors will eliminate only the heat generated in interconnections, not that inherent in transistor operation.

Superconducting materials will also reduce the weight and bulk of transformers and motors to a fraction of their present sizes. Many applications that are not even considered now because of weight limitations will suddenly become attainable. A example might be artificial limbs, with built in motors for every type of motion desired.

Transistors

One of the more recent semiconductors is gallium arsenide, GaAs, which has the same electronic structure as silicon. Studies have shown that electrons can move more freely through GaAs than through Si, resulting in less power dissipation and higher-frequency operation (up to 2 GHz). Gallium arsenide is also much less affected by high temperature and thus would appear to be a good replacement for silicon. The reason for its slow progress to date is the high cost of purification, and the difficulty of growing large crystals. Despite these problems we expect to see extensive applications in the future.

Diamond is also similar to silicon in electronic structure. Diamond has great heat resistance and excellent physical properties such as hardness and low friction surfaces. Hence it is an good candidate for fabrication of semiconductor devices. Unfortunately, the technology for the production of diamond semiconductors has not yet been achieved, though much effort is being given to its development. Anther possible application is for the manufacture of superior memory disks made of magnetic material coated with a thin layer of slippery diamond.

Computers

The progress of semiconductor memory devices has historically paralleled the development of ICs, and similar considerations hold. RAM memory in the hundreds of megabytes for personal computers is not an unreasonable expectation.

Of great future potential would be the fabrication of magnetic nonvolatile integrated circuit memories. In the early days of computers, memory was implemented by a series of tiny ferrite rings, called **cores**, one per bit, with fine wires threaded through them. The assembly process was by necessity mechanical, and the cost became prohibitive when compared to the newly developed semiconductor memories fabricated by a photolithographic process. Core memory has disappeared, but the situation would change if a method could be developed to permit production of the equivalent of magnetic cores by a lithographic process. Such memories would still be relatively slow and would require larger currents to operate, but they could be useful for ROM and would easily better the performance of hard disks. While it is difficult to conceive of hard disks with access times faster than *milli*seconds, lithographic core could operate at under one *microsec-*

ond. In addition the lack of moving parts would permit a more compact size and better resistance to shock and vibration.

It is expected that optical memory will continue to develop. At this point in time, optical recording is taking place only at the surface of the medium. When methods will be developed to record **inside** the material in three dimensions, the recording capacities might increase into the terabyte or even petabyte region.

Mainframe Computers

We often hear predictions about a "mainframe on a chip" and "mainframe in a micro," which are simply indications of the lightning fast progress of microcomputers. The principal reasons for the fast advances are the extensive market and fierce competition, resulting in gestation times for new products of perhaps only a year, This compares favorably with maybe five years development time for large computers. The latter tend to lag behind in their development.

The large computers are not likely to disappear, because they involve not only hardware, but a complex of software, service, customer confidence, tradition, status and many other elements that small computers cannot offer. More questionable is the future of minicomputer, as it suffers from competition both from mainframes and high end PCs. We predict reorganization of the market into three main classes:

1. Portable personal micros, inexpensive and with power exceeding most individual needs. They can be plugged into telephone-line networks for communications and increased computing power.

2. Desktop computers with the power of present-day mainframes and large memories, interconnected by LANs within an organization. They will fulfill most of the tasks assigned today to large computers, minis, and PCs. The LAN server will be the only major use for larger computers.

3. High-power supercomputers for science, defense, and so on. There will always be a demand for the most powerful computers available.

Communications and Networks

It is a safe bet that the most dramatic progress will be encountered in this area, since we can observe even now an extremely high slope on the application growth curve, sometimes indicating a doubling in six months! Network transmission rates of 100 Mbit/s are within sight, and a gigabit/s is conceivable in the near future. We expect computer communications to reach the same convenient, widespread use as the telephone.

Chapter 20

Mathematical Background

Throughout this book we have refrained from invoking mathematical treatments of electronic and computer circuitry, except in elementary applications. Clearly electronic engineers who design such circuits must have much more extensive and detailed methods at their command. In this chapter we explore some of these techniques. Many examples of circuit design and analysis given previously can be solved more expeditiously by the methods to be described. These techniques were not elaborated in the body of the text because it was felt that they would detract from the logical exposition of electronics. This chapter will be divided into two sections, dealing respectively with analog circuitry and with the intricacies of Boolean algebra.

Methods of Circuit Design

It must be remembered that the responses of reactances to applied voltages are time-dependent. We have emphasized this time-dependence in our definitions of impedances of reactive elements. Recall that the impedance of a capacitor is $Z_C = 1/j\omega C$, and that of an inductor is $Z_L = j\omega L$, in which j is the imaginary operator, $\sqrt{-1}$, and ω is the frequency expressed in radians per second. A number of mathematical techniques are available to simplify the analysis of such circuits.

We will consider a number of notation schemes that offer advantages in the treatment of AC circuits. We have shown earlier that an AC voltage or current can be represented as a sine wave plotted as

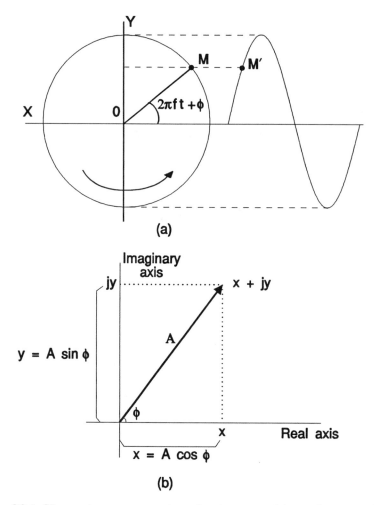

Figure 20-1 The vector representation of a sine wave: (a) as a function of time, and (b) in terms of a stationary vector, the phasor notation.

amplitude against time, or as the projection of a rotating vector. It is often simpler to consider the vector as stationary at some particular instant, with a particular amplitude and phase. Then we can plot the vector on cartesian coordinates as in Figure 20-1*b* with no loss of information compared to the sine-wave portrayal, *a* in the figure (which is identical with Figure 2-8).

Figure 20-1*b* also indicates the mathematical relationships between the polar coordinates A and ϕ, on the one hand, and the rectangular (Cartesian) coordinates, x and jy on the other. Thus a phasor can be equally well be represented as a vector by the notation

Table 20-1

Vectors for key angles

Key	Cartesian	Polar (degrees)	Polar (radians)
a	$1 + j0$	$1 \angle 0°$	$1 \angle 0$
b	$\frac{1}{\sqrt{2}} + \frac{j}{\sqrt{2}}$	$1 \angle 45°$	$1 \angle \pi/4$
c	$0 + j$	$1 \angle 90°$	$1 \angle \pi/2$
d	$-\frac{1}{\sqrt{2}} + \frac{j}{\sqrt{2}}$	$1 \angle 135°$	$1 \angle 3\pi/4$
e	$-1 + j0$	$1 \angle 180$	$1 \angle \pi$
f	$-\frac{1}{\sqrt{2}} - \frac{j}{\sqrt{2}}$	$1 \angle 225$	$1 \angle 5\pi/4$
g	$0 - j$	$1 \angle 270°$	$1 \angle 3\pi/2$
h	$\frac{1}{\sqrt{2}} - \frac{j}{\sqrt{2}}$	$1 \angle 315°$	$1 \angle 7\pi/4$

$A \angle \phi$ or as the complex number $(x + jy)$. Simple trigonometric relations permit the interconversion of the two forms:

$$x = A \cos \phi$$
$$y = A \sin \phi$$
$$x + jy = A(\cos \phi + j \sin \phi)$$
$$\phi = \arctan (y/x)$$
$$A = \sqrt{x^2 + y^2}$$

These relations are demonstrated for multiples of 45° in Figure 20-2 and Table 20-1.

The fact that AC currents and voltages can be represented by complex numbers implies that their ratio, the impedance, can also be represented by a complex number. For example, if the current I is taken as the phase reference, and the phase angle of the voltage E is found to be $+90°$, the impedance must be

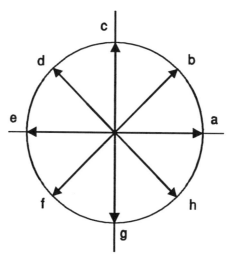

Figure 20-2 Figure to identify angles for Table 20-1.

$$Z = (E \angle 90°)/(I \angle 0) = (E / I) \angle 90°$$

The impedance in this case is a purely imaginary number. (This illustrates the rule for division or multiplication of vector quantities in polar notation: the **real coefficients**, E and I in this example, are handled as in ordinary arithmetic, but the **angular** quantities are subtracted in division or added in multiplication.)

We can perform such calculations by means of complex number notation rather than polar coordinates. Now we are interested only in phase differences. Since the current was taken as the phase reference, I can be represented by the complex $(1 + j0)$, and the voltage by $(0 + j)$, both taken from Table 20-1. Dividing these to give the impedance, we get

$$Z = \frac{0 + j}{1 + j0} = j$$

which confirms our statement that the impedance is purely imaginary in this case.

Impedances in Complex Form

The fact that impedances are complex, far from obscuring numerical relations, makes them more convenient and less likely to cause errors. The impedance of combinations of resistances, capacitances, and

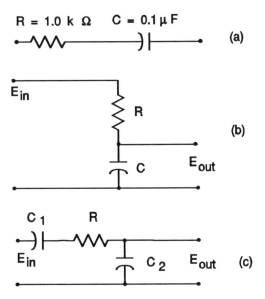

Figure 20-3 *RC* circuits to illustrate the phasor method.

inductances can be written directly if we keep in mind the following relations:

Resistances: $Z_R = R = R \angle 0$

Capacitances: $Z_C = \dfrac{1}{j\omega C} = \dfrac{1}{\omega C} \angle -90°$

Inductances: $Z_L = j\omega L = \omega L \angle 90$

Series impedances: $Z_T = Z_1 + Z_2 + Z_3 + \ldots$

Parallel impedances: $1/Z_T = 1/Z_1 + 1/Z_2 + 1/Z_3 + \ldots$

To illustrate the types of computations that can be made with complex impedances, consider the circuits shown in Figure 20-3. The impedance of the series circuit in (a) is

$$Z_T = R + \frac{1}{j\omega C}$$

Let us suppose that 100 V of AC at frequency 1 kHz is impressed on this circuit. For this frequency, the table in the Appendix tells us that Z_C = 1.6 kΩ. Impedances must be added vectorially, so we can write

$$Z_T = R + \frac{1}{j\omega C} = 1.0 \text{ k}\Omega - j1.6 \text{ k}\Omega$$

or, in polar form,

$$\sqrt{(1000)^2 + (1600)^2} \angle \arctan \frac{-1000}{1600} = 1.89 \text{ k}\Omega \ \angle(-32°).$$

From this the current can be computed as

$$I = \frac{E}{Z_T} = \frac{100 \angle 0}{1.89 \angle (-32°)} = 52.9 \angle (+32°) \text{ mA}$$

An AC ammeter in series with this circuit would indicate 52.9 mA, since such a meter cannot distinguish phase angles.[1]

Transfer Coefficients

Phasors are also applicable to transfer coefficients, defined for any circuit as the ratio of output to input, E_{out}/E_{in}. For example, the circuit in Figure 20-3b is a voltage divider, for which we can write

$$\frac{E_{out}}{E_{in}} = \frac{Z_C}{Z_R + Z_C} = \frac{\frac{1}{j\omega C}}{\frac{1}{j\omega C} + R} = \frac{1}{1 + j\omega RC}$$

---

[1] "Common sense" might predict that the total impedance would be the sum of the two impedances, 2.6 kΩ, and that the current passing would be $I = E/Z = 100/2.6 = 38.5$ mA, which is far different from the correct 52.9 m A.

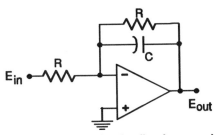

Figure 20-4   An op amp with a complex feedback network.

From this equation we can deduce the behavior of the circuit at various frequencies: at very high frequencies, the transfer coefficient approaches zero, whereas for DC it becomes unity.

Treated in a similar fashion, the more complicated circuit of Figure 20-3c, gives for its transfer coefficient

$$\frac{E_{out}}{E_{in}} = \frac{\dfrac{1}{j\,\omega\,C_2}}{\dfrac{1}{j\,\omega\,C_1} + R + \dfrac{1}{j\,\omega C_2}}$$

It is left for the reader to rewrite this expression in the conventional form $(a + jb)$, and to draw conclusions regarding its frequency response.

It is also possible to use this approach to determine the transfer coefficient of an op amp, as in Figure 20-4. It is easy to see that the transfer coefficient (the gain) is given by

$$\frac{E_{out}}{E_{in}} = -\frac{1}{(1/R + j\,\omega C)}\,\frac{1}{R} = -\frac{1}{1 + j\,\omega\,RC}$$

which, except for the negative sign, is the same as the transfer coefficient derived for the low-pass filter of Figure 20-3b.

Differential Equations

In solving the current/voltage relationships for AC currents we must deal with differential equations. We may even have to handle integro-differential equations, since a circuit with both capacitance and inductance may lead to equations containing both differentials and integrals. These can usually be solved by the traditional methods, but often the treatment is simplified by the use of **Laplace transforms**. We will solve an example both ways to demonstrate the relation between

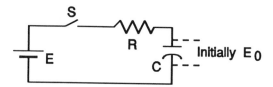

Figure 20-5 An *RC* circuit illustrating transient and steady-state conditions.

the two methods. The relations between voltages and currents for resistors, capacitors, and inductors, can be summarized as follows:

$$E = IR \qquad\qquad I = \frac{E}{R} \qquad\qquad\qquad \text{(20-1 } a,b)$$

$$E = \frac{1}{C}\int I\, dt \qquad I = C\frac{dE}{dt} \qquad\qquad \text{(20-2 } a,b)$$

$$E = L\frac{dI}{dt} \qquad\qquad I = \frac{1}{L}\int E\, dt \qquad\qquad \text{(20-3 } a,b)$$

These formulas can be combined for a given circuit to obtain the overall differential equation. For example, in the circuit shown in Figure 20-5, the capacitor is initially charged to E_0 volts, but at time zero, when the switch S is closed, it is suddenly connected to a voltage source E through a resistor R. At steady state, no current can flow, and the capacitor will be charged to the voltage E. However, a finite length of time is required for this condition to be attained, producing a **transient**[2] current. The differential equation must describe the time dependence of the transient voltage and current during the transition period. Let us apply the Kirchhoff voltage law to this circuit, obtaining

$$E = IR + \frac{1}{C}\int I\, dt + E_0 \qquad\qquad \text{(20-4)}$$

---

[2] Transients exist in reactive systems as the result of sudden changes, particularly those occurring when a signal source is connected to the system. For instance, after a sine wave is applied to the input, the output may be small in amplitude at the beginning, increasing progressively to its steady state. After a long enough time has elapsed, the transient will disappear and the current predicted by the simple treatment will be attained.

This equation can be converted to a differential equation by differentiating both sides and solving for the current, giving

$$0 = R\frac{dI}{dt} + \frac{I}{C} \quad \text{or} \quad I = -RC\frac{dI}{dt} \qquad (20\text{-}5)$$

Note that E is constant. Our immediate objective is to solve this equation for an explicit statement of the current as a function of time.

Conventional Solution

It is readily shown by the standard methods of handling differential equations that the solution for this equation is

$$I = A \exp\left(-\frac{t}{RC}\right) \qquad (20\text{-}6)$$

in which the constant A depends on the initial conditions of the circuit, specifically on the voltage E_0. At time $t = 0$, A is equal to the initial current I_0, which is given by $(E - E_0)/R$. Consequently the desired expression for the current is

$$I = \frac{E - E_0}{R} \exp\left(-\frac{t}{RC}\right) \qquad (20\text{-}7)$$

From this expression we can obtain the voltage on the capacitor as a function of time using the relation from Eq. (20-2a) and assuming an initial charge of E_0:

$$E_C = \frac{1}{C} \int I\, dt + E_0 \qquad (20\text{-}8)$$

Substituting the value of I from Eq. (20-7) into Eq. (20-8) and integrating gives

$$E_C = E\left[1 - \exp\left(-\frac{t}{RC}\right)\right] + E_0 \exp\left(-\frac{t}{RC}\right) \qquad (20\text{-}9)$$

$$E_C = E + (E_0 - E)\exp\left(-\frac{t}{RC}\right) \qquad (20\text{-}10)$$

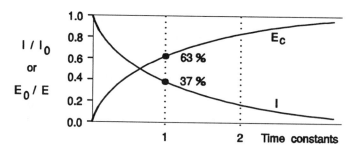

Figure 20-6 The time-behavior of the RC circuit shown in Figure 20-5.

Thus as time increases, the current approaches zero, as expected from Eq. (20-7), whereas the voltage across the capacitor approaches E, as predicted by Eq. (20-10). Similarly one can show that, for time $t = RC$ (i.e. after one time constant), the current is reduced to $1/e$ of its initial value, where e is the base of natural logarithms. Figure 20-6 gives graphs of both current and voltage against time.

Laplace Solution

To solve a differential equation by the method of Laplace transformation, we must introduce a new function, $\overline{F}(s)$. The dummy variable s is related to the former variable t by the relation

$$\overline{F}(s) = \int_0^\infty F(t) \exp(-st)\, dt \qquad (20\text{-}11)$$

It is not usually necessary to invoke this equation because corresponding pairs of functions of s and t are to be found in tabular form in many reference works; an abbreviated list is given in Table 20-2.

Taking the Laplace transform replaces the usual function $F(t)$ by a new function that is an algebraic statement in terms of the variable s, as shown in the table. The next step in solving the differential equation is to rearrange the algebraic terms into their simplest form. Then taking the inverse transform will give the solution to the equation.

Let us use this method to solve equation (20-4):

$$E = IR + \frac{1}{C} \int I\, dt + E_0 \qquad (20\text{-}12)$$

Table 20-2

Examples of Laplace transforms

	$F(t)$	$\mathcal{L}\{F(t)\} = \overline{F}(s)$
1	t	$1/s^2$
2	Step, amplitude a	a/s
3	$\exp(-at)$	$1/(s + a)$
4	$(1/a)[1 - \exp(-at)]$	$1/[s(s + a)]$
5	$t \exp(-at)$	$1/(s + a)^2$
6	$(1/a) \sin at$	$1/(s^2 + a^2)$
7	$\cos at$	$s/(s^2 + a^2)$
8	$(1/k) \exp(-at) \sin kt,$	$1/[(s^2 + 2as + b^2]$
	where $k = \sqrt{b^2 + a^2}$	

The transform of a numerical coefficient is the coefficient itself, $\mathcal{L}\{kE\}$ $= k\overline{E}$. Note that since the voltage E is connected suddenly, this transform can be considered to be a step function. Using entries from Table 20-2, we can write

$$\frac{E}{s} = R\overline{I} + \frac{\overline{I}}{sC} + \frac{E_0}{s} \qquad (20\text{-}13)$$

This relation contains only two variables, s and $\overline{I}$. It can be solved for $\overline{I}$ to give

$$\overline{I} = \frac{E - E_0}{s} \frac{1}{R + 1/sC} \qquad (20\text{-}14)$$

which can be converted to the more tractable form

$$\overline{I} = \frac{E - E_0}{R} \frac{1}{s + 1/RC} \qquad (20\text{-}15)$$

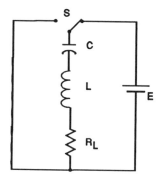

Figure 20-7 An *LC* circuit in which transients can be induced by means of a switch.

We now can use the entries in Table 20-2 in the reverse direction to give us

$$I(t) = \frac{E - E_0}{R} \exp\left(-\frac{t}{RC}\right) \qquad (20\text{-}16)$$

which is identical with Eq. (20-7) obtained by classical means. If it seems that the old method is just as good as the new, remember that this is a very simple differential equation; the Laplace procedure shows up to great advantage with more complicated equations.

As another example, consider the *LC* circuit shown in Figure 20-7. Since every inductor has appreciable resistance, it is necessary to include this explicitly in the diagram, here designated R_L. The voltage E is initially present across the capacitor, then at time $t = 0$, the switch is thrown to the left so that the capacitor can discharge through the inductance L and its resistance R_L. The current at time 0, designated as $I(0)$, is zero. Since the source of voltage vanishes at zero time, this is equivalent to a step of magnitude $-E$ volts. Again, we apply Kirchhoff's law, getting:

$$E \to 0 = \frac{1}{C} \int_0^t I\, dt + L\frac{dI}{dt} + RI \qquad (20\text{-}17)$$

The Laplace transform of the expression is

$$-\overline{E} = \frac{\overline{I}}{sC} + Ls\overline{I} + R\overline{I} \qquad (20\text{-}18)$$

solving for $\bar{I}$ gives

$$\bar{I} = -\frac{\bar{E}}{L}\frac{1}{s^2 + R s/L + 1/LC} \tag{20-19}$$

We can inversely transform this using pair No. 8 in Table 20-2, and putting $a = R/2L$ and $b = 1/\sqrt{LC}$:

$$\bar{I} = -\frac{EL}{k}\exp\left(-\frac{R}{2L}t\right)\sin\left(\frac{1}{L}kt\right) \tag{20-20}$$

where

$$k = \sqrt{L/C - R^2/4}$$

This equation indicates that the current flowing is a sine wave multiplied by a negative exponential, meaning that the amplitude of the sine wave decreases exponentially with time until it reaches zero. If the solution for this circuit is worked out with R_L omitted, the result is

$$I = -EL\sqrt{LC}\sin\left(\frac{1}{\sqrt{LC}}t\right) \tag{20-21}$$

This equation shows a continuing sine wave with no transient. Thus the effect of the resistance of the inductor is to damp out the sine wave produced by the LC combination and to introduce a minor change of frequency.

Another important property of Laplace transforms is the simplicity of the transformation of time-derivatives and time-integrals. Thus we can write

$$\pounds\{dE/dt\}) = s\bar{E} - E_0 \tag{20-22}$$

$$\pounds\left\{\int E\,dt\right\} = \frac{\bar{E}}{s} + \frac{k}{s} \tag{20-23}$$

It follows from these relations that we can write the transforms of the simple defining relations of Eqs. (20-1) to (20-3) directly as,

$$\mathcal{L}\{IR\} = \overline{I}R \tag{20-24a}$$

$$\mathcal{L}\{E/R\} = \frac{1}{R}\overline{E} \tag{20-24b}$$

$$\mathcal{L}\{\frac{1}{C}\} \int I \, dt = \frac{1}{C}(\frac{\overline{I}}{s} + \frac{k}{s}) \tag{20-25a}$$

$$\mathcal{L}\{C \frac{dE}{dt}\} = C[s\overline{E} - I(0)] \tag{20-25b}$$

$$\mathcal{L}\{L \frac{dI}{dt}\} = L[s\overline{I} - I(0)] \tag{20-26a}$$

$$\mathcal{L}\{\frac{1}{L} \int E dt\} = \frac{1}{L}(\frac{\overline{E}}{s} + \frac{k}{s}) \tag{20-26b}$$

s-Domain Impedance

We have previously defined the notion of complex impedance, which represents the ratio of steady-state AC voltage to current for reactive circuits. This approach, however, is not able to deal with transients.

A very general concept, capable of treating both steady-state and transient conditions, is the **s-domain impedance,** derived directly from the differential equation for the circuit. Consider once more a circuit containing R, L, and C components in series. The differential equation, as shown earlier, is

$$\overline{E} = IR + L\frac{dI}{dt} + \frac{1}{C} \int_0^t I \, dt \tag{20-27}$$

Laplace transforms can be written directly, denoting the resistor by R, the inductor by sL, and the capacitor by $1/sC$:

$$\overline{E} = (R + sL + \frac{1}{sC})\overline{I} \tag{20-28}$$

Since the ratio $\overline{E} / \overline{I}$ is an impedance, the quantity within the brackets meets this definition, and is called the *s*-domain impedance, or simply the impedance.

The *s*-impedance can be written directly without the need to set up the differential equation, by following these rules:

- The impedances of single elements are R, sL, or $1/sC$, as the case may be.

- Series and parallel combinations are given by the well-known rules for combinations of impedances.

For example, the RC series circuit will have an impedance given by

$$Z = R + \frac{1}{sC} = \frac{sRC + 1}{sC} \tag{20-29}$$

If a DC step voltage E is impressed upon the circuit, the current is

$$\overline{I} = \frac{\overline{E}}{\overline{Z}} = \frac{E}{s} \frac{sC}{1 + sRC} \tag{20-30}$$

(This transform was encountered earlier, with initial conditions present.) Setting $E_0 = 0$, we can use the solution given in Eq. (20-15)

$$I = \frac{E}{R} \exp\left(-\frac{t}{RC}\right) \tag{20-31}$$

This equation completely describes the transient. The steady state is obtained by setting t equal to infinity, which gives $I = 0$, a result to be expected for a DC circuit containing a capacitor.

To illustrate the use of Laplace transforms with AC systems, let us consider the case of a single capacitor. Entry 6 of Table 20-2 tells us that the transform $\pounds\{1/\omega\} \sin(\omega t)$ is $1/(s^2 + \omega^2)$. It is convenient to take an input signal of $B \sin(\omega t)$, which can be written as $\omega B[(1/\omega) \sin(\omega t)]$, with the transform $\omega B/(s^2 + \omega^2)$. Hence the transform of the current becomes

$$\frac{\overline{I}}{\overline{Z}} = \frac{\omega B/(s^2 + \omega^2)}{1/sC} \tag{20-32}$$

The inverse transform, from entry 7 of the table, gives

$$I = \omega BC \cos \omega t \tag{20-33}$$

This expression indicates that the current leads the voltage by 90° and has an amplitude that is proportional to the frequency, namely, ωBC. Note that there is no transient in this simple circuit.

Fourier Series

It was mentioned previously that if two or more sine waves of the same frequency are added together, the result will also be a sine wave. Of more interest is to examine a mixture of sinusoids of *different* frequencies. Of particular importance is a combination of waves that are simple multiples (harmonics) of a fundamental frequency. It can be shown, for example, that the summation of sine waves that are odd harmonics, with amplitudes inversely proportional with their order, will result in a close approximation of a square wave at the fundamental frequency. This can be expressed mathematically:

$$\sum_{1}^{\infty} \frac{\sin (n\omega t)}{n} \qquad \text{(for odd values of } n) \tag{20-34}$$

Similarly, a triangle wave follows the summation:

$$\sum_{1}^{\infty} \frac{\cos (n\omega t)}{n^2} \qquad \text{(for odd values of } n) \tag{20-35}$$

A third example represents the wave form of a full-wave rectified AC, which uses the summation

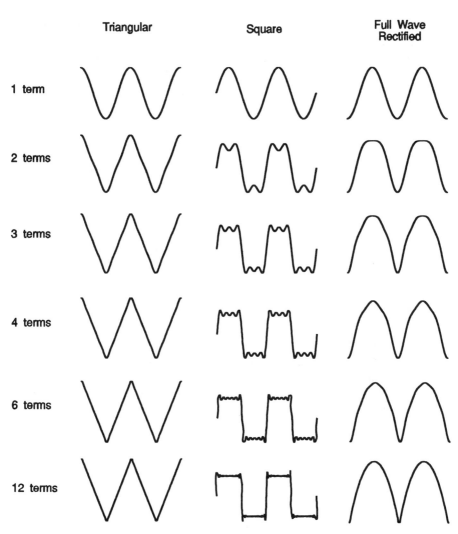

Figure 20-8 Three examples of waveforms constructed by the summation of terms according to the Fourier series of Eqs. 20−34, 20−35, and 20−36.

$$\sum_{1}^{\infty} \frac{\cos (n\omega t)}{(n/2)^2} \quad \text{(for even values of } n) \quad (20\text{-}36)$$

Figure 20-8 shows the waveforms for each of these three examples for several values of n. It will be seen that each successive term brings the sum closer to the result theoretically given by an infinite number of terms. Because of the division of terms by n or n^2, it follows that the importance of later terms drops off rapidly, so that only a few terms (e.g. to $n = 6$) are needed to give an effectively perfect representation of the desired waveform. Such summations can be generalized by an expression called the **Fourier series,** which in its fully expanded form is

$$F(\alpha) = A_0 + A_1 \cos \alpha + A_2 \cos 2\alpha + A_3 \cos 3\alpha + \cdots + A_n \cos n\alpha$$
$$+ B_1 \sin \alpha + B_2 \sin 2\alpha + B_3 \sin 3\alpha + \cdots + B_n \sin n\alpha$$
$$(20\text{-}37)$$

In order to represent a periodic voltage, the variable α is conveniently replaced by be replaced by ωt. It can be seen that this equation includes as special cases the relationships of the equations 20-34 to 20-36.

The Fourier series has three basic segments. The first is A_0, an unvarying term, which represents DC average value of the complete waveform over an integral number of cycles. The second segment consists of cosines of successive harmonics, and the third part is a similar summation of sines. There are no restrictions on the A and B coefficients, which can be of either algebraic sign, and have any numerical value including zero. In many important cases, as illustrated in the above examples, all sines or perhaps all cosines are zero.

In principle *any* repetitive waveform can be resolved into sine and/or cosine terms plus a constant (DC) term.

Boolean Algebra

In our discussion of binary logic, we have referred briefly to the use of Boolean notation. Now we will develop this approach in greater depth and demonstrate its utility in the design and optimization of logic systems.

Boolean algebra is a mathematical system useful in dealing with binary numbers since only two digits are permitted, 0 and 1. In Boolean, as in classical algebra, we can use letters to represent numbers, but instead of the familiar rules governing the operations of addition, subtraction, multiplication, and division, we are here concerned with rules controlling the four principal operators AND, OR, XOR and NOT. If these rules are followed, there will always be a

Table 20-3

Symbols of Boolean algebra

Operator	Symbol	Example
AND	•	$A \bullet B \bullet C = A$ AND B AND C
OR	+	$A + B + C = A$ OR B OR C
XOR	⊕	$A \oplus B \oplus C = $ *any one* of A, B, or C
NOT	‾	$\overline{A} = $ NOT $A = $ complement of A

correspondence between the voltages in logic gates and the results of Boolean expressions. The four basic operators and their symbols are defined in Table 20-3.

It is perhaps unfortunate that the symbols (•) and (+) have distinctly different meanings in the two algebras, but this convention is so firmly established that we can only conform. The dot symbol can be omitted if desired; thus AB is the same as $A \bullet B$. The basic operations are similar to those of standard arithmetic, but there are major differences. The AND function, for example, leads to the expressions

$$A \bullet 1 = A$$

$$A \bullet 0 = 0$$

$$A \bullet A = A$$

These relations make sense if one realizes that A can be either 0 or 1. In the third statement, for instance, if $A = 1$, then $A \bullet A$ becomes 1•1 = 1, but if $A = 0$, $A \bullet A$ becomes $0 \bullet 0 = 0$. (Translating into the context of logic gates, we can point out that a two-input AND gate will give 1 as output if both inputs are 1, or a 0 output if both inputs are 0. We say, "if both inputs are true, then the output is true.") Similarly the OR function leads to the statements

$$A + 1 = 1$$

$$A + 0 = A$$

Table 20-4

Boolean relationships

Rule No.	Relation	Remarks
1	$0 + 0 = 0$	
2	$0 \bullet 0 = 0$	
3	$1 + 1 = 1$	
4	$1 \bullet 1 = 1$	
5	$0 \bullet 1 = 0$	
6	$0 + 1 = 1$	
7	$X + X = X$	
8	$X \bullet X = X$	
9	$X \bullet \overline{X} = 0$	
10	$X + \overline{X} = 1$	
11	$X + 0 = X$	
12	$X \bullet 0 = 0$	
13	$X + 1 = 1$	
14	$X \bullet 1 = X$	
15	$X + Y = Y + X$	Commutation
16	$X \bullet Y = Y \bullet X$	Commutation
17	$(X + Y) + Z = X + (Y + Z)$	Association
18	$(X \bullet Y) \bullet Z = X \bullet (Y \bullet Z)$	Association
19	$X \bullet (Y + Z) = X \bullet Y + X \bullet Z$	Distribution
20	$X + X \bullet Y = X$	Simplification
21	$X + \overline{X} \bullet Y = X + Y$	Simplification
22	$X \bullet (X + Y) = X$	Simplification
23	$\overline{X \bullet Y} = \overline{X} + \overline{Y}$	DeMorgan
24	$\overline{X + Y} = \overline{X} \bullet \overline{Y}$	DeMorgan

$$A + A = A.$$

The NOT function is simpler; the only possibilities are

$$\bar{0} = 1$$

$$\bar{1} = 0$$

Combining the NOT function with AND or OR, to give NAND and NOR, results in two more expressions

$$A \bullet \bar{A} = 0$$

$$A + \bar{A} = 1$$

The commutative and associative rules offer no difficulties:

$$A + B = B + A$$

$$A \bullet B = B \bullet A$$

$$A + (B + C) = (A + B) + C$$

$$A \bullet (B \bullet C) = (A \bullet B) \bullet C$$

The distributive rules that permit factoring expressions are:

$$A \bullet (B + C) = (A \bullet B) + (A \bullet C)$$

$$A + (B \bullet C) = (A + B) \bullet (A + C)$$

There are so many of these rules that it is convenient to summarize them as in Table 20-4, where the letters X, Y, and Z refer to any binary variables.Other statements can be derived by combinations of the above rules. For example, the statement

$$A + (\bar{A} \bullet B) = A + B$$

can be derived as follows. We have seen in rule 13 that $X + 1 = 1$, and by rule 10 that $A + \bar{A} = 1$; hence we can write

$$
\begin{aligned}
A + (\bar{A} \bullet B) \quad &= A\,(1 + B) + (\bar{A} \bullet B) \\
&= A + A \bullet B + (\bar{A} \bullet B) \\
&= A + B\,(A + \bar{A}) \\
A + (\bar{A} \bullet B) \quad &= A + B
\end{aligned}
$$

A	B	C	P	Q	X	R	S	Y
0	0	0	1	1	0	0	0	0
0	0	1	1	1	0	0	0	0
0	1	0	1	1	0	0	0	0
0	1	1	1	1	0	0	0	0
1	0	0	1	0	1	0	1	1
1	0	1	0	1	1	1	0	1
1	1	0	1	1	0	0	0	0
1	1	1	1	1	0	0	0	0

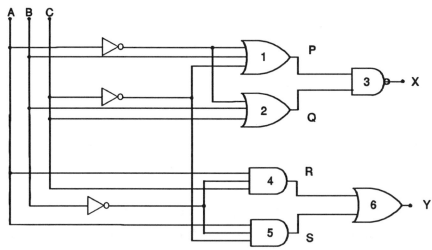

Figure 20-9   Two ways of implementing the same logic function. The truth table shows that X and Y are identical.

The DeMorgan theorem is stated in rules 23 and 24. Application of the theorem can be simplified by the use of a set of three steps:

- Replace every ($\bullet$) by ($+$) and every ($+$) by ($\bullet$).

- Place a bar over every unbarred variable and remove bars from all barred variables.

- Place a bar over the entire resultant expression.

As an example, let us derive the following statement:

$$\overline{A} \bullet (B + \overline{C}) = A + (\overline{B} \bullet C)$$

By Step 1, $\overline{A} \bullet (B + \overline{C})$ becomes $\overline{A} + (B \bullet \overline{C})$

By Step 2, $\overline{A} + (B \bullet \overline{C})$ becomes $A + (\overline{B} \bullet C)$

By Step 3, $A + (\overline{B} \bullet C)$ becomes $A + (\overline{B} \bullet C)$

Such statements permit us to transform a sum-of-products into the equivalent product-of-sums (or the reverse) as in the following example:

$$A \bullet B \bullet C + A \bullet B \bullet C = \overline{(A + B + C) \bullet (A + B + C)} \qquad (20\text{-}39)$$
$$\quad\; R \qquad\quad S \qquad\qquad P \qquad\qquad\quad Q$$

This type of relation can also be proved by the use of truth tables for the corresponding gate structures. Eq. (20-39), for instance, can be implemented by either two OR gates and an NAND gate, or by two ANDs and an OR, as shown in Figure 20-9, in which both assemblies are included. The letters R, S, P, and Q represent the intermediate outputs corresponding to the four term in Eq. (20-39). The truth table shows that output X, based on gates 1 through 3, is identical with Y, obtained using gates 4 through 6.

After a bit of practice, it will become apparent that the DeMorgan conversion can be written down on sight without explicitly passing through the several steps.

The preceeding technique can be used to simplify logical expressions so that they can be be efficiently implemented in hardware. For example, consider the expression $A(A + B)$. As it stands, this would call for two gates, connected as in Figure 20-10. However, the circuit can be simplified in the following way:

$A \bullet (A + B) = A \bullet A + A \bullet B$ (Rule 19)

$A \bullet A + A \bullet B = A + A \bullet B$ (Rule 8)

$A + A \bullet B = A \bullet (1) + A \bullet B$ (Rule 14)

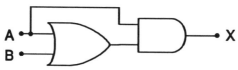

Figure 20-10 The logic equivalent of $X = A \cdot (A + B)$.

$$A \cdot (1) + A \cdot B = A \cdot (1 + B) \qquad \text{(Rule 19)}$$
$$A \cdot (1 + B) = A \cdot (1) \qquad \text{(Rule 14)}$$
$$A \cdot (A + B) = A$$

This gives the surprising result that both gates can be eliminated. It makes no difference whatever what values B may assume. This will become evident if one constructs the truth table corresponding to Figure 20-9.

Wherever the EXCLUSIVE-OR function (XOR) appears, it can be replaced by its equivalent: $A \oplus B = A \cdot \overline{B} + \overline{A} \cdot B$.

Minterms and Karnaugh Maps

An important application of Boolean algebra in our context is the simplification of systems of digital logic. The first step in this process, after establishing a working expression, is to bring it to the standard or **canonical** form. This is an equivalent expression consisting solely of the sum of terms each of which contains *all* the variables. The basic concept can be illustrated by conversion of the expression $A + B$. Each term in the canonical form must contain both A and B. The conversion is performed as follows, remembering that $X + \overline{X} = 1$, by Rule 10:

$$A + B = A \cdot (B + \overline{B}) + B \cdot (A + \overline{A})$$
$$= (A \cdot B) + (A \cdot \overline{B}) + (A \cdot B) + (\overline{A} \cdot B)$$

Each of the terms in the standard form must contain *all* the variables, and is called a **minterm**.[3] In calculating these terms, duplicates may arise (such as A • B repeated in the above expression); if this occurs, simply delete one of them. One advantage of using the canonical form is that the minterms can be represented as unique

---

[3] An alternative is to convert expressions to a product-of-sums rather than a sum-of-products. These are called **maxterms.** They are less used and will not be developed here.

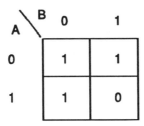

Figure 20-11 Karnaugh map for the two-variable example given in text.

binary numbers, by simply assigning the digits 1 and 0 to the letters given in sequence. An unbarred letter is given the digit 1, while a barred letter becomes a 0.

Thus for three variables there are a maximum number of eight minterms:

$$A \bullet B \bullet C + \overline{A} \bullet B \bullet C + A \bullet \overline{B} \bullet C + A \bullet B \bullet \overline{C} + \overline{A} \bullet \overline{B} \bullet C + \overline{A} \bullet B \bullet \overline{C} + A \bullet \overline{B} \bullet \overline{C} + \overline{A} \bullet \overline{B} \bullet \overline{C}$$
$$1\ 1\ 1\quad 0\ 1\ 1\quad 1\ 0\ 1\quad 1\ 1\ 0\quad 0\ 0\ 1\quad 0\ 1\ 0\quad 1\ 0\ 0\quad 0\ 0\ 0$$

The electronic significance of the minterms is that they represent all possible input combinations that give a HIGH output. A direct implementation of the above expression can be made with eight three-input AND gates, a seven-input OR gate and three inverters, but this is an unwieldy "brute-force" attack. In any given problem, only a fraction of the eight terms (and eight gates) will be needed. The circuitry can be simplified by the use of **Karnaugh maps**, also called K-maps. Each such map is a rectangular array of numbered boxes corresponding to all possible minterms. Ones and zeros are to be inserted depending on whether or not a minterm appears in the given expression.

The procedure for constructing a map is best explained with the aid of a few examples. Consider first the expression
$$X = \overline{A} + A \bullet \overline{B}$$
This expression can be converted to the equivalent canonical form as the sequence of minterms

$$X = \overline{A} \bullet B + \overline{A} \bullet \overline{B} + A \bullet \overline{B}$$
$$0\quad 1\quad\ 0\quad 0\quad\ 1\quad 0$$

BC				
A	00	01	11	10
0	0	0	1	1
1	1	0	1	1

Figure 20-12 A Karnaugh map for the three-variable example.

The 2×2 map of Figure 20-11 is filled by entering a 1 in each of the three cells coresponding to the three minterms. A 0 is inserted in the cell corresponding to 11.

By convention the individual cells in the Karnaugh are designated by the numbers at the left side followed by those at the top. Thus the upper right corner cell in the figure is 01.

Similarly for the expression

$$A \bullet B \bullet C + A \bullet B \bullet \overline{C} + A \bullet \overline{B} \bullet \overline{C} + \overline{A} \bullet B \bullet C + \overline{A} \bullet B \bullet \overline{C}$$

we can set up a map with eight cells, as in Figure 20-12 of which five will be filled with ones.

For four variables, the map must have sixteen cells. Figure 20-13 shows the K-map for the expression:

$$\overline{A} \bullet \overline{B} \bullet C \bullet \overline{D} + \overline{A} \bullet B \bullet \overline{C} \bullet \overline{D} + \overline{A} \bullet B \bullet \overline{C} \bullet D + A \bullet B \bullet \overline{C} \bullet \overline{D} +$$
$$A \bullet B \bullet \overline{C} \bullet D + A \bullet B \bullet C \bullet D + A \bullet \overline{B} \bullet C \bullet D + A \bullet \overline{B} \bullet C \bullet \overline{D}$$

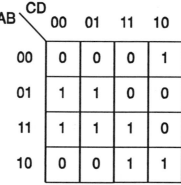

CD				
AB	00	01	11	10
00	0	0	0	1
01	1	1	0	0
11	1	1	1	0
10	0	0	1	1

Figure 20-13 A Karnaugh map for four variables.

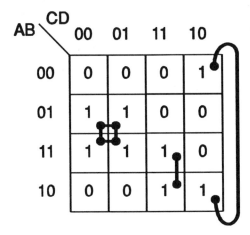

Figure 20-14 The K-map of the previous figure, with three blocks marked, one quad and two pairs.

The utility of K-maps consists of the generation of a smaller set of logical terms to express the original function in simpler form. This is done by identifying adjacent pairs, quads or octets of 1's forming "blocks." For this purpose squares are "adjacent" if they share a vertical or horizontal border. It is permissible for two squares to be functionally adjacent if they occupy opposite ends of a row or column. (Think of the map as being fitted around the surface of a cylinder so that, in Figure 20-14, for instance, cells 0010 and 1010 lie next to each other.) Notice that adjacent cells differ by only a single digit.

Once all the blocks are identified, it is posible to simplify the map by eliminating the empty cells. This is done by generating a new smaller set of terms equivalent to the original ones. Each block contributes one term to a resultant expression. Its value is determined by observing which digits (A, B, C or D) remain unchanged from one cell to another within the block; such digits are conserved (placed in the new term). The digits that change are eliminated For example (see Figure 20-14), the block containing cells 0010 and 1010 have in common the values of bits B, C and D, but not A, hence we can write a term containing each of the letters B, C and D, with bars over B and D that have the value zero, giving the term $\overline{B} \bullet C \bullet \overline{D}$. There will thus be as many terms as there are blocks marked out on the K-map. Each two-cell block uses one less variable than the set, each four-cell block uses two less, and an eight-cell block (two complete adjacent rows or columns) uses three less. The single cells that have not been paired are retained intact.

The map of Figure 20-14 shows three blocks, hence there will be three terms in the final expression. One of these as we have seen is $\overline{B} \bullet C \bullet \overline{D}$. Another, corresponding to the cells 1111 and 1011 yields the term $A \bullet C \bullet D$, because these three variables are shared, each with a 1. The quad gives the term $B \bullet \overline{C}$. Thus the expression becomes:

$$A \bullet C \bullet D + \overline{B} \bullet C \bullet \overline{D} + B \bullet \overline{C}$$

We have now the simplified equivalent of the eight-term expression we started with.

The process of first converting the given expression into its standard form (minterms), then plotting a Karnaugh map, and finally interpreting it provides an extremely powerful method of reducing a complex logic system to its simplest form.

Questions and Problems

Chapter 1

1-1 You are driving an automobile on a busy freeway. List half a dozen or so sources of noise (audible or otherwise) that you might encounter, either in the car mechanisms or in the external environment.

Chapter 2

2-1 What is the distinction between active and passive components? Is a "pot" active or passive?

2-2 Determine the maximum voltage that should be applied across resistors marked as follows: (a) 100 Ω at 2 W, (b) 10 kΩ at 0.5 W, (c) 22 MΩ at 250 mW.

2-3 (a) A square wave with an angular frequency of 10 rad/s has peak potentials of +6 and −6 V. What is the RMS voltage? What is the average voltage? (b) What are the RMS and average voltages for a triangular wave of the same amplitude and frequency?

2-4 A certain capacitor is given a charge of 10 microcoulombs (μC), resulting in a potential difference of 1.75 V. What is its capacitance?

2-5 Sketch the curve resulting from the addition of a sine wave and a cosine wave of the same frequency and amplitude.

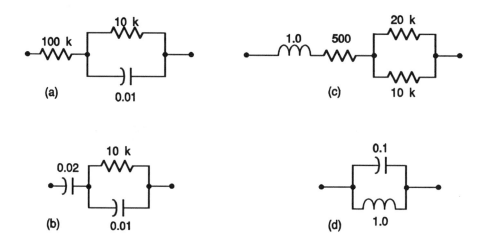

Figure P-2-9

2-6 What numerical value of ϕ will make the following expression true?

$$A \cos(\omega t) = A \sin(\omega t + \phi).$$

2-7 An amplifier with a power gain $G_p = 100$ generates an output noise of 1 mV into a load of 1000 Ω. What is its NEP figure?

2-8 A particular choke coil has an inductance of 20 mH, series resistance of 0.1 Ω, and stray capacitance of 30 pF. At what frequency will it resonate?

2-9 Compute the overall impedance (at frequency $\omega = 1000$ rad/s) of each of the circuits shown in Figure P-2-9. Capacitances are in microfarads, inductances in mH, and resistances in ohms.

2-10 What is the impedance of a 10-kΩ wire-wound resistor with an inherent inductance of 2 mH, when used at 1 kHz?

2-11 A 12-V battery is to be used to power a circuit that requires only 5 volts. To do this, a voltage divider is set up as in Figure P-2-11.
   (a) If $R_1 = 10$ kΩ, what should be the value of R_2 ?

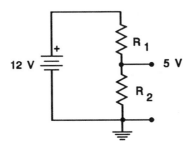

Figure P-2-11

(b) What is the maximum current that can be drawn by the load to avoid a drop of more than 0.1 volt?

2-12 A voltage divider is required that will attenuate a 10-V signal to 10 mV. The current from the source must not exceed 1 mA, and R_{out} must be less than 20 Ω. Design a circuit to meet these conditions.

2-13 In the circuits of Figure P-2-13, determine the power dissipated in each component.

2-14 Compute the power dissipated as heat in each component of the circuits of Figure P-2-14, where the power source is 1000 Hz AC at 10 V (RMS).

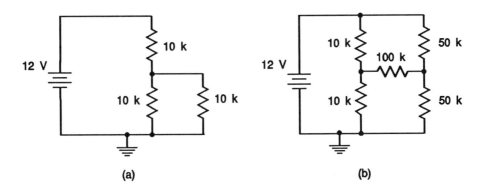

Figure P-2-13

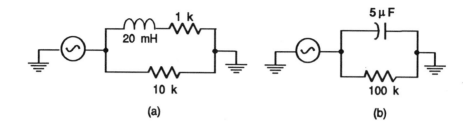

Figure P-2-14

2-15 Determine the total impedance for each of the following combinations:

(a) Resistors of 100 Ω and 151 Ω in parallel.

(b) A 33-pF capacitor in parallel with a 10-kΩ resistor, at a frequency of 50 kHz.

(c) A 100-Ω resistor in series with a 100 mH inductor of negligible resistance, measured with DC.

(d) Same as (c), but measured at 100 kHz.

2-16 A capacitor and inductor are connected in parallel and powered from an AC source, as shown in Figure P-2-16. Assume that the DC resistance of the inductor is negligible.

(a) At what frequency will the impedances of the two components be equal?

(b) What is the net impedance at that frequency?

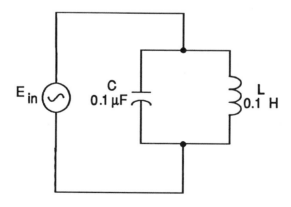

Figure P-2-16

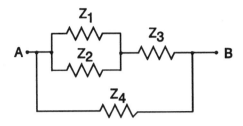

Figure P-2-19

2-17 Ten identical resistors are connected in series. Each is marked 1.0 MΩ at 0.5 W.
(a) What is the wattage rating of the string?
(b) What is the maximum voltage that can be dropped across this string without exceeding the power dissipation limits?

2-18 Answer the same questions as in Problem 2-17 for a parallel connection of the same ten resistors.

2-19 In Figure P-2-19, suppose that $Z_1 = 200$ Ω, $Z_2 = 300$ Ω, $Z_3 = 150$ Ω, and $Z_4 = 500$ Ω. What is the overall impedance of the network?

2-20 Suppose that Z_1 of Figure P-2-19 is a 0.010-μF capacitor, while all other impedances have the same values as in Problem 2-19. What will be the impedance of the circuit if measured at a frequency of 100 kHz? (Hint: The table of capacitive reactances in the Appendix may help.)

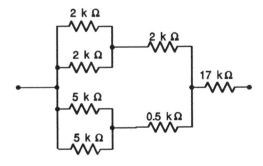

Figure P-2-21

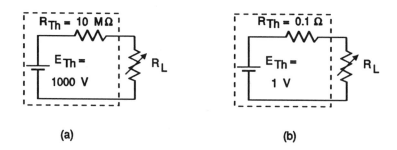

(a)   (b)

Figure P-2-22

2-21 Calculate the overall resistance of the circuit shown in Figure P-2-21.

2-22 For the circuits shown in Figure P-2-22, calculate the voltages across and the currents through the load, for R_L = 100, 1000, and 10,000 Ω. Show that the difference between sources of current and voltage is simply a matter of impedance ratios.

2-23 Obtain both the Thévenin and Norton equivalents of the circuits shown in Figure P-2-23.

2-24 The Thévenin equivalent for a particular power supply is shown in Figure P-2-24, where R_{int} is the internal resistance. The output voltage, measured with a high-impedance voltmeter, is 12.0 V. When a load resistor of 100 Ω is connected, the observed voltage is reduced to 11.5 V. What are the values of R_{int} and E?

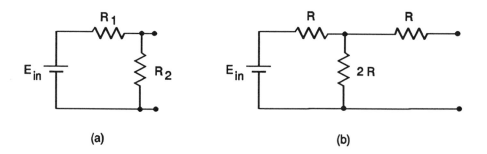

(a)   (b)

Figure P-2-23

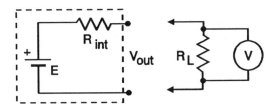

Figure P-2-24

2-25 By application of the Kirchhoff laws, calculate the current through each resistor in the circuit of Figure P-2-25, where $R_1 = R2 = R_3 = 100\ \Omega$ and $E_{in} = 2$ V DC.

2-26 An amplifier is fed from a voltage source with negligible noise. Assume the input noise of the amplifier to be solely of the Johnson type. For a given bandwidth, as an average, the noise equivalent power is 10^{-14} W. Compute the RMS voltage noise and current noise for $R_{in} = 100$, 1000, and 10,000 Ω.

2-27 An amplifier has a noise voltage, referred to the input, of $0.1\mu V/\sqrt{Hz}$ Compute the signal-to-noise ratio at the output if an input signal of 100 mV is applied, and the bandwidth of the amplifier is (a) 100 Hz, (b) 10 Hz, and (c) 1 Hz.

2-28 By dimensional analysis, justify Eq. (2-30):

$$e = \sqrt{4kTR\ \Delta f}$$

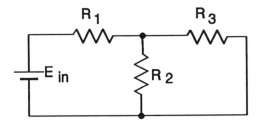

Figure P-2-25

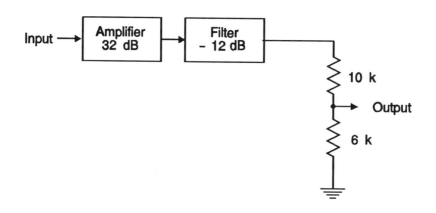

Figure P-3-3

Chapter 3

3-1 Compute the decibel values for each of the following ratios of E_{out}/E_{in}: (a) 0.01, (b) 0.1, (c) 1.0, (d) 100, (e) 2, (f) 3.142, and (g) 90.

3-2 Two circuits, A and B, are connected in series. The voltage attenuation of A is -20 dB, and that of B is -3 dB. Show mathematically that the overall attenuation is -23 dB, the sum of the individual attenuations.

3-3 Consider an amplifier with a voltage gain of 32 dB, followed by a filter with an insertion loss (attentuation) of -12 dB at the frequency of interest, followed by a voltage divider formed with 10- and 6-kΩ resistors (see Figure P-3-3). What is the overall gain if the output is taken across the 6-kΩ resistor?

3-4 Suppose that a circuit has an attenuation of -73 dB. In the decibel table in the Appendix we can find the ratios corresponding to 60, 10, and 3 dB. Show mathematically that the overall voltage ratio for the given circuit is obtained by taking the *product* of the three individual attenuations.

3-5 In the circuit shown in Figure P-3-5, the input voltage is maintained constant at 15 V. Compute the following: (a) E_{out}, I_{load}, and

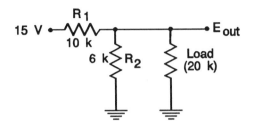

Figure P-3-5

I_{in} (b) the output impedance (c) the voltage gain in dB (d) the current gain in dB (e) the power gain in dB.

3-6 (a) In the circuit of Figure P-3-6, with the load connected, compute the input resistance and the voltage, current, and power attenuations (in decibels).

(b) What will happen to the input impedance if a number of similar "black boxes" are connected in series between source and load?

3-7 Design a low-pass filter with 1 kΩ input impedance and $f_0 = 500$ Hz. Use passive components.

3-8 A low-pass filter has an attenuation slope of 40 dB/decade and $f_0 = 3000$ Hz. The input consists of a composite signal containing a fundamental frequency of 2000 Hz (10 V RMS), with 2 V of the second harmonic and 1 V of the third harmonic. Compute the harmonic content of the output as percent harmonic distortion. You may neglect the rounding at the intersection on the Bode plot.

3-9 Design a Twin-T filter with $f_0 = 60$ Hz. Take $R_{in} = 100$ kΩ.

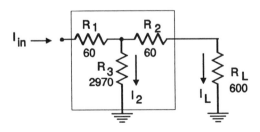

Figure P-3-6

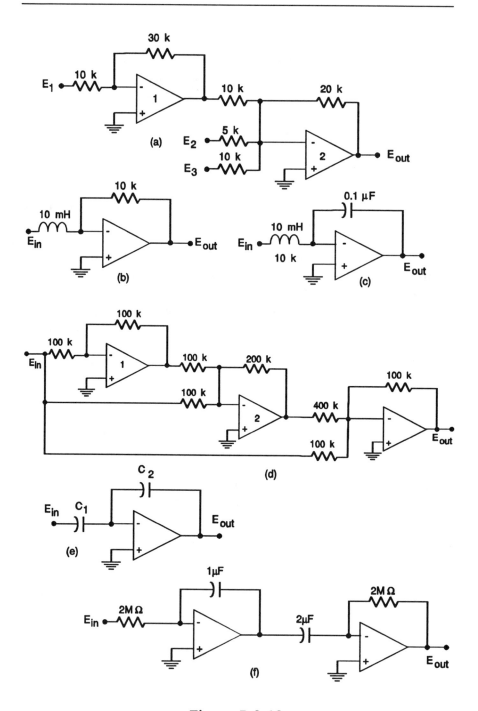

Figure P-3-13

3-10 You have on hand a few carefully matched 0.05-μF capacitors. Calculate the values of resistors needed to make up a Twin-T filter using these capacitors, to reject 400 Hz noise.

3-11 Design op amp circuits to perform each of the following data-processing tasks (The symbol t denotes time.)
  (a) $E_{out} = 5E_1 + 7E_2 - 14E_3$
  (b) $E_{out} = (10 + 5t)E_{in}$
  (c) $E_{out} = 20 \log (E_1 / E_2)$

3-12 Demonstrate mathematically that both differentiation and integration are linear operations.

3-13 Write the operating equations for each of the circuits shown in Figure P-3-13.

3-14 Write the equations for the output as a function of input for each of the two circuits shown in Figure P-3-14, in terms of the parameter k.

3-15 What voltage will be present at the output of an integrator 5 min after the start of an integration if the input resistor is 600 kΩ, the feedback capacitor is 0.5 μF, and the input voltage is 1 mV? Assume error-free operation.

3-16 What would be some disadvantages of an integrator using an inductor in the input rather than a capacitor in the feedback loop?

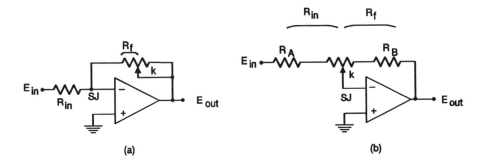

Figure P-3-14

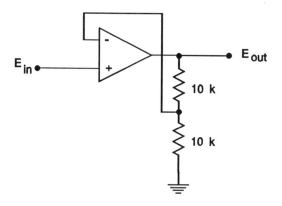

Figure P-4-1

3-17 Prove that an attenuation of 6 dB per octave is equivalent to 20 dB per decade.

Chapter 4

4-1 What is the output voltage for the circuit of Figure P-4-1, if E_{in} = +6 V? −5 V? Comment on the relation between this circuit and that of Figure 4-1b.

4-2 Consider an inverter with R_{in} = 10 kΩ and R_f = 1 MΩ, amplifying a 10-mV signal. What are the maximum permissible offset voltage and bias current, if the deviation from ideality is to be kept below 0.1 percent?

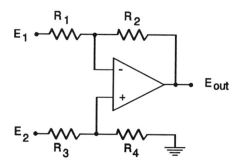

Figure P-4-3

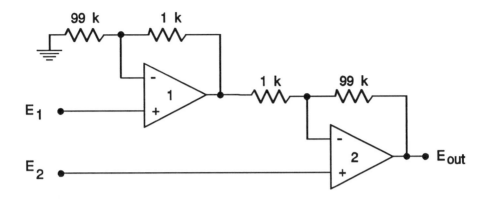

Figure P-4-5

4-3 Calculate E_{out} for the circuit of Figure P-4-3 for each of the following sets of parameters:

R1	R2	R3	R4	E1	E2
1 k	2 k	2 k	4 k	+6 V	0
1 k	2 k	2 k	4 k	0	+6 V

4-4 Carry out the calculations to derive E_{out} as a function of E_1 and E_2 for the amplifier of Figure P-4-3. (This is done in the text for the case where $R_3 = R_1$ and $R_4 = R_2$.)

4-5 The circuit of Figure P-4-5 is another version of an instrumentation amplifier. Explain how it works.

4-6 Calculate the error resulting from finite gain in the amplifier of Figure P-4-6.

4-7 Calculate the impedances of the parallel and series RC networks of Figure 4-6 in the text, and show that setting them equal to each other gives rise to the frequency expression given in the text.

4-8 For the two circuits of Figure 4-2, write a computer program that will create a table giving the output voltage as the load is changed

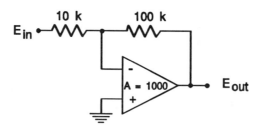

Figure P-4-6

in intervals of 20 kΩ from 50 to 500 kΩ, and intervals of 400 from 600 to 9800 kΩ. Plot these two sets of data on a single sheet of graph paper.

4-9 Explain why the sample-and-hold circuit of Figure 4-5 should make use of a FET-input amplifier for #2.

4-10 Figure P-4-10 shows a difference amplifier set up to measure current by the voltage drop caused in a 1-Ω resistor. Calculate the current corresponding to $E_{out} = 5.32$ V.

4-11 A circuit such as shown in Figure P-4-11 is often used to compensate for bias current in an op amp. R_3 should be equal to the parallel resistance of R_1 and R_2. Explain how this achieves its objective.

Chapter 5

5-1 Find in a reference book the temperature coefficients of the 1N914 and 1N4744 diodes, and show that they result in very nearly zero coefficient when connected as in Figure 5-8. See if you can find

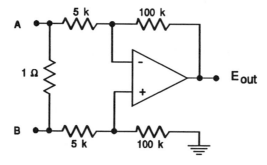

Figure P-4-10

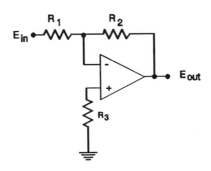

Figure P-4-11

diodes suitable to give a Zener potential of close to 15 V in a similar circuit.

5-2 In Figure 5-17, why is the MOSFET included *within* the feedback loop of the amplifier?

5-3 What would you predict for properties of a silicon crystal with equal amounts of n- and p-doping (e.g., equal atom percentages of As and Al?)

5-4 For greatest precision, a Zener diode should be operated at a constant small reverse-bias current. Can you explain why?

Chapter 6

6-1 Diodes and batteries connected as in Figure P-6-1 form a useful clipping circuit. What would be the value of E_{out} to be expected from each of the following input functions? (Neglect the forward voltage

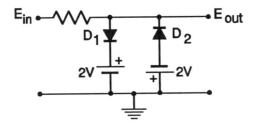

Figure P-6-1

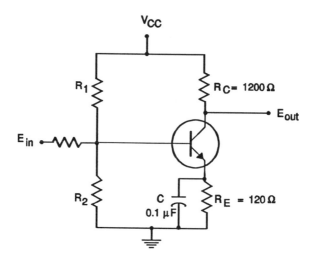

Figure P-6-2

drops of the diodes.) Sketch the outputs of (c) and (d) as functions of time, t.

(a) $E_{in} = +1.0$ V

(b) $E_{in} = -10$ V

(c) $E_{in} = 10 \sin (\omega t)$, where $\omega = 100$ rad/s

(d) $E_{in} = (10 - 2t)$ V

6-2 For the circuit of Figure P-6-2, calculate the impedance of the RC combination in the emitter lead for frequencies of 0.1 Hz and 1000 Hz. Calculate the voltage amplification, E_{out} / E_{in} for both frequencies.

6-3 Suppose that the transistor described in Figure 6-4 is to be used in the circuit of Figure P-6-2, with $C = 10$ μF. The power supply is $V_{CC} = 30$V.

(a) What should be the values of V_B and V_E to force operation at the quiescent point Q? What will be the output voltage, E_{out}?

(b) Calculate approximate values for R_1 and R_2 consistent with the selected quiescent point.

(c) Calculate the effect on E_{out} of a low-frequency AC input voltage that changes E_B by 0.1 V (peak-to-peak). Assume that V_{BE} is constant and that the effect of the emitter capacitor is negligible.

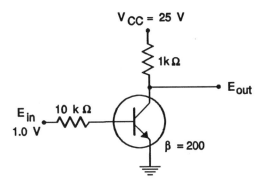

Figure P-6-4

6-4 For the circuit shown in Figure P-6-4, compute the following, neglecting the base-emitter voltage drop: (a) I_B, (b) I_C, (c) E_{out}, (d) A_V

6-5 For the circuit shown in Figure P-6-5, neglecting the base-emitter voltage, find the following quantities: (a) E_{out}; (b) the trans-impedance, dE_{out}/dI_{in}; (c) the optimal value of R_B.

6-6 Two discrete transistors are often connected as a Darlington pair, as in Figure P-6-6. Which of the two should have the higher current rating, and why?

6-7 An emitter follower, a JFET source follower, and an op amp connected as a voltage follower, all perform similar functions. Compare them with respect to the precision with which the output and input are equal.

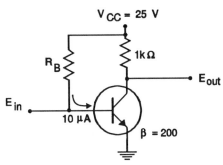

Figure P-6-5

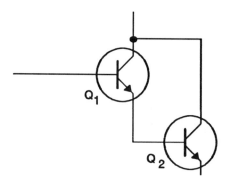

Figure P-6-6

6-8 Given a bipolar transistor with $\beta = 200$ in the emitter-follower configuration, calculate the E_{out}/E_{in} ratio. Do the same for a source follower and for an op-amp follower with $A = 10^4$.

6-9 Sketch the waveforms at the outputs of Figure P-6-9, with (a) $R_C = R_E$, (b) with $R_C = 2 \times R_E$, (c) with $R_C = 0.5 \times R_E$.

6-10 A current regulator (Figure 6-11a) is assembled using the 2N4074 (Figure 6-4). Taking $V_{EE} = 12$ V, what resistor values $(R_1, R_2,$ and $R_3)$ should be selected to give a current of 10 mA? What would be the result of doubling the load resistance, R_L?

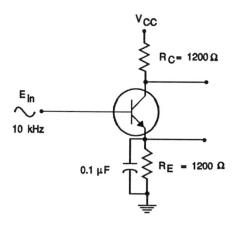

Figure P-6-9

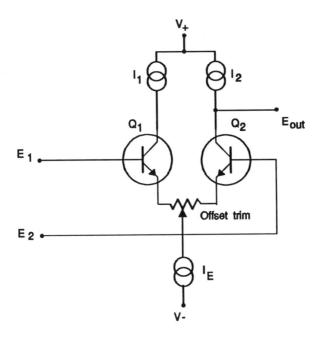

Figure P-6-12

6-11 The FET for which the characteristics are given in Figure 6-7 is connected in the circuit shown in Figure 6-11b as a current regulator. What value of resistor R_S should be selected to produce a current of 4 mA? What variation in current could be expected to result from doubling the load resistance R_L?

6-12 In the differential circuit shown in Figure P-6-12, what would be the effect on the output resulting from:
(a) Raising E_1 while holding E_2 constant?
(b) Raising E_2 while holding E_1 constant?
(c) Raising E_1 while decreasing E_2 by the same amount?
(d) Holding both E_1 and E_2 constant and varying the balance potentiometer?

6-13 Show how FETs can be used to control the "integrate," "hold," and "reset" modes of an op amp integrator. What is the distinction between an integrator with these controls and a sample-and-hold amplifier?

6-14 Devise a circuit whereby a Wien bridge oscillator can be rendered voltage-tunable by the addition of a matched pair of FETs.

6-15 Calculate the natural period of oscillation of the LC tank shown in Figure 6-16b, if $L = 300$ mH, $C_1 = 0.033$ μF and $C_2 = 0.022$ μF.

Chapter 7

7-1 A measurement not described in this chapter has to do with electrochemical cells. Electrical contact with a conductive solution is made by means of specially constructed electrodes. In one common system, current-voltage curves are plotted, their shapes providing information about the chemical make-up of the solution. Figure P-7-1 shows the appropriate electronic circuitry, making use of three electrodes, labeled Wkg (working), Ref (reference), and Aux (auxiliary), each dipping into the liquid in a cell. Amplifier #1 is referred to as a "potentiostat" because it forces the reference electrode to assume the same potential as E_{in} regardless of the current that may be flowing. Hence the current through the working electrode is determined only by the voltage difference between E_{in} and ground, and by the nature of the solution.

(a) What is the purpose of Amplifier #2?

(b) Which electrodes in the cell actually carry the current?

(c) Modify the circuit using for #1 a combination of a follower and summer. What would be the advantages?

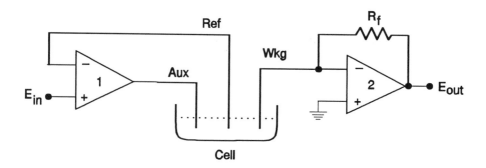

Figure P-7-1

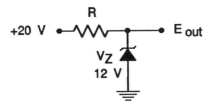

Figure P-8-1

7-2 The temperature coefficient of resistance, α, for platinum is 0.003 K^{-1}, whereas that of nickel is 0.006 K^{-1}. Which would give the larger change in resistance per degree, a coil of platinum wire with room-temperature resistance of 10 kΩ, or one of nickel with a resistance of 1 kΩ.

7-3 How could you redesign the circuits of Figure 7-3 to give a readout of conductance directly in siemens?

7-4 A photometric experiment was set up using the circuit of Figure 7-13. The dark-current pot was adjusted to give zero output with the photocell in the dark. The sensitivity pot was set at 50%. Then a shutter was opened allowing light to fall on the photodiode. The output, E_{out} was observed to change from zero to 450 mV. What was the photodiode current?

7-5 Suppose that in Figure 7-6 the feedback resistors were removed. What information would be obtainable from the output?

7-6 Explain why the photomultiplier is called a "current source," rather than a "voltage source."

7-7 In connection with Figure 7-11, it was pointed out that a photodiode can be operated at zero current or at zero voltage. What would happen if measurements were taken with a load resistor of a few kilohms? How would you draw a load-line on the curves of Figure 7-11?

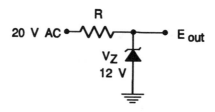

Figure P-8-2

Chapter 8

8-1 In the circuit of Figure P-8-1, compute the resistance and necessary power rating of R, and the maximum power dissipated in the diode, if the load connected from output to ground is allowed to vary from 100 to 2000 Ω. (It may be assumed that V_Z is constant over the required range of currents, and that the forward resistance of the diode is 1 Ω, the reverse resistance 1 MΩ.)

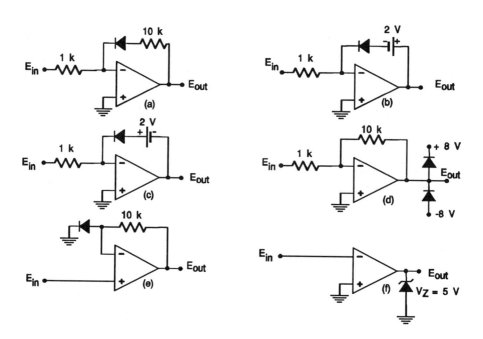

Figure P-8-3

8-2 In the half-wave rectifier shown in Figure P-8-2, the Zener diode and resistor constitute a voltage divider. Calculate the ratio E_{out}/E_{in} for both the positive and negative half-cycles of the input AC. The forward resistance of the diode is $R_{fwd} = 1\ \Omega$, the reverse resistance is $R_{rev} = 1\ M\Omega$. The series resistor $R = 1\ k\Omega$.

8-3 Describe and sketch the input-output relations for each of the circuits shown in Figure P-8-3. In each case the op amp is powered by ± 15 V. Assume all diodes are ideal and neglect the forward voltage drop.

8-4 Temperature regulator circuits are shown in Figures 7-2, 7-8, and 8-13. Compare these with respect to (1) speed of response to a changing temperature and (2) current-carrying ability.

Chapter 9

9-1 Design a variable-voltage regulated power supply to cover the range 5 to 15 V, capable of at least 100 mA, using a regulator of the 78xx series.

9-2 Consider a constant-current source consisting of a 510-V battery in series with a 2-MΩ resistor and the load.

(a) What is the current output into a 1-kΩ load? Into a 10-kΩ load?

(b) What is the percent regulation, defined as $100 \times \dfrac{\Delta I}{\Delta R}$?

(c) What is the voltage compliance?

9-3 Just as a current supply has a voltage compliance, a voltage supply can be characterized by a current compliance. Explain this, and show how it could be applied to the power supply of Figure 9-5.

Chapter 10

10-1 What are the decimal equivalents of the following binary numbers: (a) 1101101, (b) 0101000, (c) 1110, (d) 0101010, (e) 1000000?

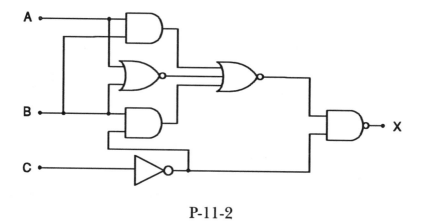

P-11-2

10-2 Write the binary equivalents of the following decimal numbers: (a) 39, (b) 65, (c) 119, (d) 250, (e) 1011.

10-3 Write down the first 20 numbers in the trinary system, base 3. Show how to convert a number from decimal to trinary, and vice versa.

Chapter 11

11-1 Design gate circuits to implement the following truth tables:

A	B	X
0	0	0
0	1	0
1	0	1
1	1	0

(a)

A	B	X
0	0	1
0	1	0
1	0	1
1	1	0

(b)

A	B	C	X
0	0	0	0
0	0	1	0
0	1	0	1
0	1	1	0

A	B	C	X
0	0	0	0
0	0	1	1
0	1	0	0
0	1	1	1

Figure P-11-5

1	0	0	1
1	0	1	0
1	1	0	1
1	1	1	0

(c)

1	0	0	0
1	0	1	1
1	1	0	0
1	1	1	0

(d)

11-2 Write the truth tables for the circuit shown in Figure P-11-2.

11-3 Implement the function: $(A$ OR $B)$ AND NOT $(C$ OR $D)$.

11-4 Design a circuit to fit the following truth table:

A	B	C	X
0	0	0	0
0	0	1	1
0	1	0	1
0	1	1	0
1	0	0	1
1	0	1	1
1	1	0	1
1	1	1	0

11-5 Show that gates shown in Figure P-11-5 perform identical functions. Include the pertinent Boolean notation.

11-6 Simplify the function $\overline{(A + B)} \bullet (B \bullet C)$, both by algebra and by a truth table.

11-7 Consider the expression $A\,\overline{BC} + AB$.
(a) Evaluate the expression for $A = B = C = 1$.

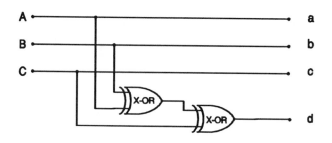

Figure P-11-12

(b) Design a logic circuit to implement this function.

11-8 Evaluate the expression $\overline{(AB + C)}$, for the following cases:
(a) $A = 1, B = 0, C = 0$
(b) $A = 1, B = 1, C = 1$

11-9 Write truth tables for the following:
(a) $X = A + (B \bullet C) + \overline{B}$
(b) $X = A \bullet (B + C)$

11-10 Construct a truth table for an interlock system that prevents a car from starting if the driver's seat is occupied but the seat belt is not fastened. Consider the case for two occupants.

11-11 Construct the equivalent of a 16-input NAND gate, using four 7420s and, if necessary, 7404 inverters.

11-12 The circuit shown in Figure P-11-12 can be used to generate a 4-digit code from one of three digits. Write its truth table.

11-13 Assume that we need to verify that a certain pattern of 1's and 0's exists in a particular 8-digit word. For example it may be essential that bits 0, 2 and 8 (counting from the right) are each "1" while it doesn't matter what the other bits are. The desired pattern can be represented by the word 1xxx x1x1, where each x is permitted to be either a 1 or a 0.
    The approach requires two steps: (a) The digits 1000 0101 are AND-ed to the existing word (requiring eight AND gates), and then (b) the result from that step is XOR-ed with the same 1000 0101. If the

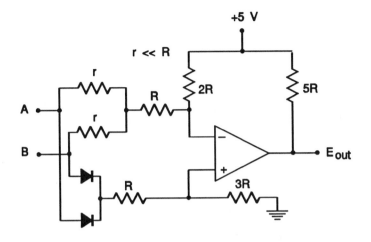

Figure P-11-17

result is now 1000 0101, it is established that the original word met the requirements. Explain the logic behind this procedure.

11-14 In the multiplexer shown in Figure 11-22, what would be the effect of replacing the OR-gate with a four input XOR, which gives a HIGH if one and only one input is HIGH.

11-15 In Figure 11-18, predict what the output lines would show in case more than one input was allowed to go high at the same time.

11-16. Draw a diagram corresponding to Figure 11-18a for an 8-to-3 encoder.

11-17 Does the comparator circuit in Figure P-11-17 constitutes a valid XOR gate?

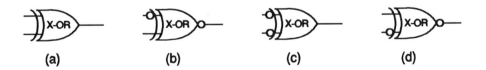

(a)   (b)   (c)   (d)

Figure P-11-18

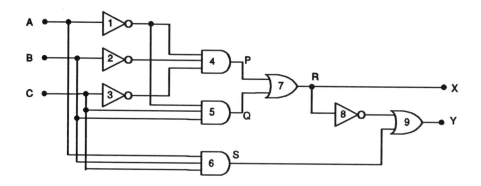

Figure P-11-19

11-18 Are all four of the gates of Figure P-11-18 equivalent?

11-19 Construct a truth table for the circuit of Figure P-11-19. How could the circuit be simplified, still giving the same logic?

Chapter 12

12-1 (a) A 555 timer is connected as an astable oscillator with R_A = R_B = 150 kΩ and C = 0.01 μF. What will be the frequency and duty cycle of the output?

(b) The output of the 555 of part (a) is connected to the clock input of a JK flip-flop. What are the frequency and duty cycle of the output of the flip-flop?

12-2 A bounceless switch can be made with NOR gates, as well as with NAND gates as in Figure 12-4. Show how the connections should be arranged for this application.

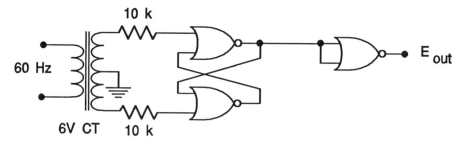

Figure P-12-3

12-3 An *RS* NOR flip-flop made with CMOS gates can be connected to a 60-Hz transformer, as in Figure P-12-3. What output would you expect?

Chapter 13

13-1 What are the advantages of using encoding followed by decoding?

13-2 Draw the block diagram for a circuit using an analog S/H amplifier to prevent glitches from appearing in a digital circuit following an A/D converter. Show a typical timing sequence for this operation.

13-3 In the staircase generator shown in Figure 13-6, what clock frequency should be selected to give a staircase with 100 steps per second?

13-4 What would be the E_{out}/E_{in} ratio for the circuit shown in Figure 13-7, if the first and fourth switches are closed while the second and third are grounded?

13-5 The clock for a certain dual-slope ADC shown in Figure 13-17 runs at 100 kHz, the reference voltage is 10.000 V, and time t_1 corresponds to 10^6 counts. If the time $(t_2 - t_1) = 15.0$ ms, what is the analog input voltage?

13-6 Explain the concept of hysteresis, and show why it is useful in the conversion of a non-square periodic signal to an equivalent square wave.

13-7 Explain the function of the two diodes at the output of the comparator used as input to the ADC circuits of Figures 13-14, -15, and -16.

13-8 Hunt up the manufacturer's literature on the LM-131, and find how this IC can be used in the inverse manner, as a frequency-to-voltage converter. You should be able to explain the operation of the circuit.

Chapter 14

14-1 Design in detail the system described in Figure 14-5

14-2 Design a simple CPU to performing, under software control, one of three operations: (a) add the 2-bit numbers A and B, (b) shift A to the right, (c) no operation.

14-3 Search in the literature and list the major types of memory both volatile and nonvolatile.

Chapter 15

15-1 Set up a sector distribution for 5:1 interleaving (Cf. Figure 15-4)

15-2 Develop the scheme of Figure 15-5, describing a complete directory/FAT setup for a hypotetical set of files.

15-3 Compare various types of mass storage media; add literature information.

15-4 Describe a system of 9 × 9 matrices for displaying letters on a character-only display; show how the word YES can be written on such a screen.

Chapter 16

16-1 Design a switching circuit to connect one of two parallel-input printers to a single computer output. Use "software" control, meaning that a line that can take 0 or 1 values is used for selecting the printer.

16-2 Repeat Problem 16-1 for a serial input.

16-3 Design a circuit to transform ASCII codes 0 through 10 into binary codes.

16-4 Design a circuit to be connected in series with an RS-232 line monitor the status each of the communication lines, by means of LEDs.

16-5 Design a device similar to the one of Problem 16-4, but able to take an instantaneous reading of the status of each line and to maintain the information on LEDs until reset (by a 1 on a separate line).

16-6 Design a centralized computer control for five different instruments in a laboratory. The system should be able to send *on, start, stop,* and *reset* logic signals to each instrument. It also should also be able to read from the analog output of each instrument, and convert the data to parallel logic. Assume that the computer is provided with an interface card that can send and receive logic words on two 12-bit sets of lines.

Chapter 17

17-1 Create a token ring variant using a double loop: line A carying the messages and line B selecting which station is active. Each station is in series with line B, and is connected in parallel to the loop of line A. One of the stations has the token, which gives an allowance of 100 ms of network control. After this time has elapsed, the station deactivates itself and sends the token to the next position by means of a $0-1-0$ pulse. The station with the token can send messages during the 100 ms alloted. The information is sent to all the stations simultaneously. The message has an address affixed to it so that only the designated receiver can processes it.

17-2 Design a bar code using three elements: thin lines, thick lines, and spaces. Implement error detection by a parity bar.

17-3 Describe in block diagrams, how you would establish a LIMS in your laboratory.

Chapter 20

20-1 Transform into cartesian coordinates: $123 \angle -22°$; $1.25 \angle 45°$; $0.03 \angle 400°$.

20-2 Transform into polar coordinates : $(5 + j5); (-3 + j0);$ $(-4 - j4); (8 - j16).$

20-3 Write the impedance in complex form for a series combination of a resistor, a capacitor and an inductor.

20-4 Calculate the impedance of a circuit element that passes a current of $0.30 \angle 60°$ A, if the voltage applied is $120 \angle 0°$ V.

20-5 Write the transfer coefficients for the diffrentiator and integrator.

20-6 Plot the frequency response of the circuit of Figure 20-3b using the phasor expression of E_{out}/E_{in}.

20-7 Derive two equations, comparable to Eq. (20-33), for an inductor driven by a sine-wave AC, first neglecting the resistance of the inductor, and then taking it into account.

20-8 Is is true that $A•B + B•C = A•B + \overline{A}•C$?

20-9 In the expression
$Q = A•B + \overline{C}•A$
determine the value of Q under the conditions:
(a) $A = B = C = 0$
(b) $A = B = C = 1$
(c) $A = 0 ; B = C = 1$

Appendixes

The ASCII Code

The ASCII convention is the following assignment:

32	(space)	56	8	80	P	104	h
33	!	57	9	81	Q	105	i
34	"	58	:	82	R	106	j
35	#	59	;	83	S	107	k
36	$	60	<	84	T	108	l
37	%	61	=	85	U	109	m
38	&	62	>	86	V	110	n
39	'	63	?	87	W	111	o
40	(	64	@	88	X	112	p
41	)	65	A	89	Y	113	q
42	*	66	B	90	Z	114	r
43	+	67	C	91	[	115	s
44	,	68	D	92	\	116	t
45	-	69	E	93	]	117	u
46	.	70	F	94	^	118	v
47	/	71	G	95	_	119	w
48	0	72	H	96	"	120	x
49	1	73	I	97	a	121	y
50	2	74	J	98	b	122	z
51	3	75	K	99	c	123	{
52	4	76	L	100	d	124	|
53	5	77	M	101	e	125	}
54	6	78	N	102	f	126	~
55	7	79	O	103	g	127	(null)

Codes 0-32 are assigned to commands. They are called control characters. On PCs they can be implemented by pressing the control key (represented here by a caret ^) together with the letter code :

Decimal value	Control key	Abbreviation	Function
0	^@	NUL	No action
1	^A	SOH	Start of Header
2	^B	STX	Start of Text
3	^C	ETX	End of Text
4	^D	EOT	End of Transm.
5	^E	ENQ	Enquire
6	^F	ACK	Acknowledge
7	^G	BEL	Rings bell
8	^H	BS	Backspace
9	^I	HT	Horizontal Tab
10	^J	LF	Line Feed
11	^K	VT	Vertical Tab
12	^L	FF	Form Feed
13	^M	CR	Carriage Return
14	^N	SO	Shift Out
15	^O	SI	Shift In
16	^P	DL	Delete
17	^Q	DC1	Device Control (1)
18	^R	DC2	Device Control (2)
19	^S	DC3	Device Control (3)
20	^T	DC4	Device Control (4)
21	^U	NAK	Not Acknowledged
22	^V	SYN	Synchronize
23	^W	ETB	End of Text Block
24	^X	CAN	Cancel
25	^Y	EM	End of Medium
26	^Z	SUB	Substitute
27	^[	ESC	Escape
28	^/	FS	File Separator
29	^]	GS	Group Separator
30	^^	RS	Record Separator
31	^_	US	Unit Separator

Of special importance is the ^C which serves to interrupt some of the ongoing tasks, and ESC, the Escape Character, which is often used to indicate the beginning of a command sequence (an "escape sequence").

The codes from 128 to 255 are subject to variation. Below is given one of the assignments, an extended character set often used in laser printers. The ASCII 218-255 are not assigned in this case.

128	Ç	161	í	194	‰
129	ü	162	ó	195	●
130	é	163	ú	196	–
131	â	164	ñ	197	—
132	ä	165	Ñ	198	°
133	à	166	ª	199	Á
134	å	167	º	200	Â
135	ç	168	¿	201	È
136	ê	169	"	202	Ê
137	ë	170	"	203	Ë
138	è	171	‹	204	Ì
139	ï	172	›	205	Í
140	î	173	¡	206	Î
141	ì	174	«	207	Ï
142	Ä	175	»	208	Ò
143	Å	176	ā	209	Ó
144	É	177	õ	210	Ô
145	æ	178	Ø	211	Š
146	Æ	179	ø	212	š
147	ô	180	œ	213	Ù
148	ö	181	Œ	214	Ú
149	ò	182	À	215	Û
150	û	183	Ã	216	Ÿ
151	ù	184	Õ	217	ß

152	ÿ	185	§	218
153	Ö	186	‡	219
154	Ü	187	†	220
155	¢	188	¶	221
156	£	189	©	222
157	¥	190	®	223
158	¤	191	™	224
159	ƒ	192	„	225
160	á	193	...	226

Any character, can be obtained by typing the appropriate numerical code on the numeric pad, while holding down the ALT key.

Decibel Table

dB	Voltage Gain	Power Gain	Voltage Loss	Power Loss
0.0	1.00	1.00	1.00	1.00
0.1	1.01	1.02	0.99	0.98
0.2	1.02	1.05	0.98	0.96
0.3	1.04	1.07	0.97	0.93
0.4	1.05	1.10	0.96	0.91
0.5	1.06	1.12	0.94	0.89
0.6	1.07	1.15	0.93	0.87
0.7	1.08	1.17	0.92	0.85
0.8	1.10	1.20	0.91	0.83
0.9	1.11	1.23	0.90	0.81
1.0	1.12	1.26	0.89	0.97
2.0	1.26	1.58	0.79	0.63
3.0	1.41	2.00	0.71	0.501
4.0	1.58	2.51	0.63	0.398
5.0	1.78	3.16	0.56	0.316
6.0	2.00	3.98	0.501	0.251
7.0	2.24	5.01	0.447	0.200
8.0	2.51	6.31	0.398	0.158
9.0	2.82	7.94	0.355	0.126
10.0	3.16	10.00	0.316	0.100
20.0	10.00	100.00	0.100	0.010

Note that the voltage ratio for 3 dB is $\sqrt{2}$, whereas that for 6 dB is 2. The corresponding power ratios are 2 and 4. If a value not listed in the table is required, one can break down the decibel number into a sum of known values, look them up in the table, and multiply them together. Thus 43.5 dB as a voltage gain can be written as $20 + 20 + 3 + 0.5$, corresponding to gains of 10, 10, 1.41, and 1.06. The net gain is given by the product of these, 149.5.

Trigonometric Table

Degrees	Radians	Sine	Cosine	Tangent
0	0	0	1	0
30	$\dfrac{\pi}{6}$	$\dfrac{1}{2}$	$\dfrac{\sqrt{3}}{2}$	$\dfrac{1}{\sqrt{3}}$
45	$\dfrac{\pi}{4}$	$\dfrac{\sqrt{2}}{2}$	$\dfrac{\sqrt{2}}{2}$	1
60	$\dfrac{\pi}{3}$	$\dfrac{\sqrt{3}}{2}$	$\dfrac{1}{2}$	$\sqrt{3}$
90	$\dfrac{\pi}{2}$	1	0	∞
120	$\dfrac{2\pi}{3}$	$\dfrac{\sqrt{3}}{2}$	$-\dfrac{1}{2}$	$-\sqrt{3}$
135	$\dfrac{3\pi}{4}$	$\dfrac{\sqrt{2}}{2}$	$-\dfrac{\sqrt{2}}{2}$	-1
150	$\dfrac{5\pi}{6}$	$\dfrac{1}{2}$	$-\dfrac{\sqrt{3}}{2}$	$-\dfrac{1}{\sqrt{3}}$
180	π	0	-1	0
$-a$		$-\sin a$	$\cos a$	$-\tan a$
$90+a$		$\cos a$	$-\sin a$	$-\operatorname{ctn} a$
$90-a$		$\cos a$	$\sin a$	$\operatorname{ctn} a$

Capacitive Impedance

The following table gives the impedance (the capacitive reactance) of selected values of capacitance calculated from the formula $Z_C = X_C = 1/(2\pi fC)$. Values less than 0.1 Ω are not likely to be useful and thus are not listed. Actual capacitors will obey this relationship only over limited frequency ranges, because of inductive and dissipative effects. All values are given in ohms.

μF	50 Hz	100 Hz	1 kHz	10 kHz	100 kHz	1 MHz
0.001	3.2 M	1.6 M	160k	16 k	1.6k	160
0.005	640k	320k	32 k	3.2 k	320	32
0.01	320k	160k	16 k	1.6 k	160	16.
0.05	64k	32k	3.2k	320	32.	3.2
0.1	32k	16k	1.6k	160	16	1.6
0.5	6.4k	3.2k	320	32	3.2	0.32
1.0	3.2k	1.6k	160	16	1.6	
5.0	640	320	32	3.2	0.32	
10.0	320	160	16	1.6	0.16	
50.0	64	32	3.2	0.32		
100.0	32.	16	1.6	0.16		
500.0	6.4	3.2	0.32			

Color Codes for Resistors and Capacitors

The values of low-wattage resistors are coded by a series of colored bands near one end of the resistor body. The first two bands from the end represent digits; the next indicates the number of zeros following the two digits. The fourth is a tolerance indicator: gold signifies 5%, silver 10%, and no band 20%. Thus a resistor banded brown-black-green-silver has a value of 1-0-00000 or 1,000,000 Ω and a 10% tolerance level. If a precision resistor has five bands, the first three indicate numerical values, the fourth is a multiplier, and the fifth is the tolerance indicator.

Color	Digit	Zeros	Tolerance
Black	0	—	
Brown	1	1	
Red	2	2	2%
Orange	3	3	
Yellow	4	4	
Green	5	5	
Blue	6	6	
Violet	7	7	
Gray	8	8	
White	9	9	
Gold	—	—	5%
Silver	—	—	10%

2   2   0000 2%
Red   Red   Yellow Red   220k 2%

2   2   4   00   1%
Red   Red   Yellow Red   Brown   22,400 1%

Capacitors sometimes use the same color code, but the first band is the temperature coefficient. The second and third bands give the digits, and the fourth the multiplier using pF units. A numerical code on a capacitor usually gives: first digit, second digit, number of zeros, tolerance (J= 5%, K = 10%, M= 20%). Thus 104K means 100,000 pF at 10%.